LONDON MATHEMATICAL SOCIETY LECTURE NOTE SERIES

Managing Editor: Professor J.W.S. Cassels, Department of Pure Mathematics and Mathematical Statistics, University of Cambridge, 16 Mill Lane, Cambridge CB2 1SB, England

The books in the series listed below are available from booksellers, or, in case of difficulty, from Cambridge University Press.

4 Algebraic topology, J.F. ADAMS
17 Differential germs and catastrophes, Th. BROCKER & L. LANDER
27 Skew field constructions, P.M. COHN
34 Representation theory of Lie groups, M.F. ATIYAH *et al*
36 Homological group theory, C.T.C. WALL (ed)
39 Affine sets and affine groups, D.G. NORTHCOTT
40 Introduction to H_p spaces, P.J. KOOSIS
42 Topics in the theory of group presentations, D.L. JOHNSON
43 Graphs, codes and designs, P.J. CAMERON & J.H. VAN LINT
45 Recursion theory: its generalisations and applications, F.R. DRAKE & S.S. WAINER (eds)
46 p-adic analysis: a short course on recent work, N. KOBLITZ
49 Finite geometries and designs, P. CAMERON, J.W.P. HIRSCHFELD & D.R. HUGHES (eds)
50 Commutator calculus and groups of homotopy classes, H.J. BAUES
51 Synthetic differential geometry, A. KOCK
54 Markov processes and related problems of analysis, E.B. DYNKIN
57 Techniques of geometric topology, R.A. FENN
58 Singularities of smooth functions and maps, J.A. MARTINET
59 Applicable differential geometry, M. CRAMPIN & F.A.E. PIRANI
60 Integrable systems, S.P. NOVIKOV *et al*
62 Economics for mathematicians, J.W.S. CASSELS
65 Several complex variables and complex manifolds I, M.J. FIELD
66 Several complex variables and complex manifolds II, M.J. FIELD
68 Complex algebraic surfaces, A. BEAUVILLE
69 Representation theory, I.M. GELFAND *et al*
74 Symmetric designs: an algebraic approach, E.S. LANDER
76 Spectral theory of linear differential operators and comparison algebras, H.O. CORDES
77 Isolated singular points on complete intersections, E.J.N. LOOIJENGA
78 A primer on Riemann surfaces, A.F. BEARDON
79 Probability, statistics and analysis, J.F.C. KINGMAN & G.E.H. REUTER (eds)
80 Introduction to the representation theory of compact and locally compact groups, A. ROBERT
81 Skew fields, P.K. DRAXL
82 Surveys in combinatorics, E.K. LLOYD (ed)
83 Homogeneous structures on Riemannian manifolds, F. TRICERRI & L. VANHECKE
84 Finite group algebras and their modules, P. LANDROCK
85 Solitons, P.G. DRAZIN
86 Topological topics, I.M. JAMES (ed)
87 Surveys in set theory, A.R.D. MATHIAS (ed)
88 FPF ring theory, C. FAITH & S. PAGE
89 An F-space sampler, N.J. KALTON, N.T. PECK & J.W. ROBERTS
90 Polytopes and symmetry, S.A. ROBERTSON
91 Classgroups of group rings, M.J. TAYLOR
92 Representation of rings over skew fields, A.H. SCHOFIELD
93 Aspects of topology, I.M. JAMES & E.H. KRONHEIMER (eds)
94 Representations of general linear groups, G.D. JAMES
95 Low-dimensional topology 1982, R.A. FENN (ed)

96 Diophantine equations over function fields, R.C. MASON
97 Varieties of constructive mathematics, D.S. BRIDGES & F. RICHMAN
98 Localization in Noetherian rings, A.V. JATEGAONKAR
99 Methods of differential geometry in algebraic topology, M. KAROUBI & C. LERUSTE
100 Stopping time techniques for analysts and probabilists, L. EGGHE
101 Groups and geometry, ROGER C. LYNDON
103 Surveys in combinatorics 1985, I. ANDERSON (ed)
104 Elliptic structures on 3-manifolds, C.B. THOMAS
105 A local spectral theory for closed operators, I. ERDELYI & WANG SHENGWANG
106 Syzygies, E.G. EVANS & P. GRIFFITH
107 Compactification of Siegel moduli schemes, C-L. CHAI
108 Some topics in graph theory, H.P. YAP
109 Diophantine Analysis, J. LOXTON & A. VAN DER POORTEN (eds)
110 An introduction to surreal numbers, H. GONSHOR
111 Analytical and geometric aspects of hyperbolic space, D.B.A.EPSTEIN (ed)
112 Low-dimensional topology and Kleinian groups, D.B.A. EPSTEIN (ed)
113 Lectures on the asymptotic theory of ideals, D. REES
114 Lectures on Bochner-Riesz means, K.M. DAVIS & Y-C. CHANG
115 An introduction to independence for analysts, H.G. DALES & W.H. WOODIN
116 Representations of algebras, P.J. WEBB (ed)
117 Homotopy theory, E. REES & J.D.S. JONES (eds)
118 Skew linear groups, M. SHIRVANI & B. WEHRFRITZ
119 Triangulated categories in the representation theory of finite-dimensional algebras, D. HAPPEL
120 Lectures on Fermat varieties, T. SHIODA
121 Proceedings of *Groups - St Andrews 1985*, E. ROBERTSON & C. CAMPBELL (eds)
122 Non-classical continuum mechanics, R.J. KNOPS & A.A. LACEY (eds)
123 Surveys in combinatorics 1987, C. WHITEHEAD (ed)
124 Lie groupoids and Lie algebroids in differential geometry, K. MACKENZIE
125 Commutator theory for congruence modular varieties, R. FREESE & R. MCKENZIE
126 Van der Corput's method for exponential sums, S.W. GRAHAM & G. KOLESNIK
127 New directions in dynamical systems, T.J. BEDFORD & J.W. SWIFT (eds)
128 Descriptive set theory and the structure of sets of uniqueness, A.S. KECHRIS & A. LOUVEAU
129 The subgroup structure of the finite classical groups, P.B. KLEIDMAN & M.W.LIEBECK
130 Model theory and modules, M. PREST
131 Algebraic, extremal & metric combinatorics, M-M. DEZA, P. FRANKL & I.G. ROSENBERG (eds)
132 Whitehead groups of finite groups, ROBERT OLIVER
133 Linear algebraic monoids, MOHAN S. PUTCHA
134 Number thoery and dynamical systems, M. DODSON & J. VICKERS (eds)
135 Operator algebras and applications, 1, D. EVANS & M. TAKESAKI (eds)
136 Operator algebras and applications, 2, D. EVANS & M. TAKESAKI (eds)
137 Analysis at Urbana, I, E. BERKSON, T. PECK, & J. UHL (eds)
138 Analysis at Urbana, II, E. BERKSON, T. PECK, & J. UHL (eds)
139 Advances in homotopy theory, B. STEER & W. SUTHERLAND (eds)
140 Geometric aspects of Banach spaces E.M.PEINADOR and A.R Usan (eds)
141 Surveys in combinatorics 1989, J. SIEMONS (ed)
142 The geometry of jet bundles, D.J. SAUNDERS

London Mathematical Society Lecture Note Series. 137

Analysis at Urbana

Volume I: Analysis in Function Spaces

Edited by
E. Berkson, T. Peck & J. Uhl
Department of Mathematics
University of Illinois

The right of the University of Cambridge to print and sell all manner of books was granted by Henry VIII in 1534. The University has printed and published continuously since 1584.

CAMBRIDGE UNIVERSITY PRESS
Cambridge
New York New Rochelle Melbourne Sydney

CAMBRIDGE UNIVERSITY PRESS
Cambridge, New York, Melbourne, Madrid, Cape Town, Singapore, São Paulo

Cambridge University Press
The Edinburgh Building, Cambridge CB2 8RU, UK

Published in the United States of America by Cambridge University Press, New York

www.cambridge.org
Information on this title: www.cambridge.org/9780521364362

First published 1989
Re-issued in this digitally printed version 2008

A catalogue record for this publication is available from the British Library

ISBN 978-0-521-36436-2 paperback

ACKNOWLEDGEMENTS

The organisers and participants gratefully acknowledge the support of the Special Year in Modern Analysis at the University of Illinois provided by the following agencies;

The Department of Mathematics, University of Illinois at Urbana-Champaign
The National Science Foundation
The Argonne Universities Association Trust Fund
The George A. Miller Endowment Fund (University of Illinois)
The Campus Research Board (University of Illinois at Urbana-Champaign)

CONTENTS

Membership of Hankel operators on planar domains in unitary ideals 1
J.Arazy

A generalised Marcel Riesz theorem on conjugate functions 41
N.Asmar and E.Hewitt

Some results in analysis related to the law of the iterated logarithm 47
R.Banuelos and C.Moore

Fourier series, mean Lipschitz spaces and bounded mean oscillation 81
P.Bourdon, J.Shapiro and W.Sledd

A remark on the maximal function associated to an analytic vector field 111
J.Bourgain

Hankel operators on H^p 133
J.Cima and D.Stegenga

Contractive projections on l_p spaces 151
W.Davis and P.Enflo

Contractive projections onto subsets of $L^1(0,1)$ 162
P.Enflo

Some Banach space properties of translation invariant subspaces of L^p 185
K.Hare and N.Tomczak-Jaegermann

Random multiplications, random coverings, and multiplicative chaos 196
J.-P.Kahane

Wavelets and operators 256
Y.Meyer

On the structure of the graph of the Franklin analysing wavelet 366
E.Berkson

Boundededness of the canonical projection for Sobolev spaces generated by finite families of linear differential operators 395
A.Pelczynski

Remarks on L^2 restriction theorems for Riemann manifolds 416
C.Sogge

PREFACE

The Special Year in Modern Analysis at the University of Illinois was devoted to the synthesis and expansion of modern and classical analysis. The program brought together analysts from around the globe for intensive lectures and discussions, including an International Conference on Modern Analysis, held March 16-19, 1987. The Special Year's success is a tribute to the outstanding merits and professional dedication of the participants. Contributions to these *Proceedings of the Special Year* were solicited from the participants in order to record and disseminate the fruits of their activities. The editors are grateful to the contributors for their response, which accurately reflects the quality and substance of the Special Year. In keeping with the wide scope of topics treated, the contents of these *Proceedings* fell naturally into two interrelated volumes, covering "Analysis in Abstract Spaces" and "Analysis in Function Spaces".

Thanks are due to the National Science Foundation, the Argonne Universities Association Trust Fund, the University of Illinois Campus Research Board, the University of Illinois Miller Endowment Fund, the University of Illinois Department of Mathematics, and J. Bourgain's Chair in Mathematics at the University of Illinois, without whose financial support the Special Year could not have taken place. Special thanks are also due to Professor Bela Bollobas, Consulting Editor at Cambridge University Press, and Mr. David Tranah, Senior Editor in Mathematical Sciences at Cambridge University Press, for the guidance and encouragement which made these *Proceedings* possible.

Earl Berkson
N. Tenney Peck
J. Jerry Uhl

Membership of Hankel operators on planar domains in unitary ideals

by

Jonathan Arazy
Department of Mathematics
University of Haifa
Haifa 31999, ISRAEL

§1. *Introduction*

A *Hankel matrix* is a matrix of the form

$$A = (c_{n+k})_{n,k \geq 0}$$

i.e., $a_{n,k} = c_{n+k}$ for $n,k \geq 0$. These matrices are naturally related to analytic functions f on the unit disk Δ via

$$f(z) = \sum_{n=0}^{\infty} \bar{c}_n z^n$$

The most efficient way of studying the properties of the Hankel matrix A as an operator on ℓ_2 is via the action of the *Hankel operator*

$$H_f = (I-P)M_{\bar{f}}P$$

on $L^2(\mathbb{T})$, where $\mathbb{T} = \partial\Delta = \{z \in \mathbb{C};\ |z| = 1\}$, P is the orthogonal projection onto the Hardy space H^2, and $M_{\bar{f}}$ is the operator of multiplication by $\bar{f}$. The connection between the two objects is that the matrix of H_f in the bases $\{z^n\}_{n\geq 0}$ in H^2 and $\{\bar{z}^k\}_{k\geq 1}$ in $L^2 \ominus H^2 = (H^2)^{\perp}$ is the Hankel matrix A.

Similar operators, also called Hankel operators, are studied in a wider context (General planar domains, non-analytic symbols, higher dimensions etc.). We discuss here the extension of the theory to planar domains.

Let Ω be a domain in the complex plane and μ a finite positive measure on Ω: Let $A^2(\mu) = A^2(\Omega,\mu)$ denote the space of all analytic functions on Ω which belong to $L^2(\mu) = L^2(\Omega,\mu)$. We assume that convergence in $A^2(\mu)$ implies uniform convergence on compact subsets of Ω. Thus $A^2(\mu)$ is closed in $L^2(\mu)$ and point evaluations are continuous lienar functionals on $A(\mu)$.

Let $P: L^2(\mu) \to A^2(\mu)$ be the orthogonal projection. For an analytic function f on Ω we consider the *Hankel operator*

$$H_f = (I-P)M_{\bar{f}}P$$

We sometimes identify H_f with its restriction to $A^2(\mu)$.

The *main theme* in studying Hankel operators is *the connection between the size of the operator* H_f — (boundness, compactness, membership in Schatten classes, S_p, etc.) and the *size of the symbol* f (rate of increase f(z) as $z \to \partial\Omega$, degree of smoothness of boundary values, rate of approximation by "nice" functions, etc.).

We would like to report here on some recent works on membership of Hankel operators in unitary ideals in the context of the unit disk [AFP2] and planar domains of finite connectivity [AFP3].

These notes are an expanded version of the talks given by the author at the University of Illinois at Urbana–Champaign to the participants of the Special Year in Modern Analysis. The author thanks Professor J. Bourgain and E. Berkson for arranging his visit.

§2. *Hankel operators in $L^2(\mathbb{T})$, Schatten p–classes, BMOA, VMOA and Besov–p spaces*

The conditions for boundedness and compactness of the Hankel operators H_f on $L^2(\mathbb{T})$, f analytic on Δ, were found by Z. Nehari and P. Hartman respectively. To formulate these results in our terminology, let

$$\varphi_a(z) = \frac{a-z}{1-\bar{a}z} \; ; \; a,z \in \Delta$$

φ_a is the *Mobius function* which preserves Δ and interchange 0 and a. An analytic function f on Δ is in the space *BMOA* (analytic functions with bounded mean oscillation) if

$$\|f\|_{BMOA} := \sup_{a\in\Delta} \|f\circ\varphi_a - f(a)\|_{H^2} < \infty$$

f belongs to *VMOA* (analytic functions of vanishing mean oscillation) if $f\in$BMOA and $\lim_{a\to\partial\Delta} \|f\circ\varphi_a - f(a)\|_{H^2} = 0$. Thus we have adopted the so–called Garsia norm to define BMOA and VMOA.

Theorem 2.1 [N] *The Hankel operator* H_f *on* $L^2(\mathbb{T})$ *is bounded if and only if* f *is in* BMOA.

Theorem 2.2 [H] *The Hankel operator* H_f *on* $L^2(\mathbb{T})$ *is compact if and only if* f *is in* VMOA.

The space of compact operators from a Hilbert space M into a Hilbert space N is denoted by $S_\infty(M,N)$, or simply S_∞ in case M and N are understood. The singular numbers of $T \in S_\infty$ are the eigenvalues of $(T^*T)^{\frac{1}{2}}$, $s_n(T) = \lambda_n((T^*T)^{\frac{1}{2}})$, arranged in a non-increasing ordering, counting multiplicity. The Schatten p-classes $S_p = S_p(M,N)$, $0 < p < \infty$, consist of all $T \in S_\infty$ for which

$$\|T\|_{S_p} := [\text{trace } (T^*T)^{p/2}]^{1/p} = (\sum_{n=1}^{\infty} s_n(T)^p)^{1/p}$$

is finite. More generally, if E is a symmetric sequence space, the associated unitary ideal $S_E = S_E(M,N)$ consists of those $T \in S_\infty$ for which $(s_n(t))_{n=1}^{\infty} \in E$, normed by

$$\|T\|_{S_E} = \|(s_n(T))\|_E$$

Clearly, $T \in S_E(M,N)$ if and only if $UTV \in S_E(M_1, N_1)$ for all unitary operators V from M_1 onto M and U from N onto N_1, and $\|T\|_{S_E} = \|UTV\|_{S_E}$. This explains the name "unitary ideals". The space S_2 (= Hilbert Schmidt operators) is a Hilbert space with respect to the inner product

$$(A,B)_{S_2} = \text{trace } (AB^*) .$$

This pairing is clearly unitarily invariant. With respect to this pairing, $S_\infty(M,N)^* = S_1(M,N)$, $S_1(M,N)^* = B(M,N)$ and

$$S_p(M,N)^* = S_q(M,N) \ ,1 < p < \infty, \frac{1}{p} + \frac{1}{q} = 1 \ .$$

See [GK] and [S] for more information on S_p and unitary ideals in general.

An analytic function f on Δ belongs to the *Besov space* B_p $(= B_{p,p}^{1/p})$, $1 < p \leq \infty$, if

$$\|f\|_{B_p} := (\int_\Delta |f'(z)|^p (1-|z|^2)^{p-2} dA(z))^{1/p}$$

is finite. Here dA is the normalized Lebesgue's measure of Δ. Thus $f \in B_p$ if and only if the Möbius–invariant derivative $(1-|z|^2)f'(z)$ is in $L^p(\mu)$, where $d\mu(z) = dA(z)/(1-|z|^2)^2$ is the Möbius–invariant measure on Δ. The space B_∞ is known also as the *Bloch space* and B_2 is the *Dirichlet space.*

For $0 < p \leq 1$ the space B_p consists of all analytic functions f on Δ for which

$$\int_\Delta (1-|z|^2)^{pn-2} \; |f^{(n)}(z)|^p dA(z) \quad < \quad \infty$$

where $1/p < n$ is an integer (It is well known that the definition is independent of the choice of n).

For $1 < p \leq \infty$, B_p are Banach spaces modulo constant functions. B_2 is Hilbert space, with a Möbins–invariant inner–product

$$(f;g)_{B_2} = \int_\Delta f'(z)\overline{g'(z)} \; dA(z) \; .$$

With respect to this pairing one has, up to an equivalent norm,

$$B_p^* = B_q; \qquad\qquad 1 < p < \infty, \; \frac{1}{p} + \frac{1}{q} = 1.$$

The space B_1 admits the equivalent description [AFP1] as those analytic functions f on Δ which admit a representation

$$f = \sum_j \lambda_j \varphi_{a_j}; \quad a_j \in \Delta \; , \quad \sum_j |\lambda_j| < \infty \; .$$

With respect to the norm

$$\|f\|_{B_1} = \inf \left\{ \sum_j |\lambda_j|; \; f = \sum_j \lambda_j \varphi_{a_j}, \quad \lambda_j \in \mathbb{C}, \quad a_j \in \Delta \right\}$$

and the B_2–inner product we have [AFP1]

$$B_1^* = B_\infty \; , \; b_\infty^* = B_1$$

where

$$b_\infty := \{f \in B_\infty; \lim_{|z|\to 1} (1-|z|^2)(f'(z) = 0\}$$

is the so–called *little Bloch space.*

See [P2] and [BL] for more information on Besov spaces.

Theorem 2.3 *Let* $0 < p < \infty$ *and let* f *be analytic on* Δ. *Then the Hankel operator* H_f *on* $L^2(\mathbb{T})$ *belongs to* S_p *if and only if* $f \in B_p$.

This result is due mainly to V.V. Peller. In [Pel1] the theorem is proved for $1 \leq p < \infty$. A little later, but independently, the result was proved for $p=1$ in [CR] and for $1 < p < \infty$ in [R1]. Peller [Pel4] and Semmes [Se] independently proved the result for $0 < p < 1$.

Let us mention also an older result of Kronecker.

Theorem 2.4: *The Hankel operator* H_f *on* $L^2(\mathbb{T})$ *is of finite rank* n *if and only if* f *is a rational function of degree* n, *with poles outside* $\overline{\Delta}$.

The theory of Hankel operators on $L^2(\mathbb{T})$ is very rich and has many applications. See [Po], [P1] and [Ni] for general surveys and extensive literature. [Pel1], [Pel4] and [PK] contain many applications of Hankel operators to approximation theory and Gaussian processes. Peller studied some other problems related to Hankel operators on $L^2(\mathbb{T})$ and unitary ideals. In [Pel4] he characterized the Hankel operators in the Lorentz ideal S_{pq}, and in [Pel2] and [Pel3] he studied the continuity properties of the averaging projection on the set of Hankel matrices.

§3. Hankel operators on weighted Bergman spaces on the unit disk

For $\alpha > -1$ let us consider the normalized weighted area measure on the unit disk Δ

$$d\sigma_\alpha(z) = (\alpha+1)(1-|z|^2)^\alpha dA(z)$$

Let $L^{2,\alpha} = L^2(\sigma_\alpha)$ and let $A^{2,\alpha} = A^2(\sigma_\alpha)$ be the subspace of analytic functions in $L^{2,\alpha}$. The spaces $A^{2,\alpha}$ are known as *weighted Bergman spaces* (for $\alpha=0$ we obtain the usual Bergman space). The reproducing kernel (*Bergman kernel*) of $A^{2,\alpha}$ is

$$K^{\alpha}(z,w) = K_w^{\alpha}(z) = (1-z\overline{w})^{-(\alpha+2)}; \quad z,w\in\Delta.$$

The orthogonal projection (*Bergman projection*)

$$P_{\alpha}: L^{2,\alpha} \longrightarrow A^{2,\alpha}$$

is given for $g\in L^{2,\alpha}$ by

$$(P_{\alpha}g)(z) = (g,K_z^{\alpha})$$

Since $\alpha > -1$ is fixed we shall write K and P instead of K^{α} and P_{α}.

If f is an analytic function on Δ, the *Hankel operator with symbol* f *on* $L^{2,\alpha}$ is

$$H_f = (I-P)M_fP$$

(with the usual convention of considering H_f as an operator from $D(H_f) \subsetneq A^{2,\alpha}$ into $(A^{2,\alpha})^{\perp}$).

The main results of [AFP2] concerning the relationship between the sizes of H_f and that of f, are the following.

Theorem 3.1 (*a*) H_f *is bounded if and only if* $f\in B_{\infty}$;
(*b*) H_f *is compact if and only if* $f\in b_{\infty}$.

Theorem 3.2 (*a*) *For* $1 < p < \infty$, $H_f \in S_p$ *if and only if* $f\in B_p$;
(*b*) $\|H_f\|_{S_2} = \|f\|_{B_2}$

The *Macaev sequence space* M is the Banach space of all sequences $x = (x_n)_{n=1}^{\infty}$ for which

$$\|x\|_M := \sup_N \frac{\sum_{n=1}^{N} x_n^*}{\sum_{n=1}^{N} \frac{1}{n}}$$

is finite. Here $\{x_n^*\}_{n=1}^{\infty}$ is the non-increasing rearrangementy of $\{|x_n|\}_{n=1}^{\infty}$. The corresponding unitary ideal S_M is called the *Macaev ideal.* Clearly $S_1 \subsetneq S_M \subsetneq S_p$ for every $p > 1$.

Theorem 3.3: *Let* $f \in B_1$, *then* $H_f \in S_M$

Theorem 3.4: S_M *is the minimal normed unitary ideal containing a non-zero Hankel operator* H_f, f *analytic.*

The case $\alpha=0$ in Theorem 3.1 was proved earlier by S. Axler [A]. Comparing Theorem 3.1 with Theorems 2.1 and 2.2, we see that in the context of the weighted Bergman spaces (i.e. the weighted *area* measure versus the Lebesgue's measure on the *circle*) *the Bloch space* B_∞ replaces BMOA and the little Bloch space b_∞ replaces VMOA. In fact it is not hard to see that

$$\|f\|_{B_\infty} \approx \sup_{a \in \Delta} \|f \circ \varphi_a - f(a)\|_{L_{2,\alpha}}$$

i.e. the usual Bloch norm is equivalent to the "Garsia–BMOA norm" with respect to the weighted area measure σ_α.

Invariance

We begin our survey of the proof of Theorems 3.1–3.4 with a few words on invariance. The *Mobius group* G = Aut (Δ) consists of all biholomorphic automorphisms of Δ. These are known to have the form

$$\psi = \lambda \varphi_a \ , \ a \in \Delta, \ \lambda \in = \partial\Delta,$$

where

$$\varphi_a(z) = \frac{a-z}{1-\bar{a}z} \ ; \ z \in \Delta$$

A Banach space X of analytic functions on Δ is *Mobius invariant* if G operates on X by compositions as a strongly–continuous group of isometries. It is not hard to see that the spaces B_p $(1 < p < \infty)$ and VMOA are Möbius–invariant, and that B_∞ and BMOA are Möbius–invariant if the continuity of the group action is understood in the w^*–topology.

Next, we define an action of the Möbius group on $L^{2,\alpha}$ via

$$V_\psi(f) = (\psi')^{1+\alpha/2}.(f\circ\psi) \quad ; \qquad \psi \epsilon G,\ f \epsilon L^{2,\alpha}$$

We have

Proposition 3.5 (*a*) *For* $\psi \epsilon G$, V_ψ *is a unitary operator on* $L^{2,\alpha}$;

(*b*) $V_{\varphi\psi} = V_\varphi V_\psi$; $\quad \varphi, \psi \epsilon G$;

(*c*) $V_{\psi_0} = I \qquad (\psi_0(z) \equiv z)$;

(*d*) $V_\psi P = PV_\varphi$; $\quad \psi \epsilon G$,

so V_φ *restricts to a unitary operator of* $A^{2,\alpha}$ *and of* $(A^{2,\alpha})^\perp$.

Proposition 3.6 *For* $\varphi \epsilon G$ *and an analytic function* f *on* Δ *holds*

$$H_{f\circ\varphi} = V_\varphi H_f V_\varphi^{-1}$$

In particular $H_{f\circ\varphi}$ *is unitarily equivalent to* H_f.

Using these elementary propositions one obtains the following important property of the map $\mathscr{H}$ which sends an analytic function f on Δ to the corresponding Hankel operator H_f. Here a "minimal" unitary ideal is a one in which the finite rank operators are dense, and a "maximal" unitary ideal is the dual of a minimal one.

Corollary 3.7: *Let* S *be a minimal unitary ideal of operator on* $L^{2,\alpha}$. *Then* $\mathscr{H}^{-1}(S)$: $=$ {f *analytic in* Δ; $H_f \epsilon S$} *is a Mobius–invariant space. If* S *is maximal, then* $\mathscr{H}^{-1}(S)$ *is Mobius–invariant with respect to its* w^**–topology.*

The proof of Theorem 3.1

One begins with the observation that for an analytic function f on Δ and an analytic function g in the domain of H_f,

$$(H_f g)(z) = \int_\Delta (\overline{f(z)} - \overline{f(w)}) K(z,w) g(w) d\sigma_\alpha(w)$$

Thus H_f is closely related to the integral operator T_{Q_f} on $L^{2,\alpha}$ with kernel

$$Q_f(z,w) := (\overline{f(z)} - \overline{f(w)}) K(z,w)$$

If $f \epsilon B_\infty$ then for all $z,w \epsilon \Delta$

$$|f(z) - f(w)| \le \|f\|_{B_\infty} d(z,w),$$

where

$$d(z,w) = \frac{1}{2} \log \left(\frac{1 + |\varphi_z(w)|}{1 - |\varphi_z(w)|} \right)$$

is the hyperbolic distance between z and w. With $u(z) = (1-|z|^2)^{-(\frac{\alpha+1}{2})}$ we see that for all $z \epsilon \Delta$

$$\int_\Delta |Q_f(z,w)| u(w) d\sigma_\alpha(w) \le C\ u(z)$$

Schur's lemma (see [HS] or [AFP2]) now implies that T_{Q_f} is bounded; hence H_f is bounded as well.

To show that boundedness of H_f implies $f \epsilon B_\infty$ we introduce a family of seminorms

$$|\!|\!| f |\!|\!|_a = \|H_f k_a\|_{L^{2,\alpha}}, \quad a \epsilon \Delta ,$$

where

$$k_a = K_a / \|K_a\|_{L^{2,\alpha}}$$

These seminorms are Möbius–invariant.

$$|\!|\!| f \circ \varphi |\!|\!|_a = |\!|\!| f |\!|\!|_{\varphi(a)} \quad ; a \epsilon \Delta \ , \ \varphi \epsilon G$$

and clearly

$$|\!|\!| f |\!|\!|_0 = \|f - f(0)\|_{L^{2,\alpha}} \ge C|f'(0)|.$$

Thus

$$\|H_f\| \geq \sup_{a\in\Delta} |||\, f \,|||_a$$
$$= \sup_{a\in\Delta} |||\, f\circ\varphi_a \,|||_0$$
$$\geq C\cdot\sup_{a\in\Delta} |(f\circ\varphi_a)'(0)|$$
$$= C\|f\|_{B_\infty}$$

The proof of part (b) of Theorem 3.1 is elementary and uses the facts that the normalized kernel k_a tends weakly to zero in $L^{2,\alpha}$ as $|a| \to 1$ and that the polynomials are dense in b_∞.

Hankel operators and monomials

The monomials $\{z^n\}$ are the eigenfunctions of the rotation subgroup of G. They play a special rõle in the theory of Hankel operators.

Lemma 3.8 (*a*) *The functions* $\left\{H_{\bar z^m}(z^n)\right\}_{n=0}^{\infty}$ *are orthogonal in* $L^{2,\alpha}$;

(*b*) *The functions* $\left\{H_{\bar z^m}(z^n)\right\}_{m=1}^{\infty}$ *are orthogonal in* $L^{2,\alpha}$

(*c*) $(H_{\bar z^m}^{*} H_{\bar z^m})^{\frac{1}{2}} z^n = w_{m,n} z^n$; $\quad m = 1,2,\ldots; \quad n=0,1,\ldots$

where, denoting $\beta_n = \|z^n\|^2$, *we have :*

$$w_{m,n}^2 = \begin{cases} \dfrac{\beta_{n+m}}{\beta_n} - \dfrac{\beta_n}{\beta_{n-m}} & ; n \geq m \\[2ex] \dfrac{\beta_{n+m}}{\beta_n} & ; n < m \end{cases}$$

(*d*) $$w_{m,n}/(\tfrac{m}{n}) \xrightarrow[n \to \infty]{} C_m > 0$$

(*e*) $$\sum_{n=0}^{\infty} w_{m,n}^2 = m.$$

Proof of Theorem 3.3

By Lemma 3.8, $\{z^n/\|z^n\|\}_{n=0}^{\infty}$ is an eigenbasis of $(H_z^* H_z)^{\frac{1}{2}}$ with the corresponding eigenvalues $\{w_{1,n}\}$. Thus $s_n(H_z) = w_{1,n} \equiv \frac{c}{n}$. If $\varphi \epsilon G$ then H_φ is unitarily equivalent to H_z, hence $s_n(H_\varphi) = w_{1,n}$ as well. Let $f \epsilon B_1$ and write $f = \sum_j \lambda_j \varphi_{a_j}$ with $\sum_j |\lambda_j| < \infty$. Then $H_f = \sum_j \bar{\lambda}_j H_{\varphi_{a_j}}$ and so

$$\sum_{n=1}^{N} s_n(H_f) \le \sum_j |\lambda_j| \sum_{n=1}^{N} s_n(H_{\varphi_{a_j}}) = \sum_j |\lambda_j| \sum_{n=1}^{N} w_{1,n}$$

Taking infimum over all representations $f = \sum_j \lambda_j \varphi_{a_j}$ we get for all N=1,2,...

$$\sum_{n=1}^{N} s_n(H_f) \le \|f\|_{B_1} \sum_{n=1}^{N} w_{1,n} \le C\|f\|_{B_1} \log(N+1)$$

and thus

$$H_f \in S_M \,. \qquad \square$$

Remark. (a) The method of the proof of 3.3 gives in the context of $L^2(\mathbb{T})$ that if $f \epsilon B_1$ then $H_f \in S_1$ and $\|H_f\|_{S_1} \le \|f\|_{B_1}$.

(b) The converse of Theorem 3.3 is false, namely there exist analytic functions f on Δ with $H_f \in S_M$ but $f \notin B_1$, see [AFP2].

Proof of Theorem 3.2 (b). $\{ z^m/\sqrt{m} \}_{m=1}^{\infty}$ is an orthonormal basis of B_2. We have $H_{z^m} \in S_M \subset S_2$ and by Lemma 3.8

$$(H_{z^m}, H_{z^k})_{S_2} = \sum_{n=0}^{\infty} \left[H_{z^m}\left[\frac{z^n}{\|z^n\|}\right], H_{z^k}\left[\frac{z^n}{\|z^n\|}\right]\right]_{L^{2,\alpha}}$$

$$= \delta_{m,k} \sum_{n=0}^{\infty} w_{m,n}^2 = \delta_{m,k} \cdot m$$

Hence $\{H_{z^m/\sqrt{m}}\}_{m=1}^{\infty}$ is an orthonormal sequence in S_2. From this it follows by general arguments that $f \in B_2$ if and only if $H_f \in S_2$ and $\|H_f\|_{S_2} = \|f\|_{B_2}$. □

Proof of Theorem 3.4

The following elementary proposition follows from standard facts in the theory of majorization, see [MO] and [LT].

Proposition 3.9 *Let* $\{a_n\}_{n=1}^{\infty}$ *be a non-increasing sequence of positive numbers satisfying* $\lim_{n\to\infty} a_n = 0$ *and* $\sum_{n=1}^{\infty} a_n = \infty$. *Then the smallest symmetric sequence space containing* $\{a_n\}_{n=1}^{\infty}$ *in its unit ball is the space* E *of all sequences* $x=(x_n)_{n=1}^{\infty}$ *for which*

$$\|x\|_E := \sup_N \frac{\sum_{n=1}^{N} x_n^*}{\sum_{n=1}^{N} a_n}$$

is finite.

Here $x^* = \{x_n^*\}_{n=1}$ is the *non-increasing rearrangement* of $\{|x_n|\}_{n=1}^{\infty}$. The main point in the proof of Proposition 3.9 is the observation that the closed unit ball of the smallest symmetric sequence space in question should be the absolute convex hull of the set of those sequences x for which $x^* = (a_n)$. It follows from Proposition 3.9 that the Macaev sequence space M, defined before the formulaton of Theorem 3.3, is the minimal symmetric space containing the sequence $(\frac{1}{n})$ in its unit ball.

Suppose now that E is a symmetric sequence space so that S_E contains a non-zero Hankel operator H_{f_0}, f_0 a non-constant analytic function on Δ. We can assume that S_E is maximal. Let $X = \mathscr{H}^{-1}(S_E) = \{f; H_f \in S_E\}$. X is a Möbius–

invariant space by Corollary 2.7. Using [AFP1] and the fact that contains a non–constant function, one gets that the identity function $\varphi_0(z) = z$ also belongs to X. That is $H_{\varphi_0} \in S_E$, and so $(s_n(H_{\varphi_0})) \in E$. But $s_n(H_{\varphi_0}) = w_{1,n} \equiv \frac{c}{n}$. Since M is the minimal symmetric sequence space containing $(\frac{1}{n})$ in its unit ball, one can conclude that $M \subseteq E$ and that $\|x\|_E \leq C \|x\|_M$ for all $x \in M$. This completes the proof of Theorem 3.4.

Interpolation, the easy cases

It remains to prove Theoprem 3.2(a), that is that for $1 < p < \infty$ an analytic function f on Δ is in B_p if and only if $H_f \in S_p$. This is proved by interpolation. Recall first that with respect to *complex interpolation* (see [BL]

$$\left.\begin{array}{l}(B_{p_0}, B_{p_1})_\theta = B_p \\ (S_{p_0}, S_{p_1})_\theta = S_p\end{array}\right\} \qquad \begin{array}{l}0 < \theta < 1 \\ 0 < p_0 < p_1 \leq \infty \\ \dfrac{1}{p} = \dfrac{1-\theta}{p_0} + \dfrac{\theta}{p_1}\end{array}$$

and

$$\left.\begin{array}{l}(B_{p_0}, b_\infty)_\theta = B_p \\ (S_{p_0}, B(H))_\theta = S_p \\ (S_{p_0}, S_\infty)_\theta = S_p\end{array}\right\} \qquad \begin{array}{l}0 < \theta < 1 \\ 0 < p_0 < \infty \\ \dfrac{1}{p} = \dfrac{1-\theta}{p_0}\end{array}$$

Let $\mathscr{H}$ be the (conjugate linear) oeprator from analytic functions on Δ to operators on $L^{2,\alpha}$ defined via $\mathscr{H}(f) = H_f$. Using the end–point results

$$\mathscr{H}: b_\infty \to S_\infty \qquad \text{(Theorem 3.1(b))}$$
$$\mathscr{H}: B_2 \to S_2 \qquad \text{(Theorem 3.2(b))}$$

we get by interpolation [BL] that

$$\mathcal{H}: B_p \to S_p \quad , 2 < p < \infty$$

as a bounded operator. We remark that we cannot use the fact that $\mathcal{H}\ B_1 \to S_\Omega$ to conclude $\mathcal{H}: B_p \to S_p$ for $1 < p < 2$, see [AFP2] (In the context of $L^2(\mathbb{T})$ and Hardy space the right result is $\mathcal{H}: B_1 \to S_1$, and so one gets $\mathcal{H}: B_p \to S_p$, $1 < p < \infty$, by interpolation).

Polarizing Theorem 3.2(b), we get

$$(H_f , H_g)_{S_2} = (g,f)_{B_2}$$

It follows by standard duality arguments that

$$H_f \in S_p \Rightarrow f \in B_p, \ 1 < p < 2.$$

Next, we need a general ___trace formula___ for operators $T:A^{2,\alpha} \to A^{2,\alpha}$ in S_1:

$$\mathrm{trace}(T) = \int_\Delta (TK_a, K_a)d\sigma_\alpha(a)$$

The proof is straightforward, and is valid in more general Hilbert spaces of analytic functions with reproducing kernel.

We introduce an *auxiliary space* Y_p, $1 < p < \infty$, consisting of all analytic functions f on Δ for which

$$\|f\|_{Y_p} := \left(\int_\Delta |||\, f\, |||_a^p \, d\mu(a)\right)^{1/p}$$

is finite. Here $|||\, f\, |||_a = \|H_f k_a\|_{A^{2,\alpha}}$, $k_a = K_a/\|K_a\|_{A^{2,\alpha}}$ and $d\mu(a) = dA(a)/(1-|a|^2)^2$ is the Möbius–invariant measure on Δ. From the definition it is clear that Y_p is a Möbius–invariant space.

As we observed in the proof of Theorem 3.1,

$$|||\, f\, |||_a \geq C|(f\circ\varphi_a)'(0)| = C(1-|a|^2)|f'(a)|$$

Thus for $1 < p < \infty$,

$$Y_p \subseteq B_p \quad \text{and} \quad C\|f\|_{B_p} \leq \|f\|_{Y_p}$$

If T is a positive operator on a Hilbert space H and $x \in H$ is normalized, then ("<u>Hölder inequality</u>")

$$(Tx,x)^\beta \le (T^\beta x,x)\ , \qquad 1 \le \beta < \infty\ ;$$
$$(Tx,x)^\beta \ge (T^\beta x,x)\ , \qquad 0 < \beta \le 1\ .$$

Combining this with the trace formula we get:

$$f \epsilon Y_p \Rightarrow H_f \epsilon S_p\ ,\ 1 < p < 2$$
$$f \epsilon Y_p \Leftarrow H_f \epsilon S_p\ ,\ 2 < p < \infty.$$

Indeed, assuming $1 < p < 2$ and $f \epsilon Y_p$, then

$$\begin{aligned}\|H_f\|_{S_p}^p &= \text{trace } (H_f^* H_f)^{p/2} \\ &= \int_\Delta ((H_f^* H_f)^{p/2} K_a, K_a) d\sigma_\alpha(a) \\ &= \int_\Delta ((H_f^* H_f)^{p/2} k_a, k_a) \|K_a\|^2 d\sigma_\alpha(a) \\ &\le \int_\Delta ((H_f^* H_f) k_a, k_a)^{p/2} \|K_a\|^2 d\sigma_\alpha(a), \qquad \text{by Hölder inequality} \\ &= (\alpha+1) \int_\Delta |||\ f\ |||_a^p\ d\mu(a) = (\alpha+1) \|f\|_{Y_p}^p\ .\end{aligned}$$

The proof for $2 < p < \infty$ is similar.

It follows that

$$Y_p = B_p = \mathscr{H}^{-1}(S_p)\ ,\ 2 < p < \infty$$

and

$$Y_p \subseteq \mathscr{H}^{-1}(S_p) \subseteq B_p,\ \ 1 < p < \infty$$

To conclude the proof of Theorem 3.2(a) it remains to prove that for $1 < p < 2$ $B_p \subseteq \mathscr{H}^{-1}(S_p)$, and it is certainly enough to show that $B_p \subseteq Y_p$.

Interpolation, the more difficult cases

We begin with $0 \le \alpha < \infty$. We transform the unit disk Δ onto the upper plane Π via the Cayley transform

$$w = \frac{1}{i} \cdot \frac{z+1}{z-1}\ ;\ z \in \Delta\ ,\ w \in \Pi\ .$$

The space Y_p gets transformed onto the space Z_p consisting of all analytic functions

g on Π so that

$$\|g\|_{Z_p}^p := \int_\Pi \left[\int_\Pi \left[\frac{|g(z)-g(w)|}{|z-\bar{w}|^{2\beta}} y^\beta v^\beta\right]^2 \frac{dxdy}{y^2}\right]^{p/2} \frac{dudv}{v^2}$$

is finite, where $\beta = 1 + \alpha/2$ and $z = x + iy$, $w = u + iv \in \Pi$. The space B_p (over Δ) gets transformed into the space B_p (over Π); see [BL] for the definition of B_p; and more generally $B^s_{p,q}$ over Π. Introducing a parameter ε in the definition of Z_p, we get a whole family of spaces $Z_{p,\epsilon}$ of analytic functions on Π, defined via the finiteness of

$$\|g\|_{Z_{p,\epsilon}}^p := \int_\Pi \left[\int_\Pi \left[\frac{|g(z)-g(w)|}{|z-\bar{w}|^{2\beta}} y^\beta v^\beta\right]^2 \frac{dxdy}{y^2}\right]^{p/2} \frac{v^{p\epsilon}dudv}{v^2}$$

The key to the rest of the proof are

Lemma 3.10 For $0 < \epsilon < \frac{1}{2}$, $B^{1-\epsilon}_{1,1} \subseteq Z_{1,\epsilon}$

Lemma 3.11 For $a \geq 0$, and $-\frac{1}{2} < \epsilon < \frac{1}{2}$, $B^{\frac{1}{2}-\epsilon}_{2,2} \subseteq Z_{2,\epsilon}$.

See [AFP2] section 7 for the proofs.

Having the lemmas, we introduce an operator

$$(Tf)(z,w) = v^\beta y^\beta \frac{f(z) - f(w)}{|z-\bar{w}|^{2\beta}}$$

$z = x + iy$, $w = u + iv \in \Pi$. Let $H = L^2\left(\Pi, \frac{dxdy}{y^2}\right)$. Then for $0 < \epsilon_0 < 1/2$ we have by Lemma 3.10

$$T: B_1^{1-\epsilon_0,1} \to L^1\left(\Pi, \frac{v^{\epsilon_0}dudv}{v^2}, H\right)$$

Also by Lemma 3.11 we get for $-1/2 < \epsilon_1 < 1/2$,

$$T: B_2^{1/2-\epsilon_1,2} \to L^2\left(\Pi, \frac{v^{2\epsilon_1}dudv}{v^2}, H\right).$$

It follows by real interpolation (see [BL]) that for $\theta \in (0,1)$, $p \in (1,2)$ which satisfy $\frac{1}{p} = \frac{1-\theta}{1} + \frac{\theta}{2}$, and for $\epsilon = (1-\theta)\epsilon_0 + \theta\epsilon$, T maps the space

$$\left[B_{1,1}^{1-\epsilon_0}, B_{2,2}^{\frac{1}{2}-\epsilon_1}\right]_{\theta,p} = B_{p,p}^{1/p-\epsilon}$$

into

$$\left[L^1\left[v^{\epsilon_0}\frac{dudv}{v^2}, H\right], L^2\left[v^{2\epsilon_1}\frac{dudv}{v^2}, H\right]\right]_{\theta,p} = L^p\left[v^{p\epsilon}\frac{dudv}{v^2}, H\right]$$

Appropriate choices of ϵ_0 and ϵ_1 produce all ϵ in the interval (1/p–1, 1/2). For $\epsilon=0$, we clearly get that $B_{p,p}^{1/p} = B_p \subseteq Z_p$ (in Π), hence $B_p \subseteq Y_p$ (in Δ).

The case $-1 < \alpha < 0$ is more complicated since for $1 < p < 2$ the space Y_p consists of constant functions only. To overcome the difficulty we need to "sharpen the singularity" of the kernel K_a, and to modify our spaces appropriately.

We define for a, $z\epsilon\Delta$

$$F_a(z) = F(z,a) = \frac{\partial}{\partial\bar{a}} K_a(z) = (\alpha+2)z/(1-\bar{a}z)^{\alpha+3}$$

This kernel reproduces derivatives

$$g'(a) = (g,F_a) \; ; \; g\epsilon A^{2,\alpha}, \; a\epsilon\Delta$$

and behaves nicely with respect to Hankel operators

$$(H_g F_a)(z) = (\overline{g(z)} - \overline{g(a)})F_a(z) - \overline{g'(a)}K_a(z)$$

For positive operator T on $A^{2,\alpha}$ we have the "modified trace formula"

$$\text{trace}\,(T) \cong \int_\Delta (T\, f_a,f_a)\, d\mu(a)$$

where $f_a = F_a\|F_a\|^{-1}_{A^{2,\alpha}}$, $d\mu(a) = (1-|a|^2)^{-2}dA(a)$,

and $\cong$ means "equivalence".

We define W_p to be the space of all analytic functions g on Δ for which

$$\|g\|_{W_p} := \left[\int_\Delta \|H_g\, f_a\|^p_{L^{2,\alpha}}\, d\mu(a)\right]^{1/p}$$

is finite.

Using the "modified trace formula" and "Hölder inequality" as above in the case $0 \leq \alpha$, $1 < p < 2$, we get

$$g\epsilon W_p \Rightarrow H_g \in S_p\,,$$

and so

$$W_p \subsetneq \mathscr{H}^1(S_p) \subsetneq B_p .$$

To prove the containment $B_p \subsetneq W_p$ (which is all we need to conclude the proof of Theorem 3.2 (a) one observes first that

$$g \epsilon W_p \iff \left[\int_\Delta \|(g-g(a))f_a\|^p_{L^{2,\alpha}} d\mu(a) \right]^{1/p} < \infty$$

From this the proof proceeds as in the case $0 \leq \alpha$ by passing to the upper half plane and introducing an auxiliary parameter ϵ, which permits us to interpolate with respect to the "smoothness parameter" and to get the desired conclusion that (back in Δ) $B_p \subsetneq W_p$.

The proof of Theorem 3.3(a) is over. Our proof yields the following characterization of the spaces B_p.

<u>*Corollary 3.12*</u> *Let* $1 < p \leq \infty$, $0 \leq \alpha$, *and let* f *be an analytic function on* Δ. *Let* $d\mu(z) = dA(z)/(1-|z|^2)^2$. *The following are equivalent:*

(*i*) $f \epsilon B_p$, *i.e.* $(1-|z|^2)f'(z) \in L^p(\mu)$

$$(ii) \quad \int_\Delta \left[\int_\Delta |f(z) - f(w)|^2 J(z,w) d\mu(z) \right]^{p/2} d\mu(w) < \infty$$

where $J(z,w) = ((1-|z|^2)(1-|w|^2)|1-z\bar{w}|^{-2})^{\alpha+2}$.

Moreover, for $1 < p < 2$ *and* $-1 < \alpha < 0$, (*i*) *is equivalent also to*

$$(iii) \quad \int_\Delta \left[\int_\Delta |f(z) - f(w)|^2 \mathfrak{J}(z,w) d\mu(z) \right]^{p/2} d\mu(w) < \infty$$

where $\mathfrak{J}(z,w) = (1-|z|^2)^{\alpha+2}(1-|w|^2)^{\alpha+4}|1-z\bar{w}|^{-2(\alpha+3)}$

Another easy consequence deals with the canonical projection onto the space of

Hankel operators. The operator $\mathcal{H}(f) = H_f$ is a conjugate–linear isometry of B_2 into S_2. Therefore $\mathcal{H}^* \mathcal{H} = I$ and $\mathcal{P} = \mathcal{H}\mathcal{H}^*$ is the orthogonal projection in S_2 onto the subspace of Hankel operators. Our results on $\mathcal{H}$ imply that $\mathcal{H}^*$ is bounded from S_p onto B_p, $1 \leq p < \infty$, and from the "small Macaev ideal" S_w into b_∞, where

$$S_w = \{T \in S_\infty,\ \|T\|_{S_w} = \sum_{n=1}^{\infty} s_n(T)/n < \infty\}.$$

These considerations give the following result.

<u>*Corollary 3.13*</u> *(a) The orthogonal projection* $\mathcal{P}$ *is bounded in* S_p, $1 < p < \infty$;
(b) $\mathcal{P}$ *is bounded from* S_1 *into the Macaev ideal* S_M;
(c) $\mathcal{P}$ *is bounded from the small Macaev ideal* S_w *into* S_∞.
(c) $\mathcal{P}$ *is unbounded on* S_∞ *or* $B(L^{2,\alpha})$.

We remark that B_∞ is injective (being isomorphic to ℓ_∞) and b_∞ is separably injective (being isomorphic to c_0). Therefore there exists a bounded projection from $B(L^{2,\alpha})$ onto $\mathcal{H}(B_\infty)$ and from $S_\infty(L^{2,\alpha})$ onto $\mathcal{H}(b_\infty)$.

We close with a word on the so-called "*reduced Hankel operator*"

$$\tilde{H}_f = \bar{P} M_f P$$

where f is analytic on Δ and $\bar{P}$ is the orthogonal projection in $L^{2,\alpha}$ onto the subspace of the conjugate–analytic functions. One observes that the matrix of $\tilde{H}_f$ with respect to the bases $\{ z^n / \sqrt{\beta_n} \}_{n=0}^{\infty}$ in $A^{2,\alpha}$ and $\{\bar{z}/\sqrt{\beta_k}\}_{k=0}^{\infty}$ in $\overline{A^{2,\alpha}}$ ($\beta_n = \|z^n\|^2$) is

$$(\tilde{H}_f)_{k,n} = \frac{\beta_{n+k}\overline{\hat{f}(n+k)}}{\sqrt{\beta_n \cdot \beta_k}}$$

Thus the study of the operators $\tilde{H}_f$ reduces to that of a variant of Hankel matrices.

Thus one obtains without much efforts the following results

Theorem 3.14 (a) $\tilde{H}_f$ *is bounded if and only if* $f \in B_\infty$;

(b) $\tilde{H}_f$ *is compact if and only if* $f \in b_\infty$;

(c) *For* $1 \leq p \leq \infty$, $H_f \in S_p$ *if and only if* $f \in B_p$.

Parts (a) and (b) for $\alpha=0$ are due to Coifman, Rochberg and Weiss [CRW]. The S_p result for all p and all α is contained (via unitary equivalence) in the family of operators considered by Peller [Pel5], Rochberg [R1] and Semmes [Se]. See also [BO].

§4. *Hankel operators on regular planar domains*

This section is devoted to a survey of the work [AFP3] on Hankel operators on regular planar domains. A domain Ω in the complex plane is called *regular* if its boundary $\Gamma = \partial\Omega$, considered as part of the Riemann sphere $\mathbb{C}^* = \mathbb{C} \cup \{\infty\}$, consists of finitely many simple closed analytic curves $\Gamma_0, \Gamma_1, \ldots, \Gamma_m$. The number m is the *connectivity* of Ω (= number of "holes" in Ω). The Γ_j are positively oriented with respect to Ω. We denote Ω_j by the interior of Ω, j=0,1,...,m, thus $\Omega = \bigcap_{j=0}^{m} \Omega_j$.

It will be explained below that our theory is conformally invariant. We shall therefore assume without less of generality that our regular domain Ω is so that $\Gamma = \partial\Omega \subseteq \mathbb{C}$, and that Ω_0 *is bounded* (thus $\infty \in \Omega_j$ for j=1,2,...,m, and Ω is bounded as well). Also, it is well known that each regular domain is conformally equivalent to a *circle domain*, i.e., a regular domain whose boundary components are circles. From time to time we shall impose this further restriction, and then take Ω_0 to be the unit disk and $\Gamma_0 = \partial\Omega_0$ the unit circle. The book [F] is a general reference to function theory on planar domains.

Let $dA(z) = \frac{1}{\pi}\, dxdy$ be the normalized area measure on Ω, and let $L^2(\Omega) =$

$L^2(\Omega,dA)$. Clearly the <u>*Bergman space*</u>

$$A^2(\Omega) = \{f \in L^2(\Omega);\ f \text{ is analytic in } \Omega\}$$

is a closed subspace, and point evaluations are continuous linear functionals. If $\{K_z^\Omega\}$ is the reproducing kernel of $A^2(\Omega)$ then the orthogonal projection P^Ω: $L^2(\Omega) \to A^2(\Omega)$ is given by $(P^\Omega f)(z) = (f,K_z^\Omega)$, $f \in L^2(\Omega)$. K^Ω and P^Ω are called the *Bergman kernel* and the *Bergman projection*, respectively. If Ω is understood, we shall omit "Ω" from the notation and write L^2, A^2 K and P for simplicity.

The <u>*Bergman metric*</u> is the function $\lambda = \lambda^\Omega$: $\Omega \to (0,\infty)$ defined by

$$\lambda(z) = \|K_z^\Omega\| = K^\Omega(z,z)^{\frac{1}{2}}; \qquad z \in \Omega.$$

The <u>*invariant measure*</u> on Ω is the measure

$$d\Sigma(z) = \lambda(z)^2 dA(z)$$

<u>*Definition 4.1*</u> *(i) for* $1 < p \le \infty$, *the space* $B_p = B_p(\Omega)$ *consists of all analytic functions* f *on* Ω *for which* $f'/\lambda \in L^p(\Omega,d\Sigma)$, *with the seminorm* $\|f\|_{B_p} = \|f'/\lambda\|_{L^p(d\Sigma)}$.

(ii) $b_\infty = b_\infty(\Omega)$ *is the subspace of* B_∞ *consisting of those functions* f *for which* $\frac{|f'(z)|}{\lambda(z)} \to 0$ *as* $d(z) \to 0$, *where* $d(z) = \inf_{w \in \Gamma} |z-w|$.

(iii) $B_1 = B_1(\Omega)$ *is the space of all analytic functions on* Ω *for which*

$$\|f\|_{B_1} = \int_\Omega |f''(z)|\,dA(z)$$

is finite.

The spaces B_p are the *Besov-p spaces.* B_∞ is the *Bloch space* and b_∞ is the *little Bloch space.* B_2 is the *Dirichlet space.*

<u>*Remark*</u>. It is a standard fact on Bergman kernels that

$$\lambda(z) \sim \frac{1}{d(z)}$$

where $d(z) = \text{dist}(z,\partial\Omega)$ and "$\sim$" means equivalence (so $C_1 \le \lambda(z)d(z) \le C_2$ for all

$z \in \Omega$). Also, $\lambda(z)$ is equivalent to the so-called Poincaré metric of Ω, see [AFP3]. The equivalence $\lambda(z) \sim \frac{1}{d(z)}$ provides a better understanding of the condition of membership in B_p. It also allows us to prove that for $1 < p \leq \infty$ and analytic function f on Ω,

$$f'/\lambda \in L^p(d\Sigma) <=> f''/\lambda^2 \in L^p(d\Sigma).$$

This provides an equivalent definition of B_p, valid for $1 \leq p \leq \infty$.

Let f be an analytic function on Ω. The associated *Hankel operator* with symbol f is $H_f = (I-P)M_fP$, considered as an operator on either $L^2(\Omega)$ or $A^2(\Omega)$. The main results of [AFP3] are the following.

Theorem 4.2 (*a*) H_f *is bounded if and only if* $f \in B_\infty$;
(*b*) H_f *is compact if and only if* $f \in b_\infty$.

Theorem 4.3 (*a*) *For* $1 < p < \infty$, $H_f \in S_p$ *if and only if* $f \in B_p$;
(*b*) *For* $f \in B_2$, $\|H_f\|_{S_2} = \|f\|_{B_2}$.

Theorem 4.4 *If* $f \in B_1$ *then* H_f *is in the Macaev ideal* S_M, *i.e.* $\sum_{n=1}^{N} s_n(H_f) = O(\log N)$.

Invariance

In general, the domain Ω need not have any non-trivial conformal (i.e., one-to-one, biholomorphic), self-map, and so the group Aut(Ω) is of very little help. Instead, we study invariance under conformal maps from one domain onto another.

Proposition 4.5: *Let* Ω *be a regular domain and let* φ *be a conformal map from* Ω *onto* D.

(*a*) *The operator* $V_\varphi f = \varphi' \cdot f\circ\varphi$ *is an isometry from* $L^2(D)$ *onto* $L^2(\Omega)$;

(*b*) $V_\varphi P^D = P^\Omega V_\varphi$;

(*c*) *If* $\psi: D \to W$ *is conformal onto, then* $V_{\psi\varphi} = V_\varphi V_\psi$. *In particular* $V_\varphi^{-1} = V_{\varphi^{-1}}$.

(*d*) *If* f *is analytic on* D *then*

$$H_{f\circ\varphi} = V_\varphi H_f V_\varphi^{-1}$$

(*e*) *The Bergman kernels of* Ω *and* $D = \varphi(\Omega)$ *are related via*

$$K^D(\varphi(z),\varphi(w))\ \varphi'(z)\overline{\varphi'(w)} = K^\Omega(z,w)$$

(*f*) *The Bergman metrics of* Ω *and* D *are related via* $\lambda^D(\varphi(z))|\varphi'(z)| = \lambda^\Omega(z)$

(*g*) *The map* $C_\varphi(f) = f\circ\varphi$ *is an isometry of* $B_p(D)$ *onto* $B_p(\Omega)$, $1 < p \leq \infty$, *and of* $b_\infty(D)$ *onto* $b_\infty(\Omega)$.

__*Localization near the boundary*__

The main idea in our study of the Hankel operators on regular domains is to "*localize near the boundary*" and then to *use the results of the simply connected case*, namely the case of the unit disk Δ. To explain this consider the __*Cauchy projections*__ C_j, j=0,1,...,m, defined via

$$(C_j(f))(z) = \frac{1}{2\pi i}\int_{\gamma_j} \frac{f(\xi)}{\xi - z}\, d\xi$$

where γ_j is a simple, closed analytic curve in Ω having the same orientation as Γ_j, separating z from Γ_j and containing all the Γ_k with $k\neq j$ in its interior. Clearly, the definition does not depend on the choice of γ_j.

__*Proposition 4.6*__: (*a*) *The operators* C_j *are bounded projections on* $A^2(\Omega)$, $C_jC_k = 0$ *for* $j\neq k$ *and* $\sum_{j=0}^{m} C_j = I$.

(*b*) $C_0(A^2(\Omega)) = A^2(\Omega_0)$, *and for* j=1,2,...,m, $C_j(A^2(\Omega))$ *is the closed subspace of those* f *in* $A^2(\Omega)$ *which extend analytically to* Ω_j *and satisfy* $f(\infty) = 0$.

(*c*) *For* $1 \leq p \leq \infty$, $f\in B_p(\Omega)$ *if and only if* $C_jf\in B_p(\Omega_j)$ *for* j=0,1,...,m.

(*d*) $f\in b_\infty(\Omega)$ *if and only if* $C_jf \in b_\infty(\Omega_j)$ *for* j = 0,1,2,...,m.

Notice that for $f\epsilon A^2(\Omega)$, hence for $f\epsilon B_p(\Omega)$ for some $1 \leq p \leq \infty$, $f - C_j f$ extends analytically across Γ_j, and is therefore bounded in a neighborhood of Γ_j. Thus the rate of increase of $f(z)$ and $C_j f(z)$ as z approaches Γ_j is the same, and so $C_j f$ can be considered as the localization of f near Γ_j.

Next, some heuristic arguments indicate that *the behavior of* $K^\Omega(z,w)$ *for* $z\epsilon\Omega$ *and* w *near* Γ_j *should be very similar to that of* $K^{\Omega_j}(z,w)$. This can be made precise in the following way. By regularity, the Dirichlet problem is solvable in Ω, see [F]. Let $G(z,w) = G^\Omega(z,w)$ be the *Green function* of Ω. Thus we can write

$$G(z,w) = \Phi(z,w) - \log|z-w|$$

where $\Phi(z,w)$ is harmonic in each variable separately in a neighborhood of $\overline{\Omega} = \Omega \cup \Gamma$. Also, $G(z,w) = G(w,z)$, $\Phi(z,w) = \Phi(w,z)$ and

$$G(z,w) = 0, \quad \Phi(z,w) = \log|z-w|; \; z\epsilon\Gamma, \; w\epsilon\Omega.$$

The Green function is related to the Bergman kernel via

$$K(z,w) = -2 \frac{\partial^2 G}{\partial z \, \partial \overline{w}} (z,w),$$

see [B] (in [B] the constant $-\frac{2}{\pi}$ replaces our constant -2 in the last formula; this is due to different normalizations of the area measures). This enables us to prove the following result.

Lemma 4.7. There are neighborhoods D_j *of* Γ_j, $j=0,1,\ldots,m$ *and a constant* C *so that*

(*a*) $D_j \cap D_k = \phi$ *for* $j\neq k$;

(*b*) $|K^\Omega(z,w) - K^{\Omega_j}(z,w)| \leq C$ *for every* $z\epsilon\Omega$ *and* $w\epsilon D_j$.

The main point in the proof is that the logarithmic singularity for $z=w$ near Γ_j cancels in the difference

$$H(z,w) = G^{\Omega}(z,w) - G^{\Omega_j}(z,w) .$$

Also, $H(z,w) = 0$ for $z \in \Omega$ and $w \in \Gamma_j$, and thus $H(z,\cdot)$ can be extended harmonically by reflection to a neighbhorhood of $\overline{\Omega} = \Omega \cup \Gamma$.

Some easy proofs

Observe first that for analytic function f on Ω, and analytic function g in the domain of definition of H_f,

$$(H_f g)(z) = \int_{\Omega} \overline{(f(z) - f(w))}\, K(z,w) g(w) dA(w);$$

thus H_f is closely related to the integral operator T_{Q_f} with kernel

$$Q_f(z,w) = \overline{(f(z) - f(w))}\, K(z,w).$$

Also, it is elementary that

$$(H_f K_w)(z) = (\overline{f(z)} - \overline{f(w)}) K_w(z) = Q_f(z,w).$$

Next, consider the *normalized kernel*

$$k_w = K_w / \|K_w\| = K_w / \lambda(w)$$

for $w \in \Omega$.

Proposition 4.8. There is a positive constant C, *depending only on* Ω, *so that if* f *is an analytic function on* Ω *and* $w \in \Omega$ *then*

$$\|H_f k_w\| \geq C |f'(w)| / \lambda(w)$$

This follows easily by estimating from below the integral over Ω by the integral over the disk with center w and radius $d(w) = \mathrm{dist}(z, \partial\Omega)$.

Definition 4.9 For $1 < p \leq \infty$, $Y_p = Y_p(\Omega)$ is the space of all analytic functions f on Ω for which

$$\|f\|_{Y_p} := \left(\int_{\Omega} \|H_f k_w\|^p d\Sigma(w)\right)^{1/p}$$

is finite, where $d\Sigma(w) = \lambda(w)^2 dA(w)$ is the invariant measure.

Clearly,

$$\|f\|_{Y_p}^p = \int_\Omega \left(\int_\Omega |f(z) - f(w)|^2 |K(z,w)|^2 dA(w)\right)^{p/2} \lambda(w)^{2-p} dA(w)$$

and Y_p is a Banach space modulo constant function. Moreover, if φ is a conformal map of Ω onto D then $C_\varphi(f) = f\circ\varphi$ is an isometry of $Y_p(D)$ onto $Y_p(\Omega)$. The auxiliary space $Y_p(\Omega)$ plays the same role played by Y_p in section 3. Our proofs show that for $1 < p \leq \infty$

$$Y_p(\Omega) = B_p(\Omega)$$

This provides an equivalent definition of the Besov spaces $Bp(\Omega)$ in terms of double integrals.

<u>*Corollary 4.10.*</u> (*a*) $Y_p \subseteq B_p$, $1 < p \leq \infty$;

(*b*) $H_f \in S_p \Rightarrow f \in Y_p$, $2 \leq p < \infty$;

(*c*) $f \in Y_p \Rightarrow H_f \in S_p$, $1 < p \leq 2$;

(*d*) *If* H_f *is bounded (or, compact) then* $f \in B_\infty$ *(respectively,* $f \in b_\infty$*).*

Part (a) follows from Proposition 4.8 and the definitions of B_p and Y_p. (d) is obvious, since $k_w \to 0$ weakly as w approaches $\partial\Omega$. Parts (b) and (c) follow as in section 3 from the trace formula

$$\text{trace}(T) = \int_\Omega (TK_w, K_w) dA(w)$$

for positive operators on A^2, and the "Hölder inequalities".

<u>*Sketch of the proof of Theorem 4.9(b)*</u>

Let f be an analytic function on Ω. By the trace formula,

$$\|H_f\|_{S_2}^2 = \int_\Omega \int_\Omega |f(z)-f(w)|^2 \, |K(z,w)|^2 \, dA(z) \, dA(w)$$

and we have to prove that this quantity equals

$$\|f\|_{B_2}^2 = \int_\Omega |f'(z)|^2 dA(z).$$

Using the properties of the Green's function ($G(z,w) = \Phi(z,w) - \log|z-w|$, the formula $K = -2\dfrac{\partial^2 G}{\partial z\,\partial\overline{w}}$ and the complex form of Green's theorem

$$\int_\Omega u\,\frac{\partial v}{\partial \overline{z}}\,dxdy = -\frac{1}{2i}\int_{\partial\Omega} uv\,d\overline{z} - \int_\Omega \frac{\partial u}{\partial \overline{z}}\,v\,dxdz\ ,$$

we get for every $w \in \Omega$

$$\int_\Omega |f(z) - f(w)|^2\ |K(z,w)|^2 dA(z) = \int_\Omega f'(z)\left(\frac{\overline{f(z)}-\overline{f(w)}}{z-w}\right)K(z,w)\ dA(z)$$

$$+\ 2\int_\Omega f'(z)\ (\overline{f(z)} - \overline{f(w)})\ K(w,z)\ \frac{\partial\Phi}{\delta\overline{w}}\ (z,w)dA(z)$$

Using the fact that $\dfrac{\partial\Phi}{\partial\overline{w}}$ (z,w) is analytic in $\overline{w}$, we get by a careful application of Fubini's theorem and the reproducing property of the Bergman kernel

$$\int_\Omega\int_\Omega |f(z)-f(w)|^2\ |K(z,w)|^2\ dA(z)\ dA(w) = \int_\Omega f'(z)\ \overline{f'(z)}\ dA(z) + 0 =$$

$$= \int_\Omega |f'(z)|^2 dA(z)$$

<u>Remark</u>: Theorems 3.2(b) and 4.3(b) can be generalized to Hankel operators in the context of quite general planar domains and measures: the Hilbert Schmidt norm of the Hankel operator is always equal to the Dirichlet norm of the symbol, see [AFJP].

<u>More information on the Besov spaces B_p</u>

The following information concerning the spaces $B_p = B_p(\Omega)$ is needed in the proofs. Here $d\Sigma(z) = \lambda^2(z)dA(z)$ is the invariant area measure.

<u>*Lemma 4.11*</u> *(a) The map* $J(f) = f'/\lambda$ *is an isometry of* B_p *into* $L^p(d\Sigma)$, $1 < p \leq \infty$, *and of* b_∞ *into* $C_0(\Omega)$;

(b) The orthogonal projection Q *from* $L^2(d\Sigma)$ *onto* JB_2 *extends to a bounded projection from* $L^p(d\Sigma)$ *onto* JB_p, $1 < p < \infty$;

(c) With respect to the pairing from B_2

$$(f,g) = \int_\Omega f'(z)\overline{g'(z)}dA(z)$$

we have $B_p^* = B_q$, $1 < p < \infty$, $\frac{1}{p} + \frac{1}{q} = 1$;

(d) The spaces B_p, $1 \leq p \leq \infty$, *form an interpolation scale with respect to complex interpolation:*

$$(B_{p_0}, B_{p_1})_\theta = B_p$$

where $0 < \theta < 1$ *and* $\frac{1}{p} = \frac{1-\theta}{p_0} + \frac{\theta}{p_1}$.

Part (a) is obvious and (b) implies (c). The proof of (d) uses (b) and the alternative description of B_p, $1 \leq p \leq \infty$: $f \epsilon Bp \iff f''/\lambda^2 \in L^p(d\Sigma)$. The main point in the proof of (b) is that for each $\beta \epsilon (0,1)$ and $z \epsilon \Omega$

$$\int_\Omega |K(z,w)|\lambda(w)^\beta dA(w) \leq C_\beta\, \lambda(z)^\beta$$

<u>*The smallness of the off-diagonal parts*</u>

Consider the Hankel operator H_f. Write $f = \sum_{j=0}^m f_j$ where $f_j = C_j f$ are the Cauchy projections of f, $0 \leq j \leq m$. Then

$$(\overline{f(z)} - \overline{f(w)})K^\Omega(z,w) =$$

$$= \sum_{j=0}^m (\overline{f_j(z)} - \overline{f_j(w)})K^{\Omega_j}(z,w)$$

$$+ \sum_{j=0}^m (\overline{f_j(z)} - \overline{f_j(w)})(K^\Omega(z,w) - K^{\Omega_j}(z,w))$$

The terms $(\overline{f_j(z)} - \overline{f_j(w)})K^{\Omega_j}(z,w)$ are the "diagonal terms" and they are closely related to the Hankel operators $H_{f_j}^{\Omega_j}$ (on $A^2(\Omega_j)$). The terms $(\overline{f_j(z)} - \overline{f_j(w)})(K^{\Omega_j}(z,w) - K^{\Omega_j}(z,w))$ are the "off-diagonal terms", and they are "perturbations" of the diagonal terms as the following result shows.

<u>*Lemma 4.12*</u> *Let* Ω *be a regular domain, let* $f \epsilon B_\infty(\Omega_0)$ *and put*

$$Q(z,w) = (\overline{f(z)} - \overline{f(w)})(K^{\Omega}(z,w) - K^{\Omega_0}(z,w))$$

$z,w \epsilon \Omega$. *Let* T *be the integral operator on* $A^2(\Omega)$ *with kernel* Q:

$$Tg(z) = \int_\Omega Q(z,w)g(w)dA(w)$$

Then

(*i*) $\|TK_z^\Omega\| \leq C|\log \lambda^\Omega(z)|$; $z\epsilon\Omega$

(*ii*) $T\epsilon S_p$ *for every* $p > 1$.

The main tool in the proof is Lemma 4.7, see [AFP3] for details.

<u>*Corollaries and the rest of the proofs of Theorems 4.2. and 4.3*</u>

<u>*Corollary 4.13*</u> *For* $1 < p < \infty$, $B_p \subseteq Y_p$. *Hence these spaces are equal and their norms are equivalent.*

<u>*Proof*</u> Let $f\epsilon B_p(\Omega_0)$, and let

$$\Phi(z) = \|H_f^\Omega K_z^\Omega\|_{L^2(\Omega)}$$

We have to show that $\Phi \; \epsilon L^p(\Omega, \lambda^{2-p}dA)$. Let T be the operator studied in Lemma 4.12. Then

$$\Phi(z) \leq \|TK_z^\Omega\|_{L^2(\Omega)} + \left[\int_\Omega |f(z) - f(w)|^2 |K^{\Omega_0}(z,w)|^2 dA(w)\right]^{1/2}$$

Since $\|Tk_z^\Omega\| \le C|\log\lambda^\Omega(z)|$, the first term is taken care of. The second term is smaller than or equal to

$$\Phi_0(z) = \left[\int_{\Omega_0} |f(z) - f(w)|^2 |K^{\Omega_0}(z,w)|^2 dA(w))\right]^{1/2}$$

which is in $L^p(\Omega_0,(\lambda^{\Omega_0})^{2-p}dA)$, since $B_p(\Omega_0) = Y_p(\Omega_0)$ by Corollary 3.12. Since Φ_0 is bounded in $\Omega \backslash D_0$ and $\lambda^\Omega(z)^{2-p}$ is area integrable there, we clearly get $\Phi_0 \epsilon L^p(\Omega, (\lambda^\Omega)^{2-p}dA)$ as well. □

In a similar fashion we get

<u>Corollary 4.14</u> *If* $f \epsilon B_\infty(\Omega)$ *then* H_f *is bounded; if* $f \epsilon b_\infty(\Omega)$ *then* H_f *is compact.*

Interpolation between $p = 2$ and $p = \infty$ yields

$$f \in B_p(\Omega) \Rightarrow H_f \in S_p\ ,\ 2 \le p < \infty$$

A standard duality argument gives from this

$$H_f \in S_p \Rightarrow f \in B_p\ ,\ 1 < p \le 2.$$

This completes the proofs of Theorems 4.2 and 4.3.

<u>B_1 and Hankel operators in the Macaev ideal</u>

The proof of Theorem 4.4, namely that if $f \epsilon B_1(\Omega)$ then $H_f \epsilon S_M$ (i.e. $\sum_{n=1}^N s_n(H_f) = O(\log N)$) is similar in principle to the proofs of Theorems 4.2 and 4.3(a). That is, one shows that for $f \in B_1(\Omega)$ the "off diagonal parts" of H_f are small (i.e. in S_M), and reduces the study of the "diagonal parts" of H_f to that of the appropriate Hankel operators on the corresponding simply connected domains. The difficulty is that Lemma 4.12 cannot be improved to yield $T \in S_M$, and so the proof requires some more efforts and new ideas.

We restrict ourselves only to the sketch of the new ingredients of the proofs.

The first result is quite standard, therefore we omit its proof.

Lemma 4.15. Let F *be a compact subset of a regular domain* Ω. *Let* $T:A^2(\Omega) \to L^2(\Omega)$ *be defined by* $Tf = f\cdot\chi_F$. *Then there exist* $0 < \alpha < 1$ *and* $0 < C$ *so that the singular numbers of T satisfy* $s_n(T) < C \cdot \alpha^n$, $n=1,2,\ldots$ *Consequently* $T \in S_p$ *for all* $0 < p$, *and in particular* $T \in S_M$.

The second tool is less standard.

Lemma 4.17. Let $(X, \mathcal{A}, \mu)$ *be a finite measure space, and let* T *be an integral operator from* $A^2(\Delta)$ *into* $L^2(X, \mathcal{A}, \mu)$ *with kernel* $L(x,\xi)$,

$$(Tf)(x) = \int_\Delta L(x,\xi)f(\xi)dA(\xi) .$$

Assume that $\sup_{\xi\in\Delta, x\in X} |L(x,\xi)| = C < \infty$.

Then $T\in S_M(A^2(\Delta), L^2(\mu))$.

Since this lemma is of independent interest, we presents its proof. We need the following result.

Lemma 4.18 [R2] For each $z\in\Delta$ *let*

$$\psi_z = (K_z^\Delta)^2/\|(K_z^\Delta)^2\|_{A^2(\Delta)}$$

Then there exist a bounded linear operator V *from* ℓ_2 *onto* $A^2(\Delta)$ *so that*

$$Ve_n = \psi_{z_n} \quad ; \quad n=1,2,\ldots$$

where $\{z_n\}_{n=1}^\infty$ *form a hyperbolic lattice in* Δ *and* $\{e_n\}_{n=1}^\infty$ *is the unit vector basis of* ℓ_2.

Proof of Lemma 4.17

Let V and the hyperbolic lattice $\{z_n\}_{n=1}^{\infty}$ be as in Lemma 4.18. We can rearrange the sequence $\{z_n\}_{n=1}^{\infty}$ so that $\{|z_n|\}$ is nondecreasing and so that the anuli

$$\Delta_\ell = \{z;\ 1 - \frac{1}{2^{\ell-1}} \le |z| < 1 - \frac{1}{2^\ell}\}$$

$\ell = 1,2,\ldots$ satisfy for some $0 < \alpha < \beta < \infty$

$$\alpha \cdot 2^\ell \le |\{n;\ z_n \epsilon \Delta_\ell\}| \le \beta \cdot 2^\ell$$

We have

$$\|(K_z^\Delta)^2\|_{A^2(\Delta)} = \left[\int_\Delta \frac{dA(\xi)}{|1-\bar{z}\xi|^8}\right]^{1/2} \approx (1-|z|)^{-3}$$

Also

$$|T(K_z^\Delta)^2(x)| \le \int_\Delta |L(x,\xi)| \frac{1}{|1-z\xi|^4}\, dA(\xi)$$

$$\le C \int_\Delta \frac{dA(\xi)}{|1-\bar{z}\xi|^4} \le C(1-|z|)^{-2}$$

Since μ is a finite measure,

$$\|T(K_z^\Delta)^2\|_{L^2(\mu)} \le C_1(1-|z|)^{-2}$$

It follows that for some constant B

$$\|T\psi_z\|_{L^2(\mu)} \le B(1-|z|)\ ;\ z \epsilon \Delta\ .$$

Thus

$$\sum_{z_n \epsilon \bigcup_{\ell=1}^N} \|T\psi_{z_n}\|_{L^2(\mu)} = \sum_{\ell=1}^N \sum_{z_n \epsilon \Delta_\ell} \|T\psi_{z_n}\|_{L^2(\mu)}$$

$$\le B \sum_{\ell=1}^N \beta \cdot 2^\ell / 2^{\ell-1} = 2B\beta N$$

From this it follows that for some constant B,

$$\sum_{n=1}^{m} \|TVe_n\|_{L^2(\mu)} \leq B \log(m+1); \quad m=1,2,\ldots$$

Thus by [AFP2] it follows that

$$TV \in S_M(\ell_2, L^2(\mu)) .$$

The open mapping theorem implies that $V|_{(\ker V)^\perp}$ is inevitable. Notice that $(\ker V)^\perp$ is infinite dimensional, being isomorphic to $A^2(\Delta)$. It follows that

$$T = T(V|_{(\mathrm{Ker}\ V)^\perp})(V|_{(\ker V)^\perp})^{-1} \in S_M(A^2(\Delta), L^2(\mu)) \qquad \square$$

§5. *Open problems and recent developments*

Our analysis of Hankel operators with analytic symbols on regular planar domain uses in an essential way the finite connectivity and the analyticity of the boundaries of these domains. A fundamental problem is to extend the theory beyond regularity and finite connectivity. Since every planar domain can be exhausted by regular domains it seems that the key problem is

Problem 1: *Investigate the dependence of the constants appearing in the theory of Hankel operators on regular planar domains on the connectivity* m. *In particular, obtain quantitative results independent of* m.

As we have indicated in Section 3, the study of the problem of membership of Hankel operators in S_p gets harder as p approaches the critical index p=1. For p=1 our result is not complete, and the following is open.

Problem 2. Describe, either in the context of the weighted Bergman spaces on Δ or in the context of area measure on a regular domain , the space of analytic function f *for which* H_f *belongs to the Macaev ideal* S_M.

We expect an affirmative answer to

Problem 3. Is S_M *the minimal unitary ideal containing a non-zero Hankel operators over a regular domain?*

We expect that many of the results presented in sections 3 and 4 generalize to quite *general nmeasure* μ *on a regular domain* Ω. Let K_z^μ be the Bergman kernel of $A^2(\Omega,\mu)$ and $\lambda(z) = \lambda_\mu(z) = K_z^\mu(z)^{\frac{1}{2}} = \|K_z^\mu\|$ the corresponding metric. Define $B_p(\Omega,\mu)$ for $1 < p \leq \infty$ as those analytic functions f on Ω for which $f'/\lambda \in L^p(\Omega,\lambda^2\mu)$. $b_\infty(\Omega,\mu)$ is defined as those analytic f for which $f'(z)/\lambda(z) \to 0$ as $z \to \partial\Omega$.

Problem 4: Let f *be an analytic function on* Ω, *and let* H_f *be the corresponding Hankel operator on* $L^2(\Omega,\mu)$. *Can the boundness, compactness and membership in* S_p $(1 < p < \infty)$ *of* H_f *be described in terms of membership of* f *in* $B_\infty(\Omega,\mu)$, $b_\infty(\Omega,\mu)$ *and* $B_p(\Omega,\mu)$ *respectively?*

The theory of Hankel operators on $L^2(\mathbb{T})$ with non-analytic symbols reduces to the case of analytic symbols. This is because the symbol f can be written as $f = f_1 + \overline{f_2}$ with f_1,f_2 analytic in Δ (via their Poisson integrals). Also, the Riesz projection is bounded in B_p, $0 < p < \infty$. This is not the case in the theory of Hankel operators over planar domains, and very little is known about *Hankel operators with non-analytic symbols.* In [AFP2] we settled a very special case.

Theorem 5.1. Let f *be a measurable function on* Δ. *Then the Hankel operator* H_f *over* $L^{2,\alpha} = L^2(\Delta,(1-|z|^2)^\alpha dA(z))$, $-1 < \alpha$ *is Hilbert Schmidt if and only if* $f = f_1$

$+ \overline{f_2} + f_3$ *where* $f_1, f_2 \in B_2(\Delta)$ and $f_3 \in L^2(\Delta, (1-|z|^2)^{-2} dA(z))$.

<u>*Problem 5.*</u> ***Study boundness, compactness and membership in S_p of Hankel operators with non–analytic symbols over planar domains.***

The theory of Hankel operators is developing in many interesting directions and is closely related to that of Toeplitz operators. We conclude with a (partial) list of references of recent works related to the subject of this survey. In [JPS] the authors investigate the action of Hankel and Toeplitz operators in the context of $\mathbb{T}$ on many functions spaces (rather than $L^2(\mathbb{T})$). The Hankel operators between two different weighted Bergman spaces on Δ are studied in [J]. Hankel forms (which in our terminology corresponding to "reduced–Hankel operators") over the Fock space $A^2(\mathbb{C}, e^{-\alpha|z|^2} dA(z))$ are studied in [JPR]. Toeplitz and Hankel operators on the Paley–Wiener space are studied in [R3]. In [JP] Hankel operators of higher weights are studied.

Toeplitz operators on the weighted Bergman spaces on Δ are closely related to "reduced Hankel operators". See [Z1] for a study of boundness and compactness and [L] for the criteria of membership in S_p.

The generalization of the theory of Hankel and Toeplitz operators to domains in $\mathbb{C}^n$ is successful in the case of the bounded symmetric domains (which are the right generalization of the unit disk). See [BCZ1], BCZ2], [BCZ3], [Z2], [Z3], [Z4] and [Ah].

References

[A] S. Axler, The Bergman space, the Bloch space, and commutators of multiplication operators, Duke Math. J. 53(1986), 315–332.

[Ah] M. Ahlman, The trace ideal criterion for Hankel operators on the weighted Bergman space $A^{\alpha 2}$ in the unit ball of $\mathbb{C}^n$, University of Lund preprint series, 1984.

[AFJP] J. Arazy, S. Fisher, S. Janson and J. Peetre, An identity for reproducing kernels in a planar domain and Hilbert Schmidt Hankel operators, in preparation.

[AFP1] J. Arazy, S. Fisher and J. Peetre, Möbius invariant function spaces, J. für reine und angewandte Math. 363(1985), 110–145.

[AFP2] J. Arazy, S. Fisher and J. Peetre, Hankel operators on weighted Bergman spaces, to appear in Amer. J. Math.

[AFP3] J. Arazy, S. Fisher and J. Peetre, Hankel operators on planar domains, in preparation.

[B] S. Bergman, "The kernel function and conformal mapping", Math Surveys No. 5, Amer. Math. Soc., Providence, R.I., 1970.

[BCZ1] C.A. Berger, L.A. Coburn and K.H. Zhu, Toeplitz operators and function theory in n–dimensions, in Pseudo–differential operators, Lecture Notes in Mathematics vol. 1256, 28–35, Springer–Verlag, 1987.

[BCZ2] C.A. Berger, L.A. Coburn and K.H. Zhu, BMO on the Bergman spaces of the classical domains, Bull. Amer. Math. Soc. 17(1987), 133–136.

[BCZ3] C.A. Berger, L.A. Coburn and K.H. Zhu, BMO in the Bergman metric on bounded symmetric domains, preprint, 1987.

[BL] J. Bergh and L. Löfström, "Interpolation Spaces, An Introduction", Springer–Verlag, Berlin, New York, 1976.

[BO] F.F. Bonsal, Hankel operators on the Bergman space for the disc, J. London Math. Soc. (2) 33(1986), 355–364.

[CR] R.R. Coifman and R. Rochberg, Representation theoremes for holomorphic and harmonic functions on L^p, Astérique 77(1980), 11–66.

[CRW] R.R. Coifman, R. Rochberg and G. Weiss, Factorization theorems for Hardy spaces in several variables, Ann. of Math. 103(1976), 611–635.

[F] S. Fisher, "Function Theory on Planar Domains," Wiley–Interscience, John Wiley & Sons, New York, 1983.

[GK] I.C. Gohberg and M.G. Krein, "Introduction to the Theory of Linear Nonselfadjoint Operators", Translations of Math. Monographs, Vol. 18, Amer. Math. Soc., Providence, R.I., 1969.

[H] P. Hartman, Completely continuous Hankel matrices, Proc. Amer. Math. Soc. 9(1958), 862–866.

[HS] P.R. Halmos and V.S. Sunder, "Bounded integral operators on L^2 spaces", Springer–Verlag, Berlin, 1978.

[J] S. Janson, Hankel operators between weighted Bergman spaces, Preprint, Uppsala University, 1987.

[JP] S. Janson and J. Peetre, A New Generalization of Hankel Operators (The Case of Higher Weights), University of Lund preprint series, 1985.

[JPR] S. Janson, J. Peetre and R. Rochberg, Hankel forms and the Fock space, Revista Matematica Iberoamericana, 3(1987), 61–138.

[JPS] S. Janson, J. Peetre and S. Semmes, On the action of Hankel and Toeplitz operators on some function spaces, Duke Math. J., 51(1984), 937–958.

[L] D.H. Luecking, Trace ideal criteria for Toeplitz operators,

[LT] J. Lindenstrauss and L. Tzafriri, "Classical Banach Spaces II, Function Spaces", Springer–Verlag, Berlin–Heidelberg–New York, 1979.

[MO] A.W. Marshall and I. Olkin, "Inequalities: Theory of Majorization and Its Applications", Academic Press, 1979.

[N] Z. Nehari, On bounded bilinear forms, Ann. of Math. 65(1957), 153–162.

[Ni] N.K. Nikolskii, Ha–plitz operators: A survey of some recent results, in Operators and Function Theory (S.C. Power Ed.); Reidel, Dordrect, 1985.

[P1] J. Peetre, Hankel operators, rational approximation and allied quesitons of analysis, Second Edmonton Conference on Approximation Theory, 1983. CMS Conference Proceedings 3, Amer. Math. Soc., Providence, R.I., 287–332.

[P2] J. Peetre, "New Thoughts on Besov spaces", Duke Univ. Math. Ser. 1, Durham, 1976.

[Pel1] V.V. Peller, Hankel Operators of Class S_p and their applications (rational approximations, Gaussian processes, the problem of majorizing operators), Math. USSR Sbornik, 41(1982), 443–479.

[Pel2] V.V. Peller, Continuity Properties of the Averaging Projection onto the set of Hankel Matrices, J. Funct. Anal., 53(1983), 74–83.

[Pel3] V.V. Peller, Metric properties of an averaging projection onto the set of Hankel matrices, Soviet Math. Dokl. 30(1984), 362–367.

[Pel4] V.V. Peller, A description of Hankel operators of class S_p for $p > 0$, an investigation of the rate of rational approximation, and other applications, Math. USSR Sbornik 50(1985), 465–494.

[Pel5] V.V. Peller, Vectorial Hankel operators, commutators and related operators on the Schatten–von Neumann class γ_p, Integral Equations and Operator Theory 5(1982), 245–272.

[PK] V.V. Peller and S.V. Khrushchev, Hankel operators, best approximations, and stationary Gaussian processes, Russian Math. Surveys 37(1982), 53–124.

[Po] S.C. Power, "Hankel operators on Hilbert spaces", Res. Notes in Math. 64, Pitman, Boston–London–Melbourne, 1982.

[R1] R. Rochberg, Trace ideal criteria for Hankel operators and commutators, Indiana Univ. Math. J. 31(1982), 913–925.

[R2] R. Rochberg, Decomposition theorems for Bergman spaces and their applications, in Operators and Function Theory (S.C. Power Ed.); Reidel, Dordrect, 1985.

[R3] R. Rochberg, Toeplitz and Hankel Operators on the Paley–Weiner Space, Integral Equations and Operator Theory.

[S] B. Simon, "Trace Ideals and their Applications", London Math. Soc. Lecture Note Ser. 35, Cambridge Univ. Press, Cambridge–London–New York–Melbourne, 1979.

[Se] S. Semmes, Trace ideal criterion for Hankel operators, $0 < p < 1$, Integral Equations and Operator Theory 7(1984), 241–281.

[Z1] K.H. Zhu, VMO, ESV and Toeplitz operatorson the Bergman space, Transactions of the Amer. Math. Soc., 302(1987), 617–646.

[Z2] K.H. Zhu, Duality and Hankel operators on the Bergman spaces of the polydisc, Preprint, 1987.

[Z3] K.H. Zhu, Multipliers of BMO in the Bergman metric with applications to Toeplitz operators, Preprint, 1987.

[Z4] K.H. Zhu, On the dual and predual of the Bergman space L_a^1 of bounded symmetric domain, Preprint, 1987.

A GENERALIZED MARCEL RIESZ THEOREM ON CONJUGATE FUNCTIONS

Nakhlé Asmar and Edwin Hewitt

1. Notation. Throughout this paper, the symbol G will denote a locally compact Abelian group with character group X. Haar measures on G and X are denoted by μ and θ, respectively. A complex-valued function f is said to be measurable on a locally compact Abelian group G if it is measurable with respect to the Haar measure on G. The class of measurable functions on G with integrable p-th power is denoted by $L_p(G)$, $1 \leq p < \infty$; the class of essentially bounded measurable functions by $L_\infty(G)$; the class of continuous functions on G with compact support will be denoted by $C_{00}(G)$. If A is a subset of G, the complement of A in G will be denoted by $G \backslash A$.

Let p be a real number such that $1 \leq p < \infty$. An essentially bounded, measurable, complex-valued function m on X is called an $L_p(G)$-multiplier if for every f in $L_1(G) \cap L_p(G)$ with compactly supported Fourier transform $\hat{f}$, there is a function g in $L_2(G) \cap L_p(G)$ such that

(i) $\hat{g} = \hat{f}\, m;$

and

(ii) $\|g\|_p \leq B \|f\|_p,$

for some number B. The smallest abmissible B is denoted by $N_p(m)$ and is called the norm of m. The linear subspace of $L_p(G)$ consisting of all f's as described above is dense in $L_p(G)$ (see Asmar and Hewitt [2], Theorem (7.1)). Therefore the linear operator $f \mapsto g$ can be extended in a unique way to a bounded linear operator on $L_p(G)$. Thus, while the multiplier m is defined on a dense subset of $L_p(G)$ by (i), we may consider it as a bounded linear operator on all of $L_p(G)$. The set of all $L_p(G)$-multipliers will be denoted by $M_p(G)$.

2. Theme of this essay. Our goal in this paper is to present a new proof of a general version of M. Riesz's theorem which contains Theorem (7.2) of Asmar and Hewitt [2], and Theorem (3.8) of Berkson and Gillespie [3], as particular cases. Our proof combines a recent characterization of Haar-measurable orders, and a well-known homomorphism theorem for multipliers. For ease of reference, we will list in this section these and other needed results.

1980 Mathematics Subject Classification (1985 Revision). 43A17, 43A22

(2.1) Theorem. (Asmar and Hewitt [2], (5.14)). *Let P be a Haar-measurable order on a locally compact Abelian group X. Let K be any compact subset of X. There are a continuous real-valued homomorphism ψ on X and a subset N of X of Haar measure zero such that:*

(i) $\psi(x) > 0$

for all $x \in (K \cap P) \setminus (N \cup \{0\})$;

(ii) $\psi(x) < 0$

for all $x \in (K \cap (-P)) \setminus (N \cup \{0\})$.

(2.2) Theorem. (The homomorphism theorem for multipliers). *Let G_1 and G_2 be locally compact Abelian groups with character groups X_1 and X_2, respectively. Let p be a real number such that $1 \le p < \infty$ and let m be a continuous $L_p(G_1)$-multiplier with multiplier norm $N_p(m)$. Let τ be a continuous homomorphism from X_2 into X_1. Then $m \circ \tau$ is an $L_p(G_2)$-multiplier with $N_p(m \circ \tau)$ less than or equal to $N_p(m)$.*

For the proof of Theorem (2.2), see Edwards and Gaudry [5], Theorem B.2.1, p. 187. Also from Edwards and Gaudry *loc. cit.* we will need Lemma B.1.2.(iii), p. 185.

(2.3) Lemma. *Suppose that $1 \le p < \infty$ and that m is an $L_p(G)$-multiplier with norm $N_p(m)$. Let k be a function in $L_1(X)$. Then the (continuous and bounded) function $k * m$ is an $L_p(G)$-multiplier, with norm $N_p(k * m)$ less than or equal to $\| k \|_1 N_p(m)$.*

We also need a certain kernel on $\mathbb{R}$, used by Berkson and Gillespie [3], Lemma (3.2).

(2.4) Lemma. *For $0 \le t \le 1$ and $n = 1, 2, \ldots$, let $k_n^{(t)} \in L_1(\mathbb{R})$ satisfy:*

(i) $k_n^{(t)} \ge 0$;

(ii) $k_n^{(t)}(x) = 0$ *for* $|x| > \frac{1}{n}$;

(iii) $\displaystyle\int_0^\infty k_n^{(t)}(x)\,dx = 1 - t$;

(iv) $\displaystyle\int_{-\infty}^0 k_n^{(t)}(x)\,dx = t$.

Now suppose that $m\colon \mathbb{R} \mapsto \mathbb{C}$ is bounded and measurable, $x_0 \in \mathbb{R}$, and that

$$\lim_{x \to x_0^+} m(x) = \alpha_1, \quad \lim_{x \to x_0^-} m(x) = \alpha_2.$$

We then have

(v) $$\lim_{n\to\infty} k_n^{(t)} * m(x_0) = t\alpha_1 + (1-t)\alpha_2.$$

Our next lemma is of independent interest. We need it to apply the homomorphism theorem for continuous multipliers to discontinuous multipliers, for example the signum function on $\mathbb{R}$.

(2.5) Lemma. *Let m be a bounded measurable function on X. Suppose that $1 \le p < \infty$ and that $\{m_n\}$ is a sequence of $L_p(G)$-multipliers such that:*

(i) $$\sup_n \|m_n\|_\infty < \infty;$$

(ii) $$\sup_n N_p(m_n) = c_p < \infty;$$

and

(iii) $$\lim_{n\to\infty} m_n(\chi) = m(\chi)$$

for all χ in X. Then m is an $L_p(G)$-multiplier with $N_p(m) \le c_p$.

Proof. Let $q = \frac{p}{p-1}$ if $1 < p < \infty$, and $q = \infty$ if $p = 1$. It is enough to show that

(1) $$\left|\int_X \hat{f}\, m\, \overline{\hat{g}}\, d\theta\right| \le c_p \|f\|_p \|g\|_q$$

for all $f \in L_p(G) \cap L_1(G)$, $g \in L_q(G) \cap L_1(G)$, for which $\hat{f}$ and $\hat{g}$ are in $C_{00}(X)$. (See Edwards and Gaudry [5], 1.2.2, (iii), p. 7.)

Hypothesis (iii) implies that the equality

$$\lim_{n\to\infty} m_n(\chi)\hat{f}(\chi)\,\overline{\hat{g}}(\chi) = m(\chi)\hat{f}(\chi)\,\overline{\hat{g}}(\chi)$$

holds for all χ in X. From (i), the inequality

$$|m_n(\chi)\hat{f}(\chi)\,\overline{\hat{g}}(\chi)| \le \sup_n \|m_n\|_\infty \|\hat{g}\|_\infty |\hat{f}(\chi)|$$

holds for all χ in X. Lebesgue's dominated convergence theorem shows that

(2) $$\lim_{n\to\infty}\left|\int_X \hat{f}\, m_n\, \overline{\hat{g}}\, d\theta\right| = \left|\int_X \hat{f}\, m\, \overline{\hat{g}}\, d\theta\right|.$$

The inequality (1) follows from (2) and the relations

$$\left|\int_X \hat{f}\, m_n\, \bar{g}\, d\theta\right| = \left|\int_G (\hat{f}\, m_n)^{\vee}\, \bar{g}\, d\mu\right|$$

$$\leq \|(\hat{f}\, m_n)^{\vee}\|_p\, \|g\|_q$$

$$\leq c_p\, \|f\|_p\, \|g\|_q. \qquad \square$$

3. Marcel Riesz's theorem on conjugate functions.

(3.1) Preliminaries. Let α_1 and α_2 be complex numbers. Consider the function m defined on $\mathbb{R}$ by

$$(1) \qquad m(s) = \begin{cases} \alpha_1 & \text{if } s > 0; \\ 0 & \text{if } s = 0; \\ \alpha_2 & \text{if } s < 0. \end{cases}$$

Riesz himself proved in Riesz [6] that m is an $L_p(\mathbb{R})$-multiplier for $1 < p < \infty$. Let $k_n^{(t)}$ be as in Lemma (2.4). From (2.4.iii, iv) and Lemma (2.3), it follows that $m{*}k_n^{(t)}$ is in $M_p(\mathbb{R})$ for all p with $1 < p < \infty$, and that

$$(2) \qquad N_p(m{*}k_n^{(t)}) \leq N_p(m)$$

for all $n = 1, 2, \ldots$. Let ψ be any continuous homomorphism from X into $\mathbb{R}$ (X need not be ordered). The homomorphism theorem (2.2) shows that $m{*}k_n^{(t)} \circ \psi$ is in $M_p(G)$ and that

$$(3) \qquad N_p(m{*}k_n^{(t)} \circ \psi) \leq N_p(m).$$

Write $m_\psi^{(t)}$ for the function on X $\lim_{n\to\infty} m{*}k_n^{(t)} \circ \psi(\chi)$, for $\chi \in X$. It is clear from (1) and (2.4.v) that:

$$(4) \qquad m_\psi^{(t)}(\chi) = \begin{cases} \alpha_1 & \text{if } \psi(\chi) > 0; \\ t\alpha_1 + (1-t)\,\alpha_2 & \text{if } \psi(\chi) = 0; \\ \alpha_2 & \text{if } \psi(\chi) < 0. \end{cases}$$

Also, standard facts about convolutions show that

$$\|m{*}k_n^{(t)} \circ \psi\|_{X,\infty} \leq \|m{*}k_n^{(t)}\|_{\mathbb{R},\infty} \leq \|m\|_\infty\, \|k_n^{(t)}\|_1 = \|m\|_\infty.$$

Thus hypotheses (ii), (iii), and (i) of Lemma (2.5) are satisfied for the sequence of $L_p(G)$-multipliers $\{m{*}k_n^{(t)} \circ \psi\}$. Accordingly the function $m_\psi^{(t)}$ is an $L_p(G)$-multiplier for all p with $1 < p < \infty$, with norm not exceeding $N_p(m)$.

We have proved the following theorem.

(3.2) Theorem. *Let G be a locally compact Abelian group with character group X. Suppose that ψ is a continuous real-valued homomorphism on X. For $0 \leq t \leq 1$, the function $m_\psi^{(t)}$ given by (3.1.4) is an $L_p(G)$-multiplier for all p such that $1 < p < \infty$, with norm not exceeding $N_p(m)$, where m is the multiplier on $L_p(\mathbb{R})$* given by (3.1.1).

We now suppose that X is ordered by a Haar-measurable order P. That is, X contains a θ-measurable subset P with the following properties:

$$P \cap (-P) = \{0\};$$

$$P \cup (-P) = X;$$

$$P + P = P.$$

We associate with P the function $m_P^{(t)}$ defined on X by

$$m_P^{(t)}(\chi) = \begin{cases} \alpha_1 & \text{if } \chi \in P \backslash \{0\}; \\ t\alpha_1 + (1-t)\alpha_2 & \text{if } \chi = 0; \\ \alpha_2 & \text{if } \chi \in (-P) \backslash \{0\} \end{cases}$$

where $0 \leq t \leq 1$, and α_1 and α_2 are arbitrary complex numbers.

A generalized version of Marcel Riesz's theorem on conjugate functions may be stated as follows.

(3.3) Theorem. *With the above notation, the function $m_P^{(t)}$ is an $L_p(G)$-multiplier for all p such that $1 < p < \infty$, with norm $N_p(m_P^{(t)})$ not exceeding $N_p(m)$.*

Proof. It suffices to show that for f in $L_p(G) \cap L_1(G)$ with $\hat{f} \in C_{00}(X)$, we have

(1) $$\|(\hat{f}\, m_P^{(t)})^{\check{}}\|_p \leq N_p(m)\, \|f\|_p.$$

Let $K = supp \hat{f}$. Let ψ be a continuous real-valued homomorphism on X such that (2.1.i) and (2.1.ii) hold. Let $m_\psi^{(t)}$ be as in (3.1.4). It is easy to verify that the equality

$$\hat{f}\, m_P^{(t)} = \hat{f}\, m_\psi^{(t)}$$

holds θ-almost everywhere on X. From Theorem (3.2) we have

$$\|(\hat{f}\, m_P^{(t)})^{\check{}}\|_p = \|(\hat{f}\, m_\psi^{(t)})^{\check{}}\|_p \leq N_p(m)\, \|f\|_p,$$

and so the proof is achieved. □

(3.4) Remarks. (a) The version of Marcel Riesz's theorem that appears in Asmar and Hewitt [2], Theorem (7.2), is a special case of Theorem (3.3) above. Simply take $\alpha_1 = -i$; $\alpha_2 = i$; and $t = \frac{1}{2}$.
(b) Note that the norms of the multipliers $m_P^{(t)}$ and $m_\psi^{(t)}$ do not depend on t, P, or ψ.
(c) We owe to Berkson and Gillespie the proof that the norm of the multiplier in the abstract version of M. Riesz's theorem on compact groups is the same as the norm of the Hilbert transform on $\mathbb{R}$. For locally compact Abelian groups, this result is contained in [2], Theorem (7.2). The proofs in [2] and [3] are different, but they are both based on a transference argument which made possible the transfer of the properties of the Hilbert transform on $\mathbb{R}$ to the conjugate function operator on a locally compact Abelian group. The homomorphism theorem has been proved recently in [1], and [4], using the transference methods. These proofs may shed more light on the validity of (b).

Acknowledgements. The authors are grateful to the editors, in particular to Professor Earl Berkson, for the opportunity they have given us to present this essay for publication in the Proceedings of the Conference. The first-named author gratefully acknowledges financial support from the Argonne Universities Association Trust Fund.

Bibliography

[1] Asmar, Nakhlé. A homomorphism theorem for multipliers (submitted).

[2] Asmar, Nakhlé, and Edwin Hewitt. Marcel Riesz's theorem on conjugate Fourier series and its descendants. Proceedings of the Analysis Conference, Singapore 1986. North Holland (to appear).

[3] Berkson, Earl, and T.A. Gillespie. The generalized M. Riesz theorem and transference. Pacific J. Math. 120 (2), 1985, 279–288.

[4] Berkson, Earl, T.A. Gillespie, and Paul Muhly. Generalized Analiticity in UMD spaces (submitted).

[5] Edwards, R.E., and G.I. Gaudry. Littlewood-Paley and multiplier theory. Berlin, Heidelberg, New York: Springer-Verlag, 1977.

[6] Riesz, Marcel. Sur les fonctions conjuguées. Math. Z. 27 (1928), 218–244.

California State University, Long Beach
Long Beach, California 90840

and

University of Washington
Seattle, Washington 98195

SOME RESULTS IN ANALYSIS RELATED TO THE LAW OF THE ITERATED LOGARITHM.

Rodrigo Bañuelos[1] and Charles N. Moore

ABSTRACT: We describe some recent results related to a law of the iterated logarithm (LIL) for analytic functions in the Bloch class, for arbitrary analytic functions in the disc, (which also holds for harmonic functions in starlike Lipschitz domains of $\mathbb{R}^n$), and for harmonic functions in the upper half space. This last result gives precise information on the relative "order of infinities" between the nontangential maximal function and the Lusin square function. Except for section 3, many of the results presented here are not new but most of the proofs are new. An effort is made to make this paper readable to those who may not be experts in harmonic analysis but who may have some interest in these type of LIL's.

§0. INTRODUCTION. Ever since shortly after Kolmogorov proved his celebrated law of the iterated logarithm (LIL) for independent random variables, analysts have worked to obtain similar results for analytic functions in the unit disc. The purpose of this paper is to describe some recent progress in this direction. Before we begin to describe the analytic LIL's, let us recall the classical theorem which, in the words of K. L. Chung [7] (p.231), "is a crowning achievement in classical probability theory."

THEOREM (Kolmogorov [15]): Let $\{X_n;\ n \geq 1\}$ be a sequence of independent random variables with mean zero and variance one. Suppose $|X_n| \leq \epsilon_n n/\sqrt{\log\log n}$ for some constants $\epsilon_n \longrightarrow 0$. Then for almost every ω,

1980 Mathematics Subject Classification (1985 Revision). 31B25, 42825,60F99.
[1]Supported by an NSF Postdoctoral Fellowship.

$$\limsup_{n\to\infty} \frac{S_n(\omega)}{\sqrt{2n \log\log n}} \equiv 1, \tag{0.1}$$

where $S_n = \sum_{k=1}^{n} X_k$.

Kolmogorov's LIL has been generalized in many directions with applications in fields ranging from statistics, number theory, differential equations, the study of Brownian motion on manifolds, dynamical systems, ergodic theory, and very recently to the study of harmonic measure. We refer the reader to the recent survey article, "variants on the law of the iterated logarithm," by N. H. Bingham [3] which contains nearly 400 references on this subject.

We now turn to the analysis LIL's. Let $D = \{z \in \mathbb{C} : |z| < 1\}$ be the unit disc in the complex plane and let $T = \partial D$ be the unit circle equipped with the probability measure $dm = d\theta/2\pi$. Almost everywhere on T will always mean with respect to m. The following beautiful result is, to the best of our knowledge, the first LIL in analysis.

THEOREM (Salem-Zygmund [21]): Let $S_n(\theta) = \sum_{k=1}^{n} \cos 2^k\theta$. Then for almost every $\theta \in T$,

$$\limsup_{n\to\infty} \frac{S_n(\theta)}{\sqrt{n \log\log n}} \le 1. \tag{0.2}$$

Generally speaking lower bounds in the LIL are much more difficult to obtain than upper bounds. They require, essentially, central limit theorem type behavior with very precise estimates on error terms. The result of Salem and Zygmund was improved by Erdös and Gál [9] who showed that in fact for almost every $\theta \in T$,

$$(0.3) \qquad \limsup_{n\to\infty} \frac{S_n(\theta)}{\sqrt{n \log\log n}} \equiv 1.$$

The trigonometric series $\sum_{k=1}^{\infty} \cos 2^k\theta$ is the classical example of a lacunary series. Salem and Zygmund, as well as Erdös and Gál also proved their results for lacunary sequences n_k and not just for 2^k. It has been shown over the years that lacunary series exhibit many of the properties of sums of independent random variables. The result of Salem-Zygmund-Erdös-Gál was extended by M. Weiss [25] to very general lacunary series. Let us recall her result. Let $S_n(\theta) = \sum_{k=1}^{n} a_k \cos n_k\theta$, where the a_k are real and the n_k are positive (not necessarily integers) with the property that $n_{k+1}/n_k > q > 1$. Let $B_n = \left[\frac{1}{2} \sum_{k=1}^{n} a_k^2\right]^{1/2}$ and $M_n = \max_{1\le k\le n} |a_k|$. Suppose that $B_n \longrightarrow \infty$ and that $M_n = o\left[B_n/(\log\log B_n)^{1/2}\right]$. Then with $S_n(\theta) = \sum_{k=1}^{n} a_k \cos n_k\theta$,

$$(0.4) \qquad \limsup_{n\to\infty} \frac{S_n(\theta)}{\sqrt{2n \log\log n}} \equiv 1$$

for almost every $\theta \in T$.

Using the invariance principle technique of W. Philipp and W. Stout [17], S. Takahashi [24] has considerably weakened the lacunary condition in

M. Weiss' results. More precisely, let $\{n_k\}$ be a sequence of positive integers such that $n_{k+1}/n_k > 1 + ck^{-\alpha}$, $c > 0$ and $0 \leq \alpha < 1/2$. Suppose $B_n = \left[\frac{1}{2}\sum_{n=1}^{n} a_n^2\right]^{1/2} \longrightarrow \infty$ and that $a_n = O\left[B_n n^{-\alpha}(\log B_n)^{-\beta}\right]$, $\beta > 1/2$. For $\theta \in T$ and $t \geq 0$, define $S(t) = S(t,\theta) = S_n(\theta)$, if $B_n^2 \leq t < B_{n+1}^2$, for $n \geq 0$, where we set $B_0 = 0$ and $S_0 = 0$. Takahashi's main result from which (0.4) immediately follows is:

(0.5) Without changing the distribution of $\{S(t),\ t \geq 0\}$ we can redefine the process $\{S(t),\ t \geq 0\}$ on a richer probability space together with standard Brownian motion $\{B(t),\ t \geq 0\}$ such that

$$S(t) = B(t) + O(t^{1/2}) \text{ a.s. as } t \longrightarrow \infty. \tag{0.6}$$

To obtain (0.4) simply apply the LIL for Brownian motion and (0.6). A consequence of (0.4) also proved in [25], is the LIL for lacunary power series in D. Let $F(z) = \sum_{k=1}^{\infty} c_k z^{n_k}$ where the c_k are complex numbers and the n_k as in M. Weiss' LIL. For $0 < \rho < 1$, set $B_\rho = \left[\sum_{k=1}^{\infty} |c_k|^2 \rho^{2\cdot n_k}\right]^{1/2}$. Then, under the same assumption on the c_k as on the a_k above,

$$\limsup_{\rho\uparrow 1} \frac{|F(\rho e^{i\theta})|}{\sqrt{B_\rho^2 \log\log B_\rho}} \equiv 1 \tag{0.7}$$

almost everywhere on T.

In the particular case of $F(z) = \sum_{k=1}^{\infty} z^{2^k}$ an easy exercise shows that

$$B_\rho^2 = \sum_{k=1}^{\infty} \rho^{2\cdot 2^k} \sim \frac{1}{\log 2} \log \frac{1}{1-\rho},$$

as $\rho \uparrow 1$, so that in this case we can write (0.7) as

$$\text{(0.8)} \qquad \limsup_{\rho\uparrow 1} \frac{|F(\rho e^{i\theta})|}{\sqrt{\log \frac{1}{1-\rho} \, \text{logloglog} \, \frac{1}{1-\rho}}} = \frac{1}{\sqrt{\log 2}}$$

almost everywhere on T.

The lacunary power series $F(z) = \sum_{k=1}^{\infty} z^{2^k}$ is an example of a function in the Bloch class. In his remarkable paper "On the distortion of boundary sets under conformal mapping," [16] N. G. Makarov has shown that the upper bound in (0.8) holds for arbitrary functions in this class. In section 1, we define this class of functions and prove Makarov's result. We shall present two different proofs. First, we show that Makarov's LIL is actually a special case of a more general LIL for arbitrary analytic functions proved in [1]. The second proof, which we learned from L. Carleson in his course on harmonic measure at UCLA, is a simple induction and an application of Green's theorem to prove Makarov's main lemma, (see Lemma 1 below). Apparently many others have independently discovered this proof. Once the lemma is proved, we proceed exactly as in the case of independent random variables. This last step is a little different than what is done in, for example, [19].

Contrary to the situation for lacunary functions, Makarov's LIL may not always have a lower bound. (Take, for example, any function in $H^{\infty}(D)$). This will be the subject of section 2. A theorem of Girela [12] says that if F belongs to B_1, a certain subclass of the Bloch class, then there is no lower bound in Makarov's LIL. Our proof of this result is an application of the Barlow-Yor [2] inequalities betweeen the square function and the local time of martingale. This is perhaps not the most efficient (nor simplest) way to obtain this result. However, we feel that the Barlow-Yor inequalities, like the Burkholder-Gundy inequalities between the square function and the maximal function, can be potentially very useful in analysis. We do not know, outside of [13], of any other applications. In section 2, we will also show, without any other assumptions, that if the L^2-means grow slowly enough we again have no lower bound in Makarov's LIL. This result is an improvement of another theorem of Girela [12].

In Section 3, we state a very recent result of I. Klemes and the authors which answers in the positive a question raised by R. F. Gundy concerning a pointwise LIL between the nontangential maximal function and the Lusin square function of an arbitrary harmonic function in the upper half space. (The proofs of the results in this section will appear elsewhere). With regards to the upper bound, this LIL is more general than any of the other LIL's discussed in this paper.

Throughout the paper, the notation C, $C_1, C_2, \ldots$ will be used to denote universal constants. By $a \sim b$ we mean there exists C_1 and C_2 such that $C_1 a \leq b \leq C_2 a$. For any positive integer n, $\log_n$ means we have iterated the log n-times.

§1. MAKAROV'S LIL. An analytic function F defined in D is said to be a Bloch function if

$$\|F\|_B = |F(0)| + \sup_{z \in D} (1 - |z|)\, |F'(z)| < \infty.$$

We denote this class by B. Let us show, for the sake of having some concrete examples, that $F(z) = \sum_{k=1}^{\infty} a_k z^{n^k} \in B$, where $\sup_k |a_k| \le 1$ and $n \ge 2$ is an integer. We write $|z| = \rho$. Then

$$\frac{1}{1-\rho} |F'(z)| \le \frac{1}{1-\rho} \sum_{m=1}^{\infty} n^m \rho^{n^m} = \frac{1}{1-\rho} \sum_{m=1}^{\infty} \left[\sum_{k=1}^{m} n^k\right]\left[\rho^{n^m} - \rho^{n^{m-1}}\right]$$

$$= \sum_{m=1}^{\infty} \left[\sum_{k=1}^{m} n^k\right]\left[\rho^{n^m} + \dots + \rho^{n^{m+1}-1}\right] \le \frac{n}{n-1} \sum_{m=1}^{\infty} n^m \left[\rho^{n^m} + \dots + \rho^{n^{m+1}-1}\right]$$

$$\le \frac{n}{n-1} \rho \sum_{j=1}^{\infty} j\, \rho^{j-1} = \frac{n}{n-1} \frac{\rho}{(1-\rho)^2}$$

from which it follows that $F \in B$.

The above computation is essentially from [18] where it is also shown that all Bloch functions are of the form $F(z) = \alpha \log g'(z)$ where $g(z)$ is a univalent function and α is some positive constant.

The following theorem is (with regards to the upper estimate) a generalization of (0.8) above.

THEOREM 1.1 (Makarov [16]). For any $F \in B$,

$$\limsup_{\rho\uparrow 1} \frac{|F(\rho e^{i\theta})|}{\sqrt{\log \frac{1}{1-\rho} \log_3 \frac{1}{1-\rho}}} \leq C \, \|F\|_B \tag{1.1}$$

for almost every $\theta \in T$.

Next we introduce the square functions. For $0 < \alpha < 1$ we define the Stoltz domain $\Gamma_\alpha(\theta)$ to be the interior of the smallest convex set containing the disc $\{|z| < \alpha\}$ and the point $e^{i\theta}$. If F is any analytic function in D we define $A_\alpha(F)(\theta)$, the Lusin square function of F, and $g_*(F)(\theta)$, the Littlewood-Paley square function of by

$$A_\alpha(F)(\theta) = \left[\int_{\Gamma_\alpha(\theta)} |F'(z)|^2 dxdy \right]^{1/2} \tag{1.2}$$

and

$$g_*(F)(\theta) = \left[\frac{1}{\pi} \int_D \log \frac{1}{|z|} P_\theta(z) |F'(z)|^2 dxdy \right]^{1/2} \tag{1.3}$$

where $z = x + iy$ (often we will write dz for $dxdy$) and

$$P_\theta(z) = \frac{1 - |z|^2}{|z - e^{i\theta}|^2} .$$

We refer the reader to E. M. Stein [22] for the basic properties of these functions. The following deep result was recently proved by A. Chang, M. Wilson, and T. Wolff.

THEOREM 1.2 ([6]). Suppose $A_\alpha(F)(\theta) \leq 1$ for almost every $\theta \in T$. Then there are constants C_α^1 and C_α^2 depending only on α such that

$$\int_T \exp\,(C_\alpha^1 |F - F(0)|^2)dm \leq C_\alpha^2. \tag{1.4}$$

What makes this result difficult is the presence of the square in the exponential. If we simply want exponential integrability this follows easily from the John-Nirenberg theorem concerning the behavior of BMO functions. (See J. Garnett [11] for properties of BMO-functions). In [1] the first author proved, using Brownian motion, a sharp version of (1.4) with the g_*-function replacing A_α from which the following theorem was obtained.

THEOREM 1.3 ([1]). Let $F(z)$ be an arbitrary analytic function in D. For $0 < \rho < 1$, let $F_\rho(z) = F(\rho z)$ and set $g_\rho = \|g_*(F_\rho)\|_{L^\infty(T)}$. Suppose $g_\rho \longrightarrow \infty$ as $\rho \uparrow 1$. Then for almost every $\theta \in T$,

$$\limsup_{\rho\uparrow 1} \frac{|F(\rho e^{i\theta})|}{\sqrt{2\, g_\rho^2 \log_2 g_\rho}} \leq 1 \tag{1.5}$$

and 1 is the smallest possible number we can have on the right-hand side of this inequality.

Random variables satisfying the estimate (1.4) are often called generalized Gaussian (also sub-Gaussian) random variables and it is well known ([23], p. 304) that such estimates lead immediately to upper half LIL's like (1.5). We observe that if we replace the numbers g_ρ by $A_\rho = \|A_\alpha(F_\rho)\|_{L^\infty(T)}$

then (1.5), with a constant on the right hand which depends only on α, will follow from (1.4).

Except for the sharpness part, Theorem 1.3 can be generalized to starlike Lipschitz domains. Let us briefly describe this extension. Let D be a bounded domain in n-dimensional Euclidean space $\mathbb{R}^n$ which is Lipschitz, that is, the boundary ∂D is described locally by a function of class Lip 1, and which is starlike with respect to the origin 0. Let $G(x,0)$ denote the Green's function for D with pole at 0. For $\xi \in \partial D$ and $x \in D$, let $K(x,\xi)$ be the Poisson kernel for D and μ_x the harmonic measure at x. If u is a harmonic function in D, define

$$(1.6) \qquad g_*^2(u)(0,\xi) = \frac{1}{K(0,\xi)} \int_D G(x,0)K(x,\xi)\,|\nabla u(x)|^2 dx.$$

Notice that for the case of the unit disc in the plane, this is exactly the same as in (1.3). Suppose u has boundary values f in $L^1(\partial D,\mu_0)$ and suppose $\|g_*(u)(0,\xi)\|_{L^\infty(\mu_0)} = a_0 < \infty$. Then by Theorem 1′ in [1] there are absolute constants C_1 and C_2 such that

$$(1.7) \qquad \int_{\partial D} \exp(C_1 |f(\xi) - u(0)|^2/a_0^2)\,d\mu_0(\xi) \le C_2.$$

Now suppose $u(0) = 0$ and let $N_\alpha(u)(\xi)$ be the nontangential maximal function of u. The size of the cone α depends on the Lipschitz constant for D. With respect to the harmonic measure μ_0 this maximal function is of weak type (1,1) and of strong type (∞,∞). (This follows from the doubling property of the harmonic measure and the fact that the Hardy-Littlewood maximal function with respect to μ_0 controls the nontangential maximal function. For full details see [4].) By interpolation we obtain for $1 < p < \infty$,

$$\|N_\alpha(u)\|^p_{L^p(\partial D,\mu_0)} \le C^p \left[\frac{p}{p-1}\right]^p \|f\|^p_{L^p(\partial D,\mu_0)}$$

$$\le C^p \|f\|^p_{L^p(\partial D,\mu_0)}, \quad p \ge 2,$$

where C is a constant depending on α, n, and the Lipschitz nature of ∂D. By summing the series for the exponential and applying (1.7) we see that

$$(1.8) \qquad \int_{\partial D} \exp(C_1^2 |N_\alpha(u)(\xi)|^2/a_0^2)\,d\mu_0(\xi) \le C_2$$

with C_1 and C_2 depending on α, n, and the Lipschitz constants of ∂D.

Now suppose u is an arbitrary harmonic function in D with $u(0) = 0$. (This restriction is irrelevant for theorem 1.4 below since square functions do not change if we subtract constants.) For $0 < \rho < 1$, let $u_\rho(x) = u(\rho x)$, $x \in D$. Define

$$(1.9) \qquad g_\rho = \|g_*(u_\rho)(0,\xi)\|_{L^\infty(\mu_0)}.$$

It is not entirely obvious that for every $0 < \rho < 1$, the numbers g_ρ should be finite. (Try it with $u = x$.) To verify this we write $g_*(u_\rho)(0,\xi)$ in terms of conditional Brownian motion (see [1], p. 656) and use the fact that the expected lifetime of this process is bounded by a constant depending on the volume of D and the Lipschitz character of ∂D, (see [5]). Once we verify that $g_\rho < \infty$, we may apply (1.8) to the functions u_ρ with $a_0 = g_\rho$ to obtain the necessary sub-Gaussian estimate and hence the following result.

THEOREM 1.4. Suppose D is a bounded Lipschitz domain in $\mathbb{R}^n$ which is starlike with respect to 0. Suppose u is a harmonic function in D and define $u_\rho(x) = u(\rho x)$, $0 < \rho < 1$. Let $N_\alpha(u_\rho)(\xi)$ be the nontangential maximal function of u_ρ and set $g_\rho = \|g_*(u_\rho)(0,\xi)\|_{L^\infty(\mu_0)}$. Suppose $g_\rho \longrightarrow \infty$ as $\rho \uparrow 1$. Then

$$\text{(1.10)} \qquad \limsup_{\rho\uparrow 1} \frac{N_\alpha(u_\rho)(\xi)}{\sqrt{g_\rho^2 \log_2 g_\rho}} \le C$$

for μ_0 almost every $\xi \in \partial D$. The constant C depends on α, n, and the Lipschitz character of ∂D.

We now show that Makarov's LIL is a special case of Theorem 1.3. Suppose $F \in B$; changing to polar coordinates we write

$$g_*^2(F_\rho)(\theta) = \frac{1}{\pi}\int_0^1\int_0^{2\pi} r \log\frac{1}{r}\, |F_\rho'(re^{i\varphi})|^2\, P_\theta(re^{i\varphi})d\varphi dr$$

$$\le \|F\|_B^2 \frac{\rho^2}{\pi}\int_0^1\int_0^{2\pi} r\log\frac{1}{r}\,\frac{1}{(1-r\rho)^2}\, P_\theta(re^{i\varphi})d\varphi dr$$

$$= 2\|F\|_B^2\, \rho^2 \int_0^1 r\log\frac{1}{r}\,\frac{dr}{(1-r\rho)^2} \le 2\,\|F\|_B^2 \log\frac{1}{1-\rho}$$

where we have used the definition of the Bloch class for the first inequality and an integration by parts for the last one. So,

(1.11) $$g_\rho^2 \le 2 \|F\|_B^2 \log \frac{1}{1-\rho}.$$

If $g_\rho \longrightarrow \infty$ as $\rho \uparrow 1$, (1.11) and Theorem 1.3 imply (1.1) with the constant $c = 2$. If g_ρ remains bounded as $\rho \uparrow 1$, then the function F has boundary values almost everywhere on T and this makes (1.1) trivially true.

We now present a second proof of Makarov's theorem. For $0 < \rho < 1$, set

$$F_\rho^*(\theta) = \sup_{0 \le r \le \rho} |F(re^{i\theta})|.$$

Without loss of generality we assume for the rest of this section that $F(0) = 0$ and $\|F\|_B \le 1$.

LEMMA 1 ([16]). There exists an absolute constant C such that

(1.12) $$\int_T (F_\rho^*(\theta))^{2k} dm \le C^k k! \left[\log \frac{1}{1-\rho}\right]^k,$$

for $k = 0,1,\ldots$.

Proof: The lemma trivially holds for $k = 0$. Suppose $k \ge 1$ and that (1.12) holds for $k - 1$. It follows from the Hardy-Littlewood maximal theorem that

$$\int_T (F_\rho^*(\theta))^{2k} dm \le C_k \int_T |F(\rho e^{i\theta})|^{2k} dm,$$

where $C_k \sim 1$ for k large. Let Δ denote the Laplace operator. Since the function F is holomorphic, $\Delta|F|^p = p^2|F|^{p-2}|F'|^2$. So applying Green's theorem we get

$$\int_T |F(\rho e^{i\theta})|^{2k} dm = \frac{4k^2}{2\pi} \int_D \log \frac{1}{|z|} |F_\rho(z)|^{2k-2} |F_\rho'(z)|^2 dz$$

$$\leq \frac{4k^2}{2\pi} \int_D \log \frac{1}{|z|} |F_\rho(z)|^{2k-2} \frac{dz}{(1 - \rho|z|)^2} .$$

Now simply change to polar coordinates and apply the induction hypothesis to obtain (1.12).

The above proof of this lemma is different from Makarov's original proof. It apparently has been discovered independently by several people.

Makarov's LIL now follows easily from (1.12). We argue as in the case of independent random variables. First we sum a series to conclude that

$$(1.13) \qquad \int_T \exp\left[C_1 (F_\rho^*)^2 / \log \frac{1}{1 - \rho}\right] dm \leq C_2$$

for universal constants C_1 and C_2. This in turn gives

$$(1.14) \qquad m\{\theta \in T : F_\rho^*(\theta) > \lambda\} \leq C_2 \exp\left[-C_1 \lambda^2 / \log \frac{1}{1 - \rho}\right]$$

for all $\lambda > 0$. Set

$$\rho_n = 1 - e^{-e^n} \quad \text{and} \quad \lambda_n = \sqrt{\frac{2}{C_1} \log \frac{1}{1 - \rho_n} \log_3 \frac{1}{1 - \rho_n}} .$$

Applying (1.14) and summing over n we conclude that

$$(1.15) \qquad \sum_{n=1}^{\infty} m\{\theta \in T : F^{*}_{\rho_n}(\theta) > \lambda_n\} \le C_2 \sum_{n=1}^{\infty} \frac{1}{n^2} < \infty.$$

It follows from (1.15) and the Borel-Cantelli lemma that for almost all $\theta \in T$, $F^{*}_{\rho_n}(\theta) \le \lambda_n$, eventually. That is, there exists an integer n_0, which may depend on θ, such that for all $n \ge n_0$, $F^{*}_{\rho_n}(\theta) \le \lambda_n$. Let $r \ge \rho_{n_0}$ and choose $n \ge n_0$ such that $\rho_n \le r < \rho_{n+1}$. Since $F^{*}_{\rho}(\theta)$ is increasing as a function of ρ, $F^{*}_{\rho}(\theta) < F^{*}_{\rho_{n+1}}(\theta) \le \lambda_{n+1} \le c\,\lambda_n \le c\sqrt{\log \frac{1}{1-r} \log_3 \frac{1}{1-r}}$, which proves (1.1).

By doing this argument a little more carefully and keeping track of the constants we get that $C = 2\sqrt{e}$ on the right hand side of (1.1). The first proof of Theorem 1.1 gives $C = 2$. We do not know whether this is best possible.

Finally we mention that yet another proof of Makarov's LIL has been given by P. Przytycki [20]. His strategy is to deduce (1.1) from W. Stout's [23] (p. 302) LIL for discrete martingales.

§2 GIRELA'S B_1-RESULT. Makarov's result raises the obvious question. Does this LIL have a lower bound? As mentioned in the introduction, this is easily shown not to be the case by taking any function in $H^{\infty}(D)$. We consider the subclass B_0 of B where the situation is not as obvious. B_0 consists of those $F \in B$ with the property that $(1 - |z|)|F'(z)| \longrightarrow 0$ as $|z| \longrightarrow 1$. The following lemma, combined with Theorem 1.3, shows that the left-hand side of (1.1) is zero almost everywhere whenever $F \in B_0$. Assume for the rest of this section that $F(0) = 0$ and $\|F\|_B \leq 1$.

LEMMA 2. Suppose $F \in B_0$. Given $\epsilon > 0$ there is a $\rho_0 = \rho_0(\epsilon)$ such that

$$g_*^2(F_\rho)(\theta) \leq \epsilon C \log \frac{1}{1-\rho}$$

for all $\rho > \rho_0$.

Proof. Since $F \in B_0$ there exists a constant ρ_1 such that if $|z| \geq \rho_1$, then $|F'(z)| \leq \frac{\epsilon}{(1-|z|)}$. Consider $\rho > \rho_1$, and compute:

$$g_*^2(F_\rho)(\theta) = \frac{1}{\pi}\rho^2 \int_0^1 \int_0^{2\pi} r \log \frac{1}{r} |F'(r\rho e^{i\varphi}|^2 P_\theta(re^{i\varphi})d\varphi dr$$

$$= \frac{1}{\pi} \int_0^\rho \int_0^{2\pi} r \log \frac{\rho}{r} |F'(re^{i\varphi})|^2 P_\theta(\frac{r}{\rho} e^{i\varphi})d\varphi dr$$

$$= \frac{1}{\pi} \int_0^{2\pi} \int_0^{\rho_1} r \log \frac{\rho}{r} |F'(re^{i\varphi})|^2 P_\theta(\frac{r}{\rho} e^{i\varphi})d\varphi dr$$

$$+ \frac{1}{\pi} \int_0^{2\pi} \int_{\rho_1}^{\rho} r \log \frac{\rho}{r} |F'(re^{i\varphi})|^2 P_\theta(\frac{r}{\rho} e^{i\varphi})d\varphi dr$$

$$= I + II.$$

By the Bloch condition and then integration by parts,

$$I \le \frac{1}{\pi}\int_0^{\rho_1}\int_0^{2\pi} r \log\frac{\rho}{r}\,\frac{1}{(1-r)^2}\,P_\theta\!\left(\frac{r}{\rho}e^{i\varphi}\right)d\varphi\,dr$$

$$= 2\int_0^{\rho_1} r\log\frac{\rho}{r}\,\frac{1}{(1-r)^2}\,dr = \frac{r}{(1-r)}\log\frac{\rho}{r}\Big|_0^{\rho_1}$$

$$+ \int_0^{\rho_1}\left(\log\frac{r}{\rho}+1\right)\frac{dr}{(1-r)}$$

$$\le \frac{C}{(1-\rho_1)} + \log\frac{1}{1-\rho_1}\,.$$

Also,

$$II \le \frac{1}{\pi}\int_0^{2\pi}\int_{\rho_1}^{\rho} r\log\frac{\rho}{r}\,\frac{\epsilon^2}{(1-r)^2}\,P_\theta\!\left(\frac{r}{\rho}e^{i\varphi}\right)d\varphi\,dr$$

$$\le C\int_0^{\rho} r\log\frac{\rho}{r}\,\frac{\epsilon^2}{(1-r)^2}\,dr \le \epsilon^2\log\frac{1}{1-\rho}\,.$$

Thus,

$$g_*^2(F_\rho)(\theta) \le \frac{C}{(1-\rho_1)} + C\log\frac{1}{1-\rho_1} + \epsilon^2\log\frac{1}{1-\rho}\,.$$

Choose ρ_0 so that $\frac{1}{(1-\rho_1)} \le \epsilon^2\log\frac{1}{(1-\rho_0)}$ to complete the proof.

Other proofs which show the nonexistence of a lower bound in the class B_0 can be found in [12] and [20]. With this result at hand the question is then: Under what conditions on F does Makarov's LIL have a lower bound on a set of positive measure? For this question the conjecture is the following:

$$\text{(2.1)} \qquad \limsup_{\rho \uparrow 1} \frac{|F(\rho e^{i\theta})|}{\sqrt{\log \frac{1}{1-\rho} \log_3 \frac{1}{1-\rho}}} = 0$$

almost everywhere if and only if

$$\text{(2.2)} \qquad \limsup_{\rho \uparrow 1} \left[\int_T |F(\rho e^{i\theta})|^2 dm / \log \frac{1}{1-\rho} \right] = 0.$$

One direction of this conjecture ((2.1) $\Rightarrow$ (2.2)) is made in [20]. We do not know who conjectured the other direction which we learned from A. Baernstein. Since $\|F_\rho(e^{i\theta})\|_2^2 = \|g_*(F_\rho)\|_2^2$, Lemma 2 shows that (2.2) holds for $F \in B_0$. In [12] Girela proved that (2.2) holds for a subclass of B larger than B_0 and that in this subclass (2.1) also holds. We now introduce this class. The class B_1 consists of those $F \in B$ with the property that if $\{z_n\} \subset D$ and $|F(z_n)| \longrightarrow \infty$ then $(1 - |z_n|)\, |F'(z_n)| \longrightarrow 0$. It is an easy exercise to show that $B_0 \subset B_1$. Examples of functions in B_1 are given by lacunary power series with coefficients tending to zero, [10]. We now present three theorems whose proofs will be given shortly.

THEOREM 2.1 ([12]). Let $F \in B_1$. Then for $0 < p < \infty$,

$$\text{(2.3)} \qquad \int_T |F(\rho e^{i\theta})|^p dm = o\left[(\log \frac{1}{1-\rho})^{p/2} \right]$$

as $\rho \uparrow 1$.

Observe that Lemma 1 ensures that for any $F \in B$ the p-moments of $|F_\rho(e^{i\theta})|$ are always $O\left[(\log \frac{1}{1-\rho})^{p/2} \right]$ as $\rho \uparrow 1$.

THEOREM 2.2. ([12]) Let $F \in B_1$. Then

$$(2.4)\qquad \limsup_{\rho\uparrow 1} \frac{|F(\rho e^{i\theta})|}{\sqrt{\log \frac{1}{1-\rho}\log_3 \frac{1}{1-\rho}}} = 0$$

for almost every $\theta \in T$.

It is also proved in [12] that if the L^2-norm of $F_\rho(e^{i\theta})$ grows slowly enough as $\rho \uparrow 1$ then, with no other assumption on F, we have the conclusion of Theorem 2.2. The following theorem generalizes this result.

THEOREM 2.3. Let $F \in B$. Suppose that for all $0 < \rho < 1$

$$\int_T |F(\rho e^{i\theta})|^2 dm \le \eta(\rho)\log \frac{1}{1-\rho}$$

where η is a function in $(0,1)$ with the property that

$$\sum_{n=2}^{\infty} \frac{\eta(1 - e^{-e^n})}{(\log n)^\beta} < \infty$$

for some $\beta \ge 0$. Then

$$(2.5)\qquad \limsup_{\rho\uparrow 1} \frac{|F(\rho e^{i\theta})|}{\sqrt{\log \frac{1}{1-\rho}\log_3 \frac{1}{1-\rho}}} = 0$$

almost everywhere on T.

Girela's result (see his Theorem 3, [12]) is the case $\eta(\rho) = \left[\log_2 \frac{1}{1-\rho}\right]^{-\alpha}$ for some $\alpha > 1$. The authors originally extended his result to the case $\beta = 0$ in the above theorem. Our proof was an application of a sharp good-λ inequality for the nontangential maximal function and the Lusin square function. A. Baernstein observed, (again for the case $\beta = 0$), that this good-λ inequality was not necessary. Our proof below for the general case of $\beta \geq 0$ is an extension of his idea.

Proof of Theorem 2.3: We may assume without loss of generality that $\beta \geq 2$ and that it is an integer. Let $\rho_n = 1 - e^{-e^n}$ and define $F^*_\rho(\theta)$ as before. From the last argument in the second proof of Theorem 1.1 it suffices to show that $m\{\theta \in T : F^*_{\rho_n}(\theta) > \epsilon\lambda_n, \text{ i.o.}\} = 0$, where $\lambda_n = \sqrt{e^n \log n}$, $\epsilon > 0$ is arbitrary and i.o. means infinitely often. From the Borel Cantelli lemma this will be the case if we can prove that

$$\sum_{n=1}^{\infty} m\{\theta \in T : F^*_{\rho_n}(\theta) > \epsilon\lambda_n\} < \infty. \tag{2.6}$$

Apply Chebychev's inequality to obtain

$$m\{\theta \in T : F^*_{\rho_n}(\theta) > \epsilon\lambda_n\} \leq \frac{1}{(\epsilon\lambda_n)^{2\beta}} \int_T (F^*_{\rho_n}(\theta))^{2\beta} dm$$

$$\leq \frac{C}{(\epsilon\lambda_n)^{2\beta}} \int_T |F(\rho_n e^{i\theta})|^{2\beta} dm$$

where we have used the Hardy-Littlewood maximal theorem. Apply Green's theorem (those who do not like Green's theorem may apply the Itô formula) to get

$$\int_T |F(\rho_n e^{i\theta})|^{2\beta} dm = C_\beta \int_D \log \frac{1}{|z|} |F_{\rho_n}(z)|^{2\beta-2} |F'_\rho(z)|^2 dz$$

$$= C_\beta \rho^2 \int_0^1 \int_0^{2\pi} r \log \frac{1}{r} |F(r\rho_n e^{i\theta})|^{2\beta-2} |F'(r\rho_n e^{i\theta})|^2 d\theta dr$$

$$\le C_\beta \rho^2 \int_0^{2\pi} \sup_{0<r<1} |F(r\rho_n e^{i\theta})|^{2\beta-2} d\theta \int_0^1 r \log\frac{1}{r} \frac{dr}{(1 - r\rho_n)^2}$$

$$\le C_\beta \left[\log \frac{1}{1-\rho_n}\right] \int_T (F^*_{\rho_n}(\theta))^{2\beta-2} dm$$

and repeating the above argument we eventually get that the previous expression is

$$\le C_\beta \left[\log \frac{1}{1-\rho_n}\right]^{\beta-1} \int_T |F(\rho_n e^{i\theta})|^2 dm$$

$$\le C_\beta \left[\log \frac{1}{1-\rho_n}\right]^{\beta} \eta(\rho_n).$$

Thus

$$m\{\theta \in T : F^*_{\rho_n}(\theta) > \epsilon\lambda_n\} \le \frac{C_\beta}{\epsilon^\beta} \frac{\eta(\rho_n)}{(\log n)^\beta}$$

and (2.6) follows from our assumption on η, finishing the proof.

For $0 < p < \infty$ and $0 < \rho < 1$ define

$$\eta_p(\rho) = \int_T |F(\rho e^{i\theta})|^p dm \Big/ \left[\log \frac{1}{1-\rho}\right]^{p/2}.$$

The proof of Theorem 2.3 above shows that if n is a positive integer then $\eta_{2n}(\rho) \le C_n \eta_2(\rho)$ and therefore $\lim_{\rho \uparrow 1} \eta_{2n}(\rho) = 0$ whenever this holds for $n = 1$. If $p > 2$ is not an integer, choose the integer n such that $2n < p < 2(n+1)$. It follows from Jensen's inequality that $\eta_p(\rho) \le (\eta_{2(n+1)}(\rho))^{\frac{p}{2(n+1)}}$. From this and the previous observation we conclude that $\eta_p(\rho) \longrightarrow 0$ as $\rho \uparrow 1$ whenever this happens for $p = 2$. If $0 < p < 2$, again apply Jensen's inequality to obtain the same conclusion. Thus, Theorem 2.1 will follow from the lemma below. In fact, this lemma also implies, as we will momentarily show, Theorem 2.2.

LEMMA 3. Let $F \in B_1$. Given $\epsilon > 0$ there exists a sequence of numbers $\rho(n,\epsilon)$, $n = 1,2,\ldots$ increasing to 1 such that with $\lambda(\rho) = \log \frac{1}{1-\rho}$,

$$\int_T (F_\rho^*(\theta))^{2n} dm \le C_1^n \epsilon^n n^n (\lambda(\rho))^n \tag{2.7}$$

for all $\rho > \rho(n,\epsilon)$. In fact,

$$\rho(n,\epsilon) = 1 - \exp(-\exp(C_2 n \log \beta(\epsilon))).$$

C_1 and C_2 are absolute constants and $\beta(\epsilon)$ depends on ϵ.

Before proving the lemma, let us deduce Theorem 2.2 from it. Recall that $\rho_n = 1 - e^{-e^n}$ and $\lambda_n = \sqrt{e^n \log n}$. From the argument already used twice it suffices to show that

$$\sum_{k=1}^{\infty} m\{\theta \in T : F_{\rho_n}^*(\theta) > \epsilon^{1/2} C \lambda_n\} < \infty \tag{2.8}$$

where C is an absolute constant. For every positive integer n let $N(n)$ be the integer part of $\log n$. Observe that if n is large enough $\rho_n > \rho(N(n),\epsilon)$. For such n's apply Chebychev's inequality and the lemma to get

$$m\{\theta \in T : F^*_{\rho_n}(\theta) > \epsilon^{1/2} C \lambda_n\}$$

$$\leq \frac{1}{\epsilon^{N(n)} C^{2N(n)} \lambda_n^{2N(n)}} \int_T (F^*_{\rho_n}(\theta))^{2N(n)} dm$$

$$\leq \frac{C_1^{N(n)} (N(n))^{N(n)}}{C^{2N(n)} (\log n)^{2N(n)}} \leq \frac{C_1^{N(n)}}{C^{2N(n)}}$$

and if we take $C = C_1^{1/2} e^{25}$ we clearly have (2.8) which proves the theorem.

<u>Proof of Lemma 3</u>. Let us first recall a few facts about local time. Let $(X_t)_{t\geq 0}$ be a continuous margingale starting at zero with square function $\langle X\rangle_t$ and with its jointly continuous family $\{L_t^a;\ t \geq 0,\ a \in \mathbb{R}\}$ of local times. The first fact we shall need is the fundamental density of occupation formula:

$$(2.9) \qquad \int_0^t f(X_s) d\langle X\rangle_s = \int_{-\infty}^{\infty} f(a) L_t^a da$$

for all bounded Borel functions $f : \mathbb{R} \longrightarrow \mathbb{R}$, [14] (p. 116). We shall also need the following beautiful result of Barlow and Yor [2]: Let $0 < p < \infty$ and set $L^* = \sup_{a\in\mathbb{R}} L_\infty^a$. There exists constants a_p and A_p depending only on p such that

$$(2.10)\qquad a_p \|L^*\|_p \le \|\langle X\rangle^{1/2}\|_p \le A_p \|L^*\|_p .$$

For $1 < p < \infty$ (2.10), together with the Burkholder-Gundy inequalitites and Doob's maximal theorem, implies that

$$(2.11)\qquad a_p \|L^*\|_p \le \|X\|_p \le A_p \|L^*\|_p$$

for some other constants a_p, A_p. Here, $X = X_\infty$.

If B_t is Brownian motion in the disc D started at the origin and $0 < \rho < 1$, let τ_ρ be the exit time from the disc $D(0,\rho) = \{z \in \mathbb{C} : |z| < \rho\}$. For $\rho = 1$, we write τ_D for τ_1. The following can be found in [8]:

$$(2.12)\qquad E\left[\int_0^{\tau_\rho} g(B_s)ds\right] = C\int_{D(0,\rho)} \log \frac{\rho}{|z|}\, g(z)dz.$$

It follows from the Hardy-Littlewood maximal theorem and Stirling's formula that it suffices to prove (2.7) with $|F(\rho e^{i\theta})|$ replacing the maximal function on the left-hand side and $n!$ replacing n^n on the right-hand side. Since $F \in B_1$, given $\epsilon > 0$ we can find $M = M(\epsilon)$ such that if $|F(z)| > M$ then $|F'(z)| \le \epsilon/(1 - |z|)$. Choose $\rho(\epsilon,1)$ such that

$$(2.13)\qquad M = \epsilon\sqrt{\lambda(\rho(1,\epsilon))}\ .$$

We notice that $\rho(1,\epsilon) = 1 - \exp\left[-\exp(2 \log \frac{M}{\epsilon})\right]$, which is of the form indicated in the statement of the lemma.

Let $\rho > \rho(1,\epsilon)$ and apply the Itô formula to get $(u = \mathrm{Re}F$ and $\Delta|F|^2 = 2^2|F'|^2 = 2^2|\nabla u|^2)$

$$\int_T |F(\rho e^{i\theta})|^2 dm = C\,E\left[\int_0^{\tau_\rho} |\nabla u(B_s)|^2 ds\right]$$

$$\leq C\,E\left[\int_0^{\tau_\rho} \chi_{\{|F|\leq M\}} |\nabla u(B_s)|^2 ds)\right] + CE\left[\int_0^{\tau_\rho} \chi_{\{|F|>M\}} |\nabla u(B_s)|^2 ds\right]$$

$$\leq C\,E\left[\int_0^{\tau_\rho} \chi_{\{|u|\leq M\}} |\nabla u(B_s)|^2 ds) + C\epsilon^2\,E\left[\int_0^{\tau_\rho} \frac{ds}{(1-|B_s|)^2}\right]\right.$$

$$= C\,E\left[\int_{-M}^{M} L^a da\right] + C\,\epsilon^2 \int_{D(0,\rho)} \log\frac{\rho}{|z|}\,\frac{dz}{(1-|z|)^2} = I + II$$

where L^a denotes the local time for the martingale $u(B_{t\wedge\tau_\rho})$ and we have used (2.9) and (2.12). Changing to polar coordinates and computing the integral we find that $II \leq C\,\epsilon^2\lambda(\rho)$. Using the obvious majorization, Jensen's inequality and (2.11) we get

$$I \leq CM\left[E(L^*)\right] \leq CM\left[E(L^*)^2\right]^{1/2}$$

$$\leq CM\left[E|u(B_{\tau_\rho})|^2\right]^{1/2}$$

$$\leq CM\left[\int_T |F(\rho e^{i\theta})|^2 dm\right]^{1/2}$$

$$\leq c\epsilon\sqrt{\lambda(\rho(1,\epsilon))}\ \sqrt{\lambda(\rho)}$$

where we have used lemma 1 to estimate the L^2-norm of $F(\rho e^{i\theta})$. Since $\rho > \rho(1,\epsilon)$,

$$\int_T |F(\rho e^{i\theta})|^2 dm \le I + II \le C\epsilon\lambda(\rho)$$

proving the lemma for $n = 1$.

We continue by induction. Suppose $n \geq 2$ and that we have found $\rho(1,\epsilon) < \rho(2,\epsilon) < \cdots < \rho(n-1,\epsilon)$ such that for all $\rho > \rho(n-1,\epsilon)$ we have

$$\int_T |F(\rho e^{i\theta})|^{2(n-1)} dm \le \epsilon^{n-1} C^{n-1} (n-1)! (\lambda(\rho))^{n-1}. \tag{2.14}$$

Let ρ be much larger than $\rho(n-1,\epsilon)$. Apply the Itô formula to obtain

$$\int_T |F(\rho e^{i\theta})|^{2n} dm = Cn^2\, E\left[\int_0^{T_\rho} |F(B_s)|^2 ds\right]$$

$$= Cn^2\, E\left[\int_0^{T_\rho} \chi_{\{|F|\le M\}} |F(B_s)|^{2n-2} |F'(B_s)|^2 ds\right]$$

$$+ Cn^2\, E\left[\int_0^{T_\rho} \chi_{\{|F|>M\}} |F(B_s)|^{2n-2} |F'(B_s)|^2 ds\right] = I + II.$$

By the same argument as above, and adopting the convention that C is a constant which may change from line to line,

$$I \le Cn^2\, M^{2n-2} M(\lambda(\rho))^{1/2}$$

$$\le Cn^2\, \epsilon^{2n-1} \left[\lambda(\rho(1,\epsilon))\right]^{n-\frac{1}{2}} \left[\lambda(\rho)\right]^{1/2}$$

$$\leq C^n n! \epsilon^n \left[\lambda(\rho)\right]^n .$$

$$II \leq Cn^2 \epsilon^2 E\left[\int_0^{T_\rho} |F(B_s)|^{2n-2} \frac{ds}{(1 - |B_s|)^2}\right]$$

$$\leq Cn^2\epsilon^2 \int_{D(0,\rho)} \log \frac{\rho}{|z|} |F(z)|^{2n-2} \frac{dz}{(1 - |z|)^2}$$

$$= Cn^2\epsilon^2 \int_0^{\rho} \int_0^{2\pi} r \log \frac{\rho}{r} |F(re^{i\theta})|^{2n-2} \frac{dr\ d\theta}{(1 - r)^2}$$

$$= Cn^2\epsilon^2 \int_0^{\rho(n-1,\epsilon)} \int_0^{2\pi} r \log \frac{\rho}{r} |F(re^{i\theta})|^{2(n-1)} \frac{dr\ d\theta}{(1 - r)^2}$$

$$+ Cn^2\epsilon^2 \int_{\rho(n-1,\epsilon)}^{\rho} \int_0^{2\pi} r \log \frac{\rho}{r} |F(re^{i\theta})|^{2(n-1)} \frac{dr\ d\theta}{(1 - r)^2}$$

$$= III + IV.$$

Integrating with respect to θ first and using the induction hypothesis we find that

$$IV \leq Cn^2 \epsilon^2 C^{n-1} \epsilon^{n-1} (n - 1)! \int_{\rho(n-1,\epsilon)}^{\rho} r \log \frac{\rho}{r} (\lambda(r))^{n-1} \frac{dr}{(1 - r)^2}$$

$$\leq C^n \epsilon^{n+1} n! n \int_0^{\rho} r \log \frac{\rho}{r} (\lambda(r))^{n-1} \frac{dr}{(1 - r)^2}$$

$$\leq C^n \epsilon^{n+1} n! (\lambda(\rho))^n .$$

For III we do not have much choice but to integrate first with respect to θ and use lemma 1. We get,

$$\text{III} \le C\, \epsilon^2 C^{n-1} n^2 (n-1)! \int_0^{\rho(n-1,\epsilon)} r \log \frac{\rho}{r} (\lambda(r))^{n-1} \frac{dr}{(1-r)^2}$$

$$= \epsilon^2 C^n n!\, (\lambda(\rho(n-1,\epsilon)))^n.$$

Thus we need to choose $\rho(n,\epsilon) > \rho(n-1,\epsilon)$ such that

$$\epsilon^2 C^n n! (\lambda(\rho(n-1,\epsilon)))^n \le \epsilon^n C^n n! (\lambda(\rho(n,\epsilon)))^n$$

or

$$\frac{1}{\epsilon^{n-2}} \left[\log \left(\frac{1}{1-\rho(n-1,\epsilon)}\right)\right]^n \le \left[\log \left(\frac{1}{1-\rho(n,\epsilon)}\right)\right]^n$$

and with $\rho(n,\epsilon) = 1 - \exp(-\exp(\beta(n,\epsilon)))$ gives

$$(1 - \frac{2}{n}) \log \frac{1}{\epsilon} \le \beta(n,\epsilon) - \beta(n-1,\epsilon).$$

So that we can take

$$\beta(n,\epsilon) \sim \log \frac{M}{\epsilon} (n - \log n) + \beta(1,\epsilon)$$

$$\sim n \log \frac{M}{\epsilon} = C\, n \log \beta(n),$$

which proves the lemma.

REMARK 1. It is natural to ask if Lemma 2 continues to hold for functions in B_1. This turns out to be false. A. Baernstein (personal communication) has found examples of functions in H^∞, hence in B_1, for which lemma 2 does not hold.

REMARK 2. P. Jones (personal communication) has shown that if $f \in B$ and if it satisfies what is essentially a uniform reverse Bloch condition, (which implies $\eta_2(\rho) > C > 0$ for all $0 < \rho < 1$, and more), then there is a lower bound in Makarov's LIL. He uses this result to study properties of the harmonic measure on what he calls "Makarov domains".

§3. AN LIL FOR HARMONIC FUNCTIONS IN $\mathbb{R}^2_+$. Let X_t be a continuous martingale with square function (quadratic variation) $\langle X\rangle_t$. Define $X^*_t = \sup_{s\leq t} |X_s|$. Since every such martingale is a time changed Brownian motion, it is easy to show [8] (p. 76) that except for sets of probability zero $\{X^*_\infty = \infty\} = \{\langle X\rangle_\infty = \infty\} = \{\lim_{t\to\infty} X_t \text{ exists}\}$. The LIL for continuous martingales [8] (p. 77), which is also obtained from the LIL for Brownian motion by time change, gives precise information on the relative "order of infinities" between X^*_∞ and $\langle X\rangle_\infty$.

$$(3.1) \qquad \limsup_{t\to\infty} \frac{X^*_t}{\sqrt{2\langle X\rangle_t \log_2 \langle X\rangle_t}} = 1$$

a.s. on $\{\langle X\rangle_\infty = \infty\}$. Notice that unlike all the previous LIL's we have discussed, the denominator is also random in (3.1). This introduces many complications which did not exist in the case when this is constant.

Now let $\mathbb{R}_+^{n+1} = \{(x,y) : x \in \mathbb{R}^n,\ y > 0\}$ and for $\alpha > 0$ and $(x,y) \in \mathbb{R}_+^{n+1}$ denote by $\Gamma_\alpha(x,y)$ the cone of aperture α, vertex at (x,y) and vertical axis, that is $\Gamma_\alpha(s,y) = \{(s,t) : |x - s| \leq \alpha(t - y)\}$. If u is harmonic in $\mathbb{R}_+^{n+1}$, define the nontangential maximal function of u, $N_\alpha(u)(x,y)$, and the Lusin area function of u, $A_\alpha(u)(x,y)$, by

$$N_\alpha(u)(x,y) = \sup_{\Gamma_\alpha(x,y)} |u|$$

and

$$A_\alpha(u)(x,y) = \left[\iint_{\Gamma_\alpha(x,y)} |\nabla u(s,t)|^2 (t - y)^{1-n} dsdt \right]^{1/2}.$$

When $y = 0$ we simply write $N_\alpha(u)(x)$ and $A_\alpha(u)(x)$. The famous results of Privalov, Marcinkiewicz and Zygmund, Spencer, Calderón, and Stein say that except for sets of Lebesgue measure zero, for all $\alpha > 0$, $\{x \in \mathbb{R}^n : N_\alpha(u)(x) = \infty\} = \{x \in \mathbb{R}^n : A_\alpha(u)(x) = \infty\} = \{x \in \mathbb{R}^n : u$ does not have nontangential limit at $x\}$, (see [22]).

The basic philosophy, which has been tremendously useful for many years now, is that X_∞^* and $\langle X \rangle_\infty^{1/2}$ behave very much like $N_\alpha(u)(x)$ and $A_\alpha(u)(x)$. Inspired by this, (3.1), and the fact that $N_\alpha(u)(x)$ and $A_\alpha(u)(x)$ are infinite on the same sets, R. F. Gundy proposed the problem of proving an LIL similar to (3.1) for N_α and A_α. Our first result below says that an upper LIL does hold, at least when $n = 1$.

THEOREM 3.1. Let u be harmonic in $\mathbb{R}^2_+$. Fix $\alpha > 1$ and $\beta > \alpha$. Suppose there is a point $(x_0, y_0) \in \overline{\mathbb{R}^2_+}$, the closure of $\mathbb{R}^2_+$, such that $N_\beta(u)(x_0, y_0) \neq \infty$. Then

$$\limsup_{y \downarrow 0} \frac{N_\alpha(u)(x,y)}{\sqrt{A^2_\beta(u)(x,y)\ \log_2 A_\beta(u)(x,y)}} \leq C \tag{3.2}$$

for almost every $x \in \{x \in \mathbb{R} : A_\beta(u)(x) = \infty\}$. Here C is a constant depending only on α and β.

As we mentioned in the introduction, this result is joint work with I. Klemes and the proof will appear elsewhere.

When we adapt this theorem to the unit disc we obtain a result more general than Theorem 1.3, (since $A_\alpha(F)(\theta) \leq C_\alpha g_*(F(\theta))$, and hence, with regards to the upper bound, more general than any of the other LIL's discussed in this paper. This also shows, together with the sharpness part of Theorem 1.3, that the constant on the right hand side of (3.2) can not be 0. (The theorem will not be very interesting if it turned out that $C \equiv 0$). At present we do not know if there is always a lower bound. The proof follows very closely the proof for discrete martingales together with a version of Theorem 1.2 for Lipschitz domains and properties of A_∞-weights.

Also, there is a similar theorem with the roles of N and A interchanged which holds in all dimensions and with less restrictions on the aperture of the cones. The proof of this result will also appear elsewhere.

THEOREM 3.2. Let u be harmonic in $\mathbb{R}^{n+1}_+$. Fix $\alpha > 0$ and $\beta > \alpha$. Suppose there is a point $(x_0, y_0) \in \overline{\mathbb{R}^{n+1}_+}$, the closure of $\mathbb{R}^{n+1}_+$, such that $A_\beta(u)(x_0, y_0) \neq \infty$. Then

$$\limsup_{y \downarrow 0} \frac{A_\alpha(u)(x,y)}{\sqrt{N_\beta^2(u)(x,y) \log_2 N_\beta(u)(x,y)}} \le C \tag{3.3}$$

for almost every $x \in \{x \in \mathbb{R}^n : N_\beta(u)(x) \equiv \infty\}$. The constant C depends only on α, β, and n.

ACKNOWLEDGEMENT: We are grateful to A. Baernstein and W. Philipp for useful conversations on the subject of this paper. We thank R. F. Gundy for his encouragement and interest in analytic LIL's.

References

1. R. Bañuelos, Brownian motion and area functions, Ind. Univ. Math. J. 35(1986), 643-668.

2. M. T. Barlow and M. Yor, Semi-martingale inequalities via the Garsia-Rodemick-Ramsey lemma, and applications to local times, J. Funct. Analy. 49(1982), 198-229.

3. N. H. Bingham, Variants on the law of the iterated logarithm, Bull. London Math. Soc. 18(1986) 433-467.

4. L. Caffarelli, E. Fabes, S. Mortola, and S. Salsa, Boundary behavior of nonnegative solutions of elliptic operators in divergence form, Ind. Univ. Math. J. 30(1981), 621-640.

5. M. Cranston, Lifetime of Conditional Brownian motion in Lipschitz domains, Z. Wahrscheinlichteitstheor. Verw. Geb., 70(1985), 335-340.

6. S. Y. A. Chang, J. M. Wilson, and T. H. Wolff, Some weighted norm inequalities concerning the Schrödinger operator, Comment. Math. Helv. 60(1985), 217-246.

7. K. L. Chung, A course in Probability Theory, 2d ed Academic Press, New York, 1974.

8. R. Durrett, Brownian Motion and Martingales In Analysis, Wadsworth, Belmont, California, 1984.

9. P. Erdös and I. S. Gál, On the law of the iterated logarithm, Nederl, Akad. Wetensch. Proc. Ser. A., 58(1955), 65-84.

10. J. L. Férnandez, On the coefficients of Bloch functions, J. London Math. Soc. 29(1984), 94-102.

11. J. Garnett, Bounded Analytic Functions, Academic Press, New York, 1980.

12. D. Girela, Integral means and radial growth of Bloch functions, Math. Zeits., (to appear).

13. R. F. Gundy, The density of the area integral, in conference on harmonic analysis in honor of Antoni Zygmund, editors, Beckner, Calderón, Fefferman, and Jones, Wadsworth, Belmont, California, 1983.

14. N. Ikeda and S. Watanabe, Stochastic Differential Equations and Diffusion Process, North Holland, Amsterdam, 1981.

15. N. Kolmogorov, Über des Gesetz des iterieten logarithmus, Math. Annalen, 101(1929), 126-136.

16. N. G. Makarov, On the distortion of boundary sets under conformal mappings, Proc. London Math. Soc. 51(1985), 369-384.

17. W. Philipp and W. Stout, Almost sure invariance principles for partial sums of weakly dependent random variables, Mem. Amer. Math. Soc., No. 161(1975).

18. Ch. Pommerenke, On Bloch functions, J. London Math. Soc. 2(1970), 689-695.

19. Ch. Pommerenke, The growth of the derivative of a univalent function, Purdue conference (1985).

20. P. Przytycki, On the law of iterated logarithm for Bloch functions, preprint.

21. R. Salem and A. Zygmund, La loi du logarithme itéré pour les séries trigonométriques lacunary, Bull. Sci. Math. 74(1950), 209-224.

22. E. M. Stein, Singular Integrals and Differentiability Properties of Functions, Princeton University Press, 1970.

23. W. F. Stout, Almost sure Convergence, Academic Press, New York, 1974.

24. S. Takahashi, Almost invariance principles for lacunary trigonometric series, Tôhoku Math. J. 31(1979), 437-451.

25. M. Weiss, The law of the iterated logarithm for lacunary series, Trans. Amer. Math. Soc. 91(1959), 444-469.

RODRIGO BAÑUELOS
DEPARTMENT OF MATHEMATICS
UNIVERSITY OF ILLINOIS
URBANA, IL 61801

CHARLES N. MOORE
DEPARTMENT OF MATHEMATICS
WASHINGTON UNIVERSITY
ST. LOUIS, MO 63130

Current address of R. Bañuelos: Department of Mathematics, Purdue University, West Lafayette, IN 47906.

FOURIER SERIES, MEAN LIPSCHITZ SPACES, AND BOUNDED MEAN OSCILLATION

Paul S. Bourdon, Joel H. Shapiro, and William T. Sledd

ABSTRACT. Using simple and direct arguments, we: (i) prove, without recourse to duality, that the mean Lipschitz spaces $\Lambda(p,1/p)$ are contained in BMO, and (ii) improve the Hardy-Littlewood $\Lambda(p,1/p)$ Tauberian theorem. Along the way we connect the Hardy-Littlewood result with a recent Tauberian theorem for BMO functions due to Ramey and Ullrich, give an exposition of the relevant classical properties of Mean Lipschitz spaces; and survey some known function theoretic applications of the spaces $\Lambda(p,1/p)$.

INTRODUCTION. We work mostly on the unit circle **T**, and study for $1 < p < \infty$ the spaces $\Lambda(p,1/p)$ consisting of functions $f \in L^p(T)$ for which $\|f - f_t\|_p = O(t^{1/p})$ as $t \to 0$, where $f_t(x) = f(x - t)$. These spaces increase with p, and while none of them consists entirely of bounded functions ($\log(1-e^{ix})$ belongs to all of them), they all lie "on the border of continuity." More precisely, if in the definition of $\Lambda(p,1/p)$ the exponent $1/p$ is replaced by anything larger, there results a space of functions, each of which, after possible correction on a set of measure zero, is continuous. Our interest in $\Lambda(p,1/p)$ derives from four sources:

(i) the observation of Cima and Petersen [5] that $\Lambda(2,1/2)$ lies inside BMO, the space of functions of bounded mean oscillation on T;

(ii) the fact that the essential range of a function of vanishing mean oscillation must be connected [26];

(iii) a 1928 result of Hardy and Littlewood which states that $\Lambda(p,1/p)$ is a Tauberian condition relating Cesàro and ordinary summability. More precisely, the Fourier series of $f \in \Lambda(p,1/p)$, if

---

1980 Mathematics Subject Classification (1985 Revision). 42A20, 46E36.
Research supported in part by the National Science Foundation

(C,1) summable at a point of T, must actually converge there ([15], Theorem 1, page 613);

(iv) a recent theorem of Ramey and Ullrich [23] asserting that BMO is a Tauberian condition relating Abel summability and differentiability of indefinite integrals.

The original proofs of (i) and (iii) require complicated preliminaries. That of (i) in [5] uses the deepest part of the Fefferman-Stein duality theorem: the characterization of BMO functions by Carleson measures ([14], Chapter 6, Theorem 3.4, page 240; [10]); while Hardy and Littlewood prove (iii) by interpolating between ordinary Cesàro summability, and Cesàro boundedness of negative orders. Mysteriously, neither Hardy and Littlewood nor their successors in the literature appear to have considered the question of whether <u>Abel</u> summability at a point implies summability for the Fourier series of a $\Lambda(p,1/p)$ function.

This paper addresses both these complaints, and treats some additional topics suggested by the connection between the spaces $\Lambda(p,1/p)$ and BMO. First, we give a direct proof that $\Lambda(p,1/p) \subset$ BMO for all finite p. There is a corresponding containment between the "little oh" space $\lambda(p,1/p)$ and VMO, the space of functions of vanishing mean oscillation, which along with (ii) above shows that functions in $\lambda(p,1/p)$ must have connected essential range. We then extend the Hardy-Littlewood $\Lambda(p,1/p)$ Tauberian theorem to Abel summability by proving that at each point of the circle the sequence of Fourier partial sums of each $\Lambda(p,1/p)$ function is slowly oscillating. The fact that Abel summability of the Fourier series implies summability then follows from a standard Tauberian theorem. Our argument is considerably simpler than that of Hardy and Littlewood in that it avoids interpolation arguments and negative order Cesaro means. We discuss the connection between the Hardy-Littlewood Tauberian theorem and the Ramey-Ullrich theorem mentioned in (iv) above.

Both Hardy and Littlewood ([15], Lemma 12, page 620) and Cima and Petersen ([5], Theorem 2.1 and Cor. 2.2) noted that a function belongs to $\Lambda(2,1/2)$ whenever its Fourier coefficients decay like $O(1/n)$. Thus our $\Lambda(p,1/p)$ Tauberian theorem can be viewed as an extension of Littlewood's $O(1/n)$ Tauberian theorem. By duality, the $O(1/n)$ sufficient condition for membership in $\Lambda(2,1/2)$ can also be regarded as a generalization of Hardy's inequality from functions in the Hardy space H^1 to functions in a somewhat larger space.

Several classical papers extend the work of Hardy and Littlewood on Fourier series of mean Lipschitz functions (see [11] for further references, and for a unified treatment of some of this) and there is a vast literature about various generalizations of these spaces (see [4], [7], [12], [13], [17], [22], [25], [31] - [33] for the flavor of some of these generalizations, and for further references). Thus it seems quite possible that, while we have not yet come across them, our "new" results may already be known. However, much of the literature of mean Lipschitz spaces deals with settings far more complicated than ours, and is therefore not always as accessible as it should be to researchers in one variable function theory. For this reason we feel that an account of our work placed within the context of a detailed discussion of the relevant classical properties of mean Lipschitz spaces may, in any case, be of interest to function theorists.

Accordingly, we adhere to the following ground rules. We state without proof, but with detailed references: (i) function theoretic facts that can be found in Duren's book [8] on H^p theory, or Rudin's text [24]; (ii) basic facts about BMO such as are set out in Garnett's book [14], and (iii) classical Tauberian theorems for numerical series. On the other hand we give detailed proofs of all prerequisites on mean Lipschitz spaces that do not occur in these sources.

Here is an outline of the rest of the paper. Section 1 contains definitions, notation, and first properties of the mean Lipschitz spaces. Here we review the characterization of these spaces via Poisson integrals, and their resulting self-conjugacy. In the second section we discuss containments among the mean Lipschitz spaces, and prove that $\Lambda(p,1/p) \subset \mathrm{BMO}$. The proof of our Tauberian theorem, as well as the discussion of the Ramey-Ullrich theorem, occupies section 4. This proof depends on a well-known characterization, presented in section 3, of $\Lambda(p,1/p)$ by the degree to which its members can be approximated by Fourier partial sums. The approximation theorem follows from a Littlewood-Paley type dyadic decomposition theorem, as used in [27]. The dyadic decomposition leads to a characterization of the Fourier coefficients of functions in $\Lambda(2,1/2)$, which in turn yields the previously mentioned "$O(1/n)$" sufficient condition for membership in $\Lambda(2,1/2)$. In the fifth and final section we comment how this sufficient condition generalizes Hardy's inequality for H^1, and survey a few other situations in function theory where the spaces $\Lambda(p,1/p)$ occur.

1. PRELIMINARIES. $L^p(T)$ $(1 \leq p < \infty)$ denotes the space of (equivalence classes of) 2π-periodic measurable functions f on the real line for which

$$\|f\|_p^p = \frac{1}{2\pi}\int_{-\pi}^{\pi} |f(t)|^p \, dt$$

while $L^\infty(T)$ is the corresponding space of essentially bounded measurable functions. The translate f_t of a function f on the real line by the real number t is defined by: $f_t(x) = f(x - t)$ for all real numbers x.

1.1. MEAN LIPSCHITZ SPACES $\Lambda(p,\alpha)$. For $1 \le p \le \infty$ and $0 < \alpha \le 1$ we define $\Lambda(p,\alpha)$ to be the collection of $f \in L^p(T)$ for which there exists a constant $C < \infty$ such that:

$$\|f_t - f\|_p \le C|t|^\alpha \text{ for all } t \in [-\pi,\pi] .$$

If $p = \infty$ we write Λ_α instead of $\Lambda(\infty,\alpha)$. This is the usual Lipschitz space for the exponent α. More precisely, $f \in \Lambda_\alpha$ if and only if f coincides a.e. with a 2π periodic function F for which:

$$|F(x-t) - F(x)| \le |t|^\alpha \text{ for all } x, t \in [-\pi,\pi].$$

It is not difficult to show that the norm

$$\|f\|_{p,\alpha} = \|f\|_p + \sup\{t^{-\alpha}\|f_t - f\|_p : t \in [-\pi,\pi]\}$$

turns $\Lambda(p,\alpha)$ into a Banach space.

Clearly the spaces $\Lambda(p,\alpha)$ decrease as either p or α increases (with the other index held fixed). In the next section we examine the containments between these spaces in more detail. The key to these results, as well as to much of our subsequent work, is a useful characterization of mean Lipschitz spaces in terms of Poisson integrals.

1.2. POISSON INTEGRALS. Let U denote the open unit disc of the complex plane. For $f \in L^p(T)$, we denote by P[f] the Poisson integral of f:

$$P[f](re^{i\theta}) = \frac{1}{2\pi}\int_{-\pi}^{\pi} P_r(\theta - t)\, f(t)\, dt ,$$

where $P_r(t) = \mathrm{Re}\{(1+ re^{it})/(1- re^{it})\}$, and $re^{i\theta} \in U$. It is well known that $u = P[f]$ is harmonic in U, that the integral means

$$M_p(u,r) = \left\{\frac{1}{2\pi}\int_{-\pi}^{\pi} |u(re^{it})|^p dt\right\}^{1/p}$$

are bounded, and that the radial limit

$$u^*(e^{i\theta}) = \lim u(re^{i\theta}) \quad (r \to 1-)$$

exists and equals $f(\theta)$ for a.e. real θ ([8], Chapter 1). Conversely, if u is any harmonic function on U for which the integral means $M_p(u,r)$ are bounded, then $f = u^*$ exists a.e., belongs to $L^p(T)$, and $u = P[f]$. In short, if h^p denotes the class of harmonic functions u for which $M_p(u,r)$ is bounded, then the radial limit map $u \to u^*$ establishes an isometric isomorphism between h^p, taken in the natural norm imposed by its definition, and $L^p(T)$. The next result identifies the class of harmonic functions that corresponds in the same manner with $\Lambda(p,\alpha)$.

1.3. PROPOSITION. Suppose u is harmonic in U, $u = \mathrm{Re}\, F$ where F is holomorphic in U, $1 \leq p \leq \infty$, and $0 < \alpha < 1$. Then following conditions are equivalent:

(a) $u = P[f]$ for some $f \in \Lambda(p,\alpha)$,

(b) $M_p(F',r) \leq C(1-r)^{\alpha-1}$ for all $0 \leq r < 1$.

When (b) holds, the functional

$$f \to |u(0)| + \sup \{M_p(F',r)(1-r)^{1-\alpha}: 0 \leq r < 1\}$$

is a norm on $\Lambda(p,\alpha)$ that is equivalent to $\|\cdot\|_{p,\alpha}$.

With F in place of u, this result was proved by Hardy and Littlewood ([15], Theorem 3, page 625). It can also be found in [8] (Theorem 5.4, page 78). The proof is exactly the one given in these references, except that in proving (a) → (b) one represents F by a completed Poisson integral of f, rather than a Cauchy integral.

It follows easily from Proposition 1.3 that, as we mentioned in the Introduction, the function $\log|1 - e^{i\theta}|$ is in $\Lambda(p,1/p)$.

Higher dimensional versions of this result, for more general spaces, can be found in [28], Chapter 5, and [31].

It follows from the M. Riesz theorem that the classes $\Lambda(p,\alpha)$ are self-conjugate if $1 < p < \infty$. However the result above gives a more elementary proof, valid even if $p = 1$ or ∞.

1.4. COROLLARY. If $1 \leq p \leq \infty$, $0 < \alpha < 1$, and $f \in \Lambda(p,\alpha)$, then so is its conjugate function $\tilde{f}$.

In [15] Hardy and Littlewood also prove Corollary 1.4 for the case $p > 1$, $0 < \alpha \leq 1$; and the case $p = 1$, $0 < \alpha < 1$ (Lemma 13, page 621). The result is false for $p = \alpha = 1$. Indeed, $f \in \Lambda(1,1)$ if and only if f coincides a.e. with a function of bounded variation ([15], Lemma 9, page 619), and the class of functions of bounded variation is not self-conjugate.

2. CONTAINMENTS AMONG THE SPACES $\Lambda(p,\alpha)$. We observed in the last section that the spaces $\Lambda(p,\alpha)$ decrease as either p or α increases. The next result gives more precise information.

2.1 PROPOSITION ([15] Theorem 5, p. 627). Suppose $1 \leq p \leq q \leq \infty$, $0 < \alpha < 1$, and $\delta = p^{-1} - q^{-1}$. Then $\Lambda(p,\alpha) \subset \Lambda(q,\alpha-\delta)$.

PROOF. By the self-conjugacy of the spaces in question, it is enough to prove the theorem for $f \in \Lambda(p,\alpha)$ with Fourier transform vanishing on the negative integers. Then $F = P[f]$ is holomorphic in U, and by Proposition 1.3:

$$M_p(F',r) \leq C_p \|f\|_{p,\alpha}(1-r)^{\alpha-1} \qquad (0 < r < 1).$$

The Hardy-Littlewood theorem on comparative growth of means ([15], Theorem 2, page 623; [8], Theorem 5.9, page 84) asserts that if $g \in H^p$, with p, q, and δ as above, then

$$M_q(g,r) \leq C_{p,q} \|g\|_p (1-r)^{-\delta}.$$

We apply this result to the dilated function $g = (F')_r$ to obtain

$$M_q(F',r^2) \leq C_{p,q} M_p(F',r)(1-r)^{-\delta}.$$

Thus:

$$M_q(F',r^2) \leq C_{p,q} \|f\|_{p,\alpha}(1-r)^{\alpha-\delta-1},$$

which, along with another application of Proposition 1.3, proves the desired result. ///

Upon respectively setting $\alpha = 1/p$ and $q = \infty$, we justify the claims made in the Introduction that the spaces $\Lambda(p,1/p)$ increase with p, and lie on the border of continuity.

2.3. COROLLARY ([15], Theorems 5 and 6, pp. 627-8).

(a) If $1 < p < q < \infty$, then $\Lambda(p,1/p) \subset \Lambda(q,1/q)$.

(b) If $\alpha > 1/p$, then $\Lambda(p,\alpha) \subset \Lambda_{\alpha-(1/p)}$, hence each $f \in \Lambda(p,\alpha)$ coincides a.e. with a continuous function.

2.4. BOUNDED MEAN OSCILLATION. Suppose $f \in L^1(T)$. If I is a subinterval of $[0,2\pi]$, let $|I|$ denote its length, and write

$$f_I = \frac{1}{|I|}\int_I f(t)dt .$$

Set

$$\|f\|_* = \sup\left\{\frac{1}{|I|}\int_I |f(t)-f_I|dt : I \text{ a subinterval of } [0,2\pi]\right\}.$$

The space BMO of functions of *bounded mean oscillation* is the collection of $f \in L^1(T)$ for which $\|f\|_* < \infty$. The John-Nirenberg theorem ([18]; [14], Chapter VI, Theorem 2, page 230) implies that $BMO \subset L^p(T)$ for all $p < \infty$, and that the same space, with equivalent norm, results if we redefine $\|f\|_*$ by:

$$\|f\|_*^p = \sup\left\{\frac{1}{|I|}\int_I |f(t)-f_I|^p dt: I \text{ a subinterval of } [0,2\pi]\right\}.$$

As we mentioned in the Introduction, Cima and Petersen [5] used deep results about BMO to show that $\Lambda(2,1/2) \subset BMO$. Here is a generalization (since the spaces $\Lambda(p,1/p)$ increase with p), for which we give a direct proof.

2.5. THEOREM. For $1 \leq p < \infty$, $\Lambda(p,1/p) \subset BMO$.

PROOF. By the translation-invariance of both spaces, it is enough to show that for each $f \in \Lambda(p,1/p)$ there exists $C < \infty$ such that for each $0 < \delta < \pi/2$, upon letting $I = [-\delta,\delta]$ we have:

$$\frac{1}{2\delta}\int_{-\delta}^{\delta} |f(t)-f_I|^p \, dt \leq C .$$

Suppose $f \in \Lambda(p,1/p)$ with $\|f\|_{p,1/p} \leq 1$, and fix $0 < \delta < \pi/2$. Then for all real t:

$$2\pi t \geq \int_{-\pi}^{\pi} |f(s-t) - f(s)|^p \, ds .$$

We integrate both sides of this inequality over the interval [-2δ,2δ]; then successively use Fubini's theorem and change variables on the resulting inner integral to obtain:

$$8\pi\delta^2 \geq \int_{-2\delta}^{2\delta}\int_{-\pi}^{\pi} |f(s-t) - f(s)|^p \, ds\, dt$$

$$= \int_{-\pi}^{\pi}\int_{-2\delta}^{2\delta} |f(s-t) - f(s)|^p \, dt\, ds$$

$$= \int_{-\pi}^{\pi}\int_{s-2\delta}^{s+2\delta} |f(t) - f(s)|^p \, dt\, ds$$

$$\geq \int_{-\delta}^{\delta}\int_{-\delta}^{\delta} |f(t) - f(s)|^p \, dt\, ds$$

where the last line follows from the fact that if $-\delta \leq s,t \leq \delta$, then $s - 2\delta \leq t \leq s + 2\delta$. Now divide both sides of the above inequality by $4\delta^2$ and apply Jensen's convexity theorem:

$$2\pi \geq \frac{1}{2\delta}\int_{-\delta}^{\delta}\left[\frac{1}{2\delta}\int_{-\delta}^{\delta} |f(t) - f(s)|^p \, dt\right] ds$$

$$\geq \frac{1}{2\delta}\int_{-\delta}^{\delta}\left|\frac{1}{2\delta}\int_{-\delta}^{\delta}[f(t) - f(s)]\, dt\right|^p ds$$

$$= \frac{1}{2\delta}\int_{-\delta}^{\delta}|f(s) - f_I|^p \, ds ,$$

where $I = [-\delta,\delta]$. Since $0 < \delta < \pi/2$ is arbitrary, we have achieved the desired result. ///

2.6. "LITTLE OH." The collection $\lambda(p,\alpha)$ of $f \in L^p(T)$ for which $\|f - f_t\|/t^\alpha \to 0$ as $|t| \to 0$ is a closed subspace of $\Lambda(p,\alpha)$, and the previously stated results all have "little oh" analogues with $\lambda(p,\alpha)$ replacing $\Lambda(p,\alpha)$. In particular, $\lambda(p,1/p) \subset$ VMO, the space of functions of vanishing mean oscillation. It has recently been shown that every VMO function has connected essential range [26], hence,

PROPOSITION. If $1 < p < \infty$, and $f \in \lambda(p,1/p)$, then the essential range of f is connected.

2.7. GENERALIZED MEAN LIPSCHITZ SPACES. The following generalization of the spaces $\Lambda(p,\alpha)$ occurs frequently in the literature ([31] - [Ta3]; [28] Chapter V, section 5; [12]). If $1 < p, q < \infty$, and $0 < \alpha < 1$, say a function $f \in L^p(T)$ belongs to $\Lambda(p,q,\alpha)$ if

$$\int_0^\infty \| f - f_t \|_p^q \, t^{-1-\alpha q} \, dt < \infty$$

(note that the convergence of the integral depends only upon the behavior of the integrand for t near 0). Our spaces $\Lambda(p,\alpha)$ correspond to the limiting case $q = \infty$ here. These generalized mean Lipschitz spaces play no role in this paper because of the containment: $\Lambda(p,q,\alpha) \subset \lambda(p,\alpha)$ ([12], page 125).

3. PARTIAL SUMS AND DYADIC BLOCKS. For $f \in L^1(T)$ and n an integer, let $\hat{f}(n)$ denote the nth Fourier coefficient of f:

$$\hat{f}(n) = \frac{1}{2\pi} \int_{-\pi}^{\pi} f(t) e^{int} \, dt \; .$$

For n positive, write

$$s_n f(\theta) = \sum_{|k| \leq n} \hat{f}(k) e^{ik\theta}$$

for the nth partial sum of the Fourier series of f, and setting

$$I(n) = \{k \text{ an integer}: 2^n < |k| \leq 2^{n+1}\},$$

write

$$\Delta_n f(\theta) = \sum_{k \in I(n)} \hat{f}(k) e^{ik\theta} ,$$

for the nth dyadic block of that series. A standard argument involving the M. Riesz theorem shows that for each $1 < p < \infty$ there exists a constant $C_p < \infty$ for which:

$$\|s_n f\|_p \leq C_p \|f\|_p, \text{ and } \|\Delta_n f\|_p \leq C_p \|f\|_p,$$

for all $f \in L^p(T)$, and all positive integers n. The first of these inequalities, along with the density of trigonometric polynomials in $L^p(T)$, shows that $\|s_n f - f\|_p \to 0$ for each $f \in L^p(T)$ $(1 < p < \infty)$. These matters are discussed, for example, in [24], Ch. 17, Problem 25.

In this section we survey some known results which relate the mean smoothness of f with the speed at which it is approached by $s_n f$. The main result is:

3.1. THEOREM. Suppose $1 \leq p < \infty$ and $0 < \alpha < 1$. Then for $f \in L^p(T)$, the following three conditions are equivalent.

(a) $f \in \Lambda(p,\alpha)$.

(b) $\|\Delta_n f\|_p = O(2^{-n\alpha})$ as $n \to \infty$.

(c) $\|f - s_n f\|_p = O(n^{-\alpha})$ as $n \to \infty$.

The equivalence of (a) and (c) is mentioned without proof in Hardy and Littlewood's paper [15]. That of (a) and (b) often serves as the basis for the definition of spaces generalizing mean Lipschitz spaces in various settings (see [4], [13], [27], for example). In the next section, Theorem 3.1 will play an important role in the proof of our $\Lambda(p,1/p)$ Tauberian theorem. It also has notable appeal in the case $p = 2$, where through the Parseval identity it characterizes the Fourier coefficients of functions in $\Lambda(2,1/2)$.

3.2. COROLLARY. Suppose $f \in L^2(T)$ and $0 < \alpha < 1$. Then the following three conditions on f are equivalent:

(a) $f \in \Lambda(2,\alpha)$,

(b) $\sum_{k \in I(n)} |\hat{f}(k)|^2 = O(2^{-2n\alpha})$ as $n \to \infty$,

(c) $\sum_{|k| \geq n} |\hat{f}(k)|^2 = O(n^{-2\alpha})$ as $n \to \infty$.

3.3. COROLLARY (see [15], Lemma 12; and [5], Corollary 2.2). If $f \in L^1(T)$ and $|\hat{f}(n)| = O(1/|n|)$ as $n \to \infty$, then $f \in \Lambda(2,1/2)$.

The proof of Theorem 3.1 requires some preliminary estimates. If f is a trigonometric polynomial:

$$f(\theta) = \sum a_n e^{in\theta} \quad \text{(finite sum)},$$

let us write

$$f'(\theta) = \sum n a_n e^{i(n-1)\theta},$$

and for $r \geq 0$,

$$f_r(\theta) = \sum a_n r^{|n|} e^{in\theta}.$$

The lemma below generalizes to $1 \leq p \leq \infty$ some estimates that are obvious if $p = 2$.

3.4. LEMMA. Suppose $0 \leq N \leq M < \infty$, and

$$f(\theta) = a_N e^{iN\theta} + \dots + a_M e^{iM\theta}.$$

Then for $1 \leq p \leq \infty$:

(a) $r^M \|f\|_p \leq \|f_r\|_p \leq r^N \|f\|_p \quad (0 \leq r \leq 1)$,

and

(b) $N\|f\|_p \leq \|f'\|_p \leq M\|f\|_p$.

PROOF. Both inequalities follow from a standard fact about convolutions: $\|K*f\|_p \le \|K\|_1 \|f\|_p$ whenever $K \in L^1(T)$ and $f \in L^p(T)$. To prove (a), observe first that for any trigonometric polynomial f, if $0 \le r < 1$, then $f_r = P_r*f$, where P_r is the Poisson kernel. Now for f as in the hypothesis of the lemma, let $h(\theta) = e^{-iN\theta} f(\theta)$ and $g(\theta) = e^{-iM\theta} f_r(\theta)$. It is easy to check that:

(1) $f_r(\theta) = r^N e^{iN\theta} h_r(\theta)$,

and

(2) $r^M f(\theta) = e^{iM\theta} g_r(-\theta)$.

From (1):

$$\|f_r\|_p = r^N \|h_r\|_p = r^N \|P_r*h\|_p \le r^N \|P_r\|_1 \|h\|_p = r^N \|f\|_p$$

This proves the second inequality of (a). For the first one we use (2) above, along with the same convolution inequality to obtain:

$$r^M \|f\|_p = \|g_r\|_p \le \|g\|_p = \|f_r\|_p \,,$$

which completes the proof of (a).

Part (b) is proved similarly. Let $g(\theta) = e^{-i(N-1)\theta} f'(\theta)$, and

$$K(\theta) = \lim_{n\to\infty} \sum_{k=-n}^{n} \frac{1}{|k|+N} e^{ik\theta} \quad .$$

Since the coefficient sequence for K is symmetric about 0, and convex for non-negative n, the series on the right converges whenever $e^{i\theta} \neq 1$; moreover the resulting function is positive and integrable on T ([8], Theorem 4.5, page 64). In particular:

$$\|K\|_1 = \hat{K}(0) = 1/N.$$

Now $e^{-iN\theta} f(\theta) = K*g(\theta)$, so

$$\|f\|_p \le \|K\|_1 \|g\|_p = N^{-1} \|f'\|_p ,$$

which proves the first inequality of (b).

For the second inequality, write:

$$g(\theta) = Ma_M + (M-1)a_{M-1}e^{i\theta} + \ldots + Na_N e^{i(M-N)\theta} ,$$

and

$$K(\theta) = \sum_{|n|< M} (M - |n|)e^{in\theta} .$$

Then $g = K*h$, where $h(\theta) = e^{-iM\theta} f(-\theta)$. Thus:

$$\|f'\|_p = \|g\|_p = \|K*h\|_p \le \|K\|_1 \|h\|_p = \|K\|_1 \|f\|_p ,$$

and the proof is completed by observing that K is just M times the M-1st Fejer kernel, so $\|K\|_1 = M$. ///

The example $f(\theta) = e^{iN\theta}$ shows that the inequalities of Lemma 3.4 cannot be improved. The second inequality in part (a) is a version of Bernstein's inequality ([19], page 17, Problem 12).

3.5. PROOF OF THEOREM 3.1. First some notation. If F is holomorphic in U, let s_nF and Δ_nF denote respectively the nth partial sum and nth dyadic block of the Taylor expansion of F about the origin. Since the spaces $\Lambda(p,\alpha)$ are self-conjugate for the indices considered, it suffices to prove the theorem for functions f whose Fourier transform vanishes on the negative integers, i.e. for f of "power series type". In the arguments below, "C" always denotes a finite positive constant which may vary from line to line, but never depends on the parameters n or r.

(1) → (2). Suppose $f \in \Lambda(p,\alpha)$ is of power series type, so its Poisson integral $F = P[f]$ is holomorphic on U. Fix a positive integer n. Then for $0 \le r < 1$ we have:

$$M_p(\Delta_n(zF'),r) \le C\,M_p(zF'),r) \le C(1 - r)^{\alpha-1}$$

where the first inequality follows from the M. Riesz theorem (since $1 < p < \infty$), and the second from Proposition 1.3, the characterization of mean Lipschitz spaces by Poisson integrals. Now set $r = 1-2^{-n}$, and use successively both left-hand inequalities in Lemma 3.4:

$$
\begin{aligned}
2^{n(1-\alpha)} &\geq C\,M_p(\Delta_n(zF'),r) \\
&\geq C r^{2^{n+1}} \|(\Delta_n f)'\|_p \qquad \text{(notation as in Lemma 3.4)} \\
&\geq C\, 2^n \|\Delta_n f\|_p .
\end{aligned}
$$

Thus $\|\Delta_n f\|_p \leq C^{-1} 2^{-n\alpha}$, as desired.

(2) → (3). Suppose $f \in L^p(T)$ obeys (2), and n is a fixed positive integer. Choose $N = 2^{j-1}$ so that $N < n \leq 2N$. By the M. Riesz theorem,

$$
\begin{aligned}
\|f - s_n f\|_p &\leq C\|f - s_N f\|_p \\
&\leq C \sum_{k \geq j-1} \|\Delta_k f\|_p \\
&\leq C \sum_{k \geq j-1} 2^{-k\alpha} \\
&\leq C\,2^{-j\alpha} \\
&\leq C n^{-\alpha},
\end{aligned}
$$

which is (3).

(3) → (1). Suppose $f \in L^p(T)$ satisfies (3), i.e.

$$\|s_n f - f\|_p \leq C n^{-\alpha} \qquad (n = 1,2,\ldots).$$

Then by the M. Riesz theorem:

$$\|\Delta_n f\|_p \leq C\,2^{-n\alpha}$$

for each positive integer n, so using both right-hand inequalities in Lemma 3.4, we obtain:

$$
\begin{aligned}
M_p(zF',r) &\le \sum M_p(\Delta_n(zF'),r) \\
&\le \sum r^{2^n} \| (\Delta_n f)' \|_p \qquad \text{(by right-hand inequality of 3.4(a))} \\
&\le \sum r^{2^n} 2^{n+1} \| \Delta_n f \|_p \qquad (\text{" \quad " \quad " \quad " \quad " } 3.4(b)) \\
&\le C \sum r^{2^n} 2^{n(1-\alpha)} ,
\end{aligned}
$$

where in each line the range of summation is $0 \le n < \infty$. Since

$$
r^{2^n} 2^{n(1-\alpha)} \le \sum_{k=2^{n-1}+1}^{2^n} r^k k^{-\alpha} ,
$$

the previous estimate gives

$$
M_p(F',r) \le C \sum k^{-\alpha} r^k \le C/(1-r)^{1-\alpha} ,
$$

which, by Proposition 3.1, shows that $f \in \Lambda(p,\alpha)$, and completes the proof of the Theorem. ///

4. TAUBERIAN NATURE OF $\Lambda(p,1/p)$. Here we give our simple proof of the fact that the Fourier series of a function $f \in \Lambda(p,1/p)$ converges at a point whenever it is Abel summable there. As noted in the Introduction, $\Lambda(1,1)$ is essentially the space of functions of bounded variation, so we always assume without further mention that $1 < p < \infty$. We begin with a review of some summability matters, and hope the reader will not be offended that we begin at the beginning.

4.1. SUMMABILITY. Let $(a_n: 0 \le n < \infty)$ be a sequence of complex numbers. The series $\sum a_n$ is said to be:

(i) *summable (to S)* if the sequence of partial sums

$$
s_n = a_0 + a_1 + \dots + a_n
$$

converges (to S);

(ii) *Cesaro summable (to S)* if the sequence of arithmetic means

$$\sigma_n = (s_0 + s_1 + \ldots + s_n)/(n+1)$$

converges (to S); and

(iii) *Abel summable (to S)* if

$$\lim_{r \to 1-} \sum_{n=0}^{\infty} a_n r^n = S ,$$

where the tacit assumption is, of course, that the series on the left converges for all $0 \leq r < 1$.

If $f \in L^1(T)$ and θ is real, then we say the Fourier series of f converges in one of these modes to S if the corresponding numerical series $\sum a_n$ does, where $a_0 = \hat{f}(0)$, and

$$a_n = \hat{f}(-n)e^{-in\theta} + \hat{f}(n)e^{in\theta} \qquad (n > 0).$$

In particular, the Fourier series of f converges at θ if and only if the sequence $(s_n f(\theta): n \geq 0)$ of symmetric partial sums defined in section 3 converges, and it is Abel summable at θ if and only if

$$\sum \hat{f}(n) r^{|n|} e^{in\theta} = P[f](re^{i\theta})$$

converges as $r \to 1-$.

As everyone knows: summability ⇒ Cesaro summability ⇒ Abel summability, but in general, not conversely. Here is the main result of this section.

4.2. $\Lambda(p,1/p)$ TAUBERIAN THEOREM. If the Fourier series of $f \in \Lambda(p,1/p)$ is Abel summable at θ, then it is summable at θ.

As we mentioned in the Introduction, Hardy and Littlewood proved this result under the stronger hypothesis that the Fourier series be Cesàro summable. Their proof, which involved interpolation between ordinary Cesàro summability and negative order Cesàro boundedness, is considerably more complicated than the one we will present for the stronger result stated above.

However their proof does give additional information: it implies that the Fourier series is Cesaro convergent for certain negative orders.

The crucial Tauberian concept in our proof is that of *slow oscillation.*

4.3. DEFINITION. A sequence $(a_n: n \geq 0)$ is said to be *slowly oscillating* if for every $\varepsilon > 0$ there exists $\lambda = \lambda(\varepsilon) > 1$ and a positive integer $N = N(\varepsilon)$ such that :

$$\max\{|a_j - a_k|: n \leq j, k \leq \lambda n\} < \varepsilon$$

whenever $n \geq N$.

For example, if $|a_n| = O(1/n)$, then the corresponding sequence of partial sums is slowly oscillating. This shows that the next result is a generalization of Littlewood's $O(1/n)$ Tauberian theorem.

4.4. SASZ'S TAUBERIAN THEOREM. Suppose (a_n) is a numerical sequence whose partial sums (s_n) form a slowly oscillating sequence. If the series Σa_n is Abel summable, then it is summable.

This result originally appeared in [29]. An eminently readable proof is presented in [30]. Theorem 4.2 follows immediately from Sasz's theorem and:

4.5. THEOREM. For every $f \in \Lambda(p, 1/p)$ and real θ, the numerical sequence $(s_n f(\theta): n \geq 0)$ is slowly oscillating.

PROOF. According to the definition of slow oscillation, our goal is to prove that:

$$(1)\quad \overline{\lim_{n\to\infty}} \max_{n<m\leq\lambda n} \left| s_m f(\theta) - s_n f(\theta) \right| \to 0 \text{ as } \lambda \to 1+ .$$

To this end, fix $n < m$ and let

$$Q_{nm}(\theta) = \sum_{n<k\leq m} e^{ik\theta} .$$

Then:

$$(2)\quad s_n f(\theta) - s_m f(\theta) = Q_{nm} * f(\theta) = Q_{nm} * (f - s_n f)(\theta),$$

where the last equality follows from the fact that $s_n f * Q_{nm} \equiv 0$. Upon applying Holder's inequality and the translation invariance of Lebesgue measure to (2) we obtain:

$$(3)\quad \left| s_n f(\theta) - s_m f(\theta) \right| \leq \| f - s_n f \|_p \, \| Q_{nm} \|_{p'}$$

where p' is the index conjugate to p. Now

$$\left| Q_{nm}(\theta) \right| \leq 2 K_{m-n-1}(\theta),$$

where

$$K_m(\theta) = \left| 1 + e^{i\theta} + \ldots + e^{im\theta} \right|$$

$$= \left| \frac{\sin (m+1)(\theta/2)}{\sin (\theta/2)} \right|$$

Thus

$$\| Q_{nm} \|_{p'} \leq 2 \| K_{m-n-1} \|_{p'} \leq C(m-n)^{1/p} ,$$

where C is a constant that does not depend on m or n. The last inequality is a straightforward computation based on the closed-form expression for K_m and a change of variable. We leave the details to the reader. From inequality (3) above and Theorem 3.1:

$$\left| s_n f(\theta) - s_m f(\theta) \right| \leq C(m - n)^{1/p} n^{-1/p} ,$$

so for $\lambda > 1$:

$$\sup_{n<m\leq\lambda n} \left| s_m f(\theta) - s_n f(\theta) \right| \leq C(\lambda - 1)^{1/p} .$$

This inequality yields (1), and completes the proof of the Theorem.

4.6. THE TAUBERIAN NATURE OF BMO. Recall from section 2 that $\Lambda(p,1/p) \subset$ BMO. However BMO, or even VMO, cannot replace $\Lambda(p,1/p)$ in Theorems 4.2 or 4.5, since there exist continuous functions on T whose Fourier series diverge at a given point. Nevertheless, Ramey and Ullrich [23] have recently shown that BMO is a Tauberian condition linking Abel and various other methods of summability. They work on the real line **R**, instead of the circle. To state their result efficiently we need the notion of normalized dilate. If $K \in L^1(\mathbf{R})$ and $y > 0$, let $K_y(x) = y^{-1}K(x/y)$. Thus $K \in L^1(\mathbf{R})$, and $\|K_y\|_1 = \|K\|_1$. Ramey and Ullrich prove the following Tauberian theorem for BMO(**R**).

THEOREM ([23], Theorem 4.4). Suppose $f \in$ BMO(**R**), $x \in \mathbf{R}$, and $P[f](x,y) \to L$ as $y \to 0+$. Then also $f*K_y(x) \to L$ whenever $K \in L^1(\mathbf{R})$ obeys the following additional conditions:

(i) $\left|K(x)\right| <$ constant$(1 + x^2)^{-1}$ for all real x, and

(ii) $$\int_{-\infty}^{\infty} K(x)\,dx = 1 .$$

The proof of this result involves an elegant mixture of functional analysis (weak* convergence in BMO(**R**)), and function theory (normal families), along with the crucial observation that the BMO(R) norm is dilation invariant. Actually Ramey and Ullrich work in higher dimensions, but state their result only for K the characteristic function of a ball. However their proof gives the full result stated above. Letting K be respectively the characteristic function of the interval [-1/2,1/2], and the Fejer (Cesàro) kernel:

$$K(x) = \frac{1}{\pi}\left[\frac{\sin x}{x}\right]^2,$$

this result yields:

COROLLARY. If the Fourier integral of $f \in BMO(R)$ is Abel summable to L at $x \in R$, then it is Cesaro summable to L at x, and also:

$$\lim_{h \to 0+} \frac{1}{2h}\int_{x-h}^{x+h} f(t)\,dt = L .$$

This Corollary can be transferred to the unit circle by means of the Poisson summation formula, and the observation that if a 2π-periodic function is in BMO(T), then it is in BMO(R). Thus: *the Corollary above remains true if R is replaced by T and "Fourier integral" is replaced by "Fourier series".*

Since $\Lambda(p,1/p) \subset BMO(T)$, this result for the circle, along with Hardy and Littlewood's original Tauberian theorem for $\Lambda(p,1/p)$, give another proof of Theorem 4.2. In [15] (Theorem 1, page 613) Hardy and Littlewood also state that the convergence of the integral average in the above Corollary is a necessary and sufficient condition for the Fourier series of $f \in \Lambda(p,1/p)$ to converge at the point x. This also follows from the considerations above: If the averages converge to L, then as is well known ([24],

Theorem 11.2, page 257), $P[f](re^{i\theta}) \to L$ as $r \to 1-$. Thus by Theorem 4.2, $s_n f(\theta) \to L$. Conversely, if $s_n f(\theta) \to L$, then since $\Lambda(p,1/p) \subset$ BMO, the circle version of the Ramey-Ullrich Corollary above asserts that the integral averages converge to L.

5. $\Lambda(p,1/p)$ AND FUNCTION THEORY.

In this final section we survey some function theoretic situations in which the spaces $\Lambda(p,1/p)$ arise naturally.

5.1. HARDY'S INEQUALITY.

That the coefficient condition $|\hat{f}(n)| = O(1/|n|)$ suffices for $f \in$ BMO seems to be part of the folklore. It is usually obtained by duality from Hardy's inequality:

$$\sum \frac{|\hat{f}(n)|}{n+1} \leq \pi \|f\|_1$$

for $f \in H^1$ (see [2], page 25, for example). However, following [5], Corollary 3.3 gives the result directly, and therefore by duality provides a different proof of Hardy's inequality (although with a less precise constant).

In fact, Corollary 3.3 gives a result more general than Hardy's inequality. Flett [12] has shown that for $1 < p < \infty$, $\Lambda(p,\alpha)$ is the dual space of the space $X(p',1-\alpha)$ consisting of functions u harmonic on U for which

$$\|u\|_X = \int_{-\pi}^{\pi} M_{p'}(u,r)(1-r)^{1-\alpha} dr < \infty ,$$

and the two spaces are paired by integration on the circle. Now the same duality argument that lead from Hardy's inequality to the $O(1/n)$ sufficient condition for BMO can be reversed to give a version of Hardy's inequality for $X(2,1/2)$. Since $\Lambda(2,1/2) \subset$ BMO, it follows easily that the reverse inequality is true of the preduals: $\mathrm{Real}\{H^1\} \subset X(2,1/2)$. More directly, this last inclusion follows from a result of Hardy and Littlewood ([8], Theorem 5.11, p. 87]). Thus, the $O(1/n)$ sufficient condition for membership in $\Lambda(2,1/2)$ leads to an improvement of Hardy's inequality.

5.2. BLASCHKE PRODUCTS. Suppose (z_n) is a sequence of points in the unit disc, arranged in order of increasing modulus. Let $d_n = 1 - |z_n|$, and suppose $\Sigma d_n < \infty$. Then the infinite product

$$B(z) = \prod_{n=0}^{\infty} \omega_n \frac{z_n - z}{1 - \bar{z}_n z}$$

where $\omega_n = \bar{z}_n/|z_n|$, converges uniformly on compact subsets of U to a bounded holomorphic function B, called *the Blaschke product with zeros* (z_n). Moreover, B is an inner function, that is, its radial limit

$$B^*(e^{i\theta}) = \lim B(re^{i\theta}),$$

which exists a.e. by Fatou's theorem, has modulus 1 a.e (see [8], Chapter 2, or [24], Sections 15.21-15.24, pages 333-336).

Suppose $B(z) = \Sigma\, \hat{B}(n) z^n$ is the Taylor expansion of B about the origin. In 1962 D.J. Newman and H.S. Shapiro [21] proved a number of interesting results about the Taylor coefficients of Blaschke products, and more generally, of inner functions. In the first place, they showed among inner functions, only the Blaschke products with finitely many factors have Taylor coefficients which tend to zero like $o(1/n)$. This can be seen in modern terms as follows. It is known [Ste], [26] that among inner functions, only the finite Blaschke products can have boundary function belonging to VMO, and we know from section 2 that $\lambda(2,1/2) \subset$ VMO. Now the $o(1/n)$ coefficient condition for an inner function f implies that

$$(*) \qquad n \sum_{|k|>n} |\hat{f}(k)|^2 \to 0 \quad \text{as } n \to \infty,$$

which, by the "little oh" version of Corollary 3.2, is necessary and sufficient for $f^* \in \lambda(2,1/2)$. Note that this argument actually improves the Newman-Shapiro result; it shows that any inner function whose Taylor coefficients obey (*) above must be a finite Blaschke product.

Newman and Shapiro also constructed Blaschke products B with coefficients $|B(n)| = O(1/n)$. In fact they showed that any Blaschke product for which

(1) $\sup_n d_{n+1}/d_n < 1$

has this property. In our language: condition (1) implies $B^* \in \Lambda(2,1/2)$. In the converse direction, Ahern [1] showed in 1979 that (for Blaschke products) necessary and sufficient for

(2) $d_n = O(a^n)$ (some $0 < a < 1$)

is

(3) $$\sum_{k=0}^{N} \sum_{n=k}^{\infty} |\hat{B}(n)|^2 = O(\log N)$$

([A], Lemma 3.1 and Theorem 3.3, pages 327 - 331). Now if $B^* \in \Lambda(2,1/2)$, then by our Corollary 3.2, the estimate (3) above holds, hence the zeros of B must satisfy (2). Finally, in 1982 Verbitskii [34] obtained the complete result:

THEOREM. For B a Blaschke product, the following four conditions are equivalent:

(a) The zeros of B can be decomposed into finitely many sequences, each of which satisfies (1) above,

(b) $B^* \in \Lambda(2,1/2)$,

(c) $B^* \in \Lambda(p,1/p)$ for some $1 < p < \infty$,

(d) $|\hat{B}(n)| = O(1/n)$.

5.3. UNIVALENT FUNCTIONS. S is the class of analytic, univalent functions f on U for which $f(0) = 0$ and $f'(0) = 1$. The associated logarithmic function $g(z) = \log[f(z)/z]$ was shown by Baernstein [3] to belong to BMOA, the class of functions in H^2 with boundary function in BMO, and an alternate proof was later given by Cima and Schober [6]. Thus it is natural to ask if g must actually belong to $\Lambda(2,1/2)$. However Hayman [16] has constructed an example which shows that this is not the case. On the other hand, in [6] Cima and Schober show that it *is* the case if f is a support point of S. This raises the apparently open question, first asked by Allen Shields, of whether $g \in \Lambda(2,1/2)$ whenever f is an *extreme point* of S (for more details see [9]). Duren and Leung [9] have shown that $g \in \Lambda(2,1/2)$ whenever the modulus of f dominates a positive multiple of $(1 - |z|)^{-2}$ on some sequence that tends to the boundary.

Of course, smoothness classes of analytic functions show up in many other contexts. For example, applications to approximation theory and operator theory can be found in the appendix by Hruschev and Peller to Nikolskii's book [20].

BIBLIOGRAPHY

[1] P. Ahern, "The mean modulus and the derivative of an inner function", Indiana Univ. Math. J. 28 (1979), 311 - 347.

[2] A. Baernstein II, "Analytic functions of bounded mean oscillation," in *Aspects of Contemporary Complex Analysis*, Conference Proceedings, edited by D.A. Brannan and J.G. Clunie, Academic Press, New York, 1980.

[3] ________________, "Univalence and bounded mean oscillation," Michigan Math. J. **23** (1976), 217 - 223.

[4] J. Boman and H.S. Shapiro, "Comparison theorems for a generalized modulus of continuity," Arkiv för Mat. **9** (1971), 91-116.

[5] J.A. Cima and K.E. Petersen, "Some analytic functions whose boundary values have bounded mean oscillation," Math. Z. **147** (1976), 237-247.

[6] _________ and G. Schober, "Analytic functions with bounded mean oscillation and logarithms of H^p functions," Math. Z. **151** (1976) 295 - 300.

[7] R.A. DeVore and R.C. Sharpley, "Maximal functions measuring smoothness," Memoirs Amer. Math. Soc. **47**, #293, (1984) 1-115.

[8] P.L. Duren, *Theory of H^p Spaces*, Academic Press, New York, 1970.

[9] _________ and Y.J. Leung, "Logarithmic coefficients of univalent functions," J. d'Analyse Math. 36 (1979), 36 - 43.

[10] C. Fefferman and E.M. Stein, "H^p spaces of several variables," Acta Math. **129** (1972) 137-193.

[11] T.M. Flett, "Some more theorems concerning absolute summability of Fourier series and power series," Proc. London Math. Soc. (3) **8** (1958), 357-387.

[12] ________, "Lipschitz spaces of functions on the circle and the disc," J. Math. Anal. & App. **39** (1972), 125-158.

[13] M. Frazier and B. Jawerth, "Decomposition of Besov spaces," Indiana Univ. Math. J. **34** (1985), 777-799.

[14] J. Garnett, *Bounded Analytic Functions,* Academic Press, New York, 1981.

[15] G.H. Hardy and J.E. Littlewood, "A convergence criterion for Fourier series," Math. Z. **28** (1928), 612 - 634.

[16] W. Hayman, "The logarithmic derivative of multivalent functions," Michigan Math. J. **27** (1980), 149 - 179.

[17] S. Janson, "Generalization of Lipschitz spaces and an application to Hardy spaces and bounded mean oscillation," Duke Math. J. **47** (1980), 959-982.

[18] F. John and L. Nirenberg, "On functions of bounded mean oscillation," Comm. Pure Appl. Math. **14** (1961), 415-426.

[19] Y. Katznelson, *Introduction to Harmonic Analysis,* J. Wiley & Sons, New York, 1968.

[20] N.K. Nikolskii, Treatise on the Shift Operator. Springer-Verlag, New York, 1985.

[21] D.J. Newman and H.S. Shapiro, "The Taylor coefficients of inner functions," Michigan Math. J. **9** (1962), 249-255.

[22] J. Peetre, *New Thoughts on Besov Spaces,* Duke Univ. Math. Series **1**, Dept. of Math. Duke Univ. Durham, N.C. 1976.

[23] W. Ramey and D. Ullrich, "On the behavior of harmonic functions near a boundary point," preprint 1987.

[24] W. Rudin, *Real and Complex Analysis, 2nd ed.* McGraw Hill, New York, 1974.

[25] H. S. Shapiro, "A Tauberian theorem related to approximation theory," Acta Math. **120** (1968), 279 - 292.

[26] J. H. Shapiro, "Cluster set, essential range, and distance estimates in BMO," Michigan Math. J. **34** (1987), 323 - 336.

[27] W. T. Sledd, "Some results about spaces of analytic functions introduced by Hardy and Littlewood," J. London Math. Soc. (2) **9** (1974), 328 - 336.

[28] E. M. Stein, *Singular Integrals and Differentiability Properties of Functions.* Princeton Univ. Press, Princeton, N.J., 1970.

[29] O. Szász, "Converse theorems of summability for Dirichlet series," Trans. Amer. Math. Soc. **39** (1936) 117 - 130.

[30] ________, *Introduction to the Theory of Divergent Series*, Lecture notes, Univ. of Cincinnatti, Cincinnatti, Ohio 1944.

[31] - [33] M. H. Taibleson, "On the theory of Lipschitz spaces of distributions on Euclidean n-space; I. Principal properties; II. Smoothness and integrability of Fourier transforms; III. Translation invariant operators, duality, and interpolation," J. Math. Mech. **13** (1964), 407 - 479; **14** (1965), 821 - 839; **15** (1966) 973 - 981.

[34] I.E. Verbitskii, "On Taylor coefficients and L^p moduli of continuity of Blaschke products", Zap. Nauchn. Sem. Leningrad. Otdel. Inst. Steklov. (LOMI) **107** (1982), 27 - 35 (Russian, English summary). MR 84d:30059.

DEPARTMENT OF MATHEMATICS
MICHIGAN STATE UNIVERSITY
EAST LANSING, MICHIGAN 48824

A REMARK ON THE MAXIMAL FUNCTION ASSOCIATED TO AN ANALYTIC VECTOR FIELD

J. Bourgain[(*)]

1. Introduction

We consider the planar case Let thus $\Omega \subset \mathbb{R}^2$ be a bounded open set and $v: \Omega' \to S^1$ or, more generally, $v: \Omega' \to \mathbb{R}^2$ be a vectorfield defined on a neighborhood Ω' of the closure $\bar{\Omega}$ of Ω.

For $\epsilon < \epsilon_0$ (taken small enough), consider the averages (along v)

$$A_\epsilon f = \frac{1}{\epsilon} \int_0^\epsilon f(x+tv(x))\,dt \tag{1.1}$$

where f is a priori a bounded measurable function

The differentiation problem f: $\lim_{\epsilon \to 0} A_\epsilon f$ alsmore sure (a.s.) leads naturally to estimating the corresponding naximal function

$$M_v f = Mf: \sup_{\epsilon < \epsilon_0} |A_\epsilon f| ,$$

although our interest in this object here will be rather the purely harmonic analysis aspects.

Already the boundedness of $A_\epsilon f$ (as an operator on L^2 say) requires hypothesis on v, as shown by the example of the Nikodym set (see [G]). The differentiation problem has been solved affirmatively in the analytic case but is still open for $\mathcal{C}^\infty$ vector fields. More precisely, estimates on v were obtained when v has non-vanishing curvature (see [N-S-W]), say

$$\inf_{x \in \Omega} \det[Dv(x)\,v(x), v(x)] > 0 . \tag{1.2}$$

(*) IHES, University of Illinois.

Assume v normalized and $\det[Dv(x)v(x), v(x)] > 0$ on Ω. Then an L^2-bound on M_v may in fact be formulated in terms of the quantity

$$\frac{\sup_{\Omega} \det[Dv(x)v(x), v(x)]}{\inf_{\Omega} \det[Dv(x)v(x), v(x)]} \tag{1.3}$$

thus the ratio between maximum and minimum curvature of the integral curves of v. In the context of "non-vanishing curvature", this theory has been generalized and developed in several directions (for instance, the higher dimensional case and maximal functions associated to certain Raden-transforms, of [Chr], [Ph-St.]).

It has been observed that for a constant curve $\Gamma_x = \Gamma$, the maximal function

$$M_\Gamma f(x) = \sup_{\epsilon > 0} \frac{1}{\epsilon} \int_0^\epsilon |f(x + \Gamma(t))| dt \tag{1.4}$$

requires geometric conditions on Γ in order to have non-trivial boundedness properties. However, these restrictions on Γ did not so far necessary conditions on the integral curves in the vectorfield case. Possibly M_v is bounded as soon as v is $\mathcal{C}^1$.

2. Statement of a Condition and Verification in the Analytic Case

The letters $c, C > 0$ will stand for various constants.

The result proved here is obtained using known ideas and techniques. As such, a condition on v will be imposed, expressing "how well v turns". This condition will permit us to prove the L^2-boundedness of M_v and will apply also in certain flat cases, even including straight lines among the integral waves of v.

For $x \in \Omega$ and t small enough, define the function

$$\omega_x(t) = \omega(t) = |\det[v(x + tv(x)), v(x)]| \ . \qquad (2.1)$$

We require a uniform estimate of the type

$$\text{mes}\{t \in [-\varepsilon,\varepsilon] : \omega(t) < \tau \sup_{-\varepsilon \le t \le \varepsilon} \omega(t)\} \le C\tau^c\varepsilon \qquad (2.2)$$

valid for all $0 < \tau < 1$, $0 < \varepsilon < \varepsilon_0$, where $0 < c, C < \infty$ are constants independent of the point $x \in \Omega$.

Theorem 1: If v is $\mathcal{C}^1$ and satisfies a condition (2.2), then M_v is bounded on $L^2(\Omega)$.

Combining the argument with some additional interpolation, the analogue statement for $p > 1$ may be obtained. We do not carry the details out in this paper. It is also likely that similar ideas permit us to prove boundedness results on the Hilbert transform along v

$$H_v f(x) = \text{Pv} \int_{-\varepsilon_0}^{\varepsilon_0} \frac{f(x + tv(x))}{t} \, dt \ .$$

In case of positive curvature, (2.2) holds with $c = 1$ and for C the expression (1.3). The main additional application is the following strengthening of the differentiation result for real analytic vectorfields.

Theorem 2: Let v be real analytic on Ω'. Then for $\varepsilon_0 > 0$ chosen small enough, M_v is bounded on $L^2(\Omega)$.

In order to verify (2.2), consider first the polynomial case. Clearly, for fixed x,

$$pt() = \det[v(x + tv(x)), v(x)]$$

is a polynomial in t of degree bounded independently of x.

Now on polynomials $p = p(t)$ of a given degree d, there is a uniform estimate (2.2) In fact, one has for $p \neq 0$,

$$\Big\{\frac{1}{\epsilon}\int_{-\epsilon}^{\epsilon} |p(t)|^{-\rho}\, dt\Big\}^{1/\rho} \sup_{|t| < \epsilon} |p(t)| < C(\rho,d) \ , \quad \rho < \frac{1}{d} \tag{2.3}$$

which is easily checked by factorization of p and using Hölder's inequality.

Assume now $v = v(x_1, x_2)$ real analytic in a neighborhood of the closure $\bar{\Omega}$ of Ω. If each point $x_0 \in \bar{\Omega}$ has a neighborhood on which (2.2) holds, then (2.2) will be valid on $\bar{\Omega}$, hence Ω, by compactness.

Since v is real analytic, we may write for $|x-x_0| < \delta$

$$F(x,t) = \sum_{k \geq 1} f_k(x-x_0)\, t^k$$

where for some constant C,

$$|\hat{f}_k(\alpha)| \leq C^{k+|\alpha|} \ , \ \alpha \in \mathbb{N}^1 = \mathbb{N} \times \mathbb{N} \ . \tag{2.4}$$

We use the following

Lemma 2.3: There is $\nu > 0$ such that for $|x| < \nu$,

$$|f_k(x)| \leq c \cdot C^k \max_{k \leq k_0} |f_k(x)| \tag{2.5}$$

holds, where k_0 is some integer.

It will then clearly suffice to have a uniform estimate (2.2) for real analytic functions $\omega(t) = \sum_{k>0} \hat{\omega}(k)t^k$ satisfying

$$\sup_{k>0} |\hat{\omega}(k)| = \sup_{k \leq k_0} |\hat{\omega}(k)| = 1 . \tag{2.6}$$

Such ω may be written on a (fixed) neighborhood of 0 as a product $p \cdot \omega_1$ where p is a polynomial of bounded degree (depending on k_0) and $|\hat{\omega}_1(0)| > \delta = \delta(k_0)$. Verification of (2.2) then reduces to p, i.e., the polynomial case.

Proof of Lemma 2.3: We will invoke a quantitation version of the division theorem for convergent power series, which may be found in [Br] (see Th. II). Take thus $\ell_1, \ell_2 > 0$ linearly independent over $\mathbb{Z}$ and consider the Banach algebra $K[\rho]$ of formal power series f in x_1, x_2 such that

$$\|f\| = \sum_{j_1, j_2 \geq 0} \rho_1^{j_1} \rho_2^{j_2} |\hat{f}(j_1, j_2)| < \infty$$

denoting $\rho = (\rho_1, \rho_2)$, $\rho_1 = \tau\eta^{\ell_1}$, $\rho_2 = \tau\eta^{\ell_2}$ and where $0 < \tau, \eta < 1$ will be suitably chosen. Given pairs $E_1, \ldots, E_p$ in $\mathbb{N}^2$, denote $\Delta = \bigcup_{1 \leq i \leq p} (E_1 + \mathbb{N}^2)$ and $R(\Delta, \rho)$ the subspace of elements $f = \sum_\alpha \hat{f}(\alpha) x^\alpha$ in $K[\rho]$ such that $\hat{f}(\alpha) = 0$ for $\alpha \in \Delta$. For appropriate $\tau > 0$, by (2.5), the sequence $\{f_k\}$ is contained in $K[\rho]$ and $\|f_k\| \leq c^k$. Let $\mathfrak{a}$ be the ideal generated by $\{f_k\}$ in $K[\rho]$.

It is proved in [Br] (see Th. II) that there this $\tau, \eta > 0$, a finite generating sequence $F_1, \ldots, F_p$ in $\mathfrak{a}$ and a set $\Delta \subset \mathbb{N}^2$ as above, such that the map φ given by

$$\varphi(g_1,\dots,g_p,h) = \sum_{i=1}^{p} g_i F_i + h$$

maps $H[\rho]$ onto $K[\rho]$, denoting

$$H[\rho] = K[\rho]^p \oplus R(\Delta,\rho)$$

the Banach space with norm

$$\|(g_1,\dots,g_p,h)\| = \sum_{i=1}^{p} \rho^{E_i}\|g_i\| + \|h\| .$$

Moreover, every element f in $K[\rho]$ is equivalent up to an element of $\mathcal{Q}$ with a unique element of $R(\Delta,\rho)$. Thus for $f \in \mathcal{Q}$, $\|f\| \leq 1$, there is a representation $f = \sum_{1 \leq i \leq p} g_i F_i$ for some $g_1,\dots,g_p \in K[\rho]$, $\|g_i\| \leq B$ $(1 \leq i \leq p)$. In particular, for some $\nu > 0$,

$$|f(x)| \leq C \max|F_i(x)| \quad \text{if} \quad |x| < \nu . \tag{2.7}$$

Since the $F_i \in \mathcal{Q}$, there is an integer k_0 such that the right member of (2.7) is bounded by $C \max_{k \leq k_0} |f_k(x)|$. Applying (1.11) to the functions $f = C^{-k} f_k$ completes the proof.

<u>Remark</u>: The author is grateful to P. Milman for pointing out the reference [Br] to him, simplifying an earlier argument.

We now come back to the proof of Theorem 1. The basic idea is the following. To the vectorfield v, we associate a "natural" system of rectangles, say $\mathcal{C}$, for which the corresponding maximal operator $M_{\mathcal{C}}$ satisfies a weak-type estimate and hence is bounded on L^2. The difference between M_v and M_ℓ is then taken care of by "best" estimates, possible because of condition (2.2).

In the next section, a general estimate is proved which is later exploited using (2.2).

3. A Local Estimate on the A_ϵ-Averages

Assume $|\nabla v| < B$ and choose $\epsilon < \epsilon_0 < \frac{1}{100B}$. It then easily follows from a change of variable that

$$\|A_\epsilon f\|_2 \le 2\|f\|_2 \tag{3.1}$$

where, for convenience, $A_\epsilon f$ is redefined as

$$A_\epsilon f(x) = \int f(x + \epsilon t v(x))\alpha(t)dr \tag{3.2}$$

taking for α a fixed, positive C^∞-function supported by [-1,7], $\alpha(-t) = \alpha(t)$, $\int \alpha(t)dt = 1$.

Since M_v is a positive operator, we may as well put

$$M_v f(x) = \sup_{\epsilon < \epsilon_0} |A_\epsilon f|$$

where $A_\epsilon f$ is defined by (3.2).

Take a point $x_0 \in \Omega$ and assume $v(\alpha_0) \neq 0$. Let R be the rectangle with center x_0, orientation $v(x_0)$, length $\epsilon|v(x_0)|$ (in direction $v(x_0)$) and width

$$\delta = \epsilon \cdot \sup_{|t| < \epsilon} |\det[v(x_0 + tv(x_0)), \frac{v(x_0)}{|v(x_0)|}]| \tag{3.3}$$

which we assume non-zero.

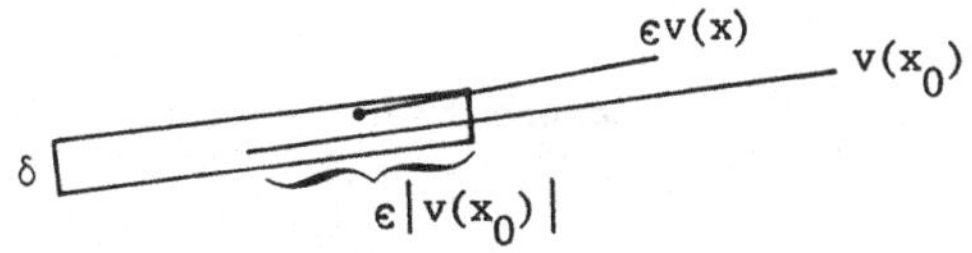

Since $\epsilon < \epsilon_0 < \frac{1}{100B}$, $x + \epsilon v(x)$ will clearly lie in a 2δ-neighborhood of $x_0 + \mathbb{R}v(x_0)$, hence in the doubled rectangle R_1 of R, for each $x \in R$.

Let T be a positive number (practically $T > \delta^{-1}$) and assume $\operatorname{supp} \hat{f} \subset B(0,2T)\backslash B(0,T)$. Let φ, ψ be a pair of positive C^∞-functions on $\mathbb{R}^2$ satisfying

$$\int \varphi(x)\,dx = 1 = \int t(\alpha)\,dx \tag{3.4}$$

$$\varphi, \hat{\psi} \text{ are supported by the disc } B(0,1)\ . \tag{3.5}$$

Denote as usual $\varphi_\delta(x) = \delta^{-2}\varphi(\delta^{-1}x)$, hence supported by $B(0,\delta)$. Thus, denoting R_3 a doubling of R_2, $\chi_{R_3} * \varphi_\delta = 1$ on R_2 and by previous considerations, also

$$A_\epsilon f|_R = A_\epsilon f_1|_R \quad \text{where} \quad f_1 = f(\chi_{R_3} * \varphi_\delta)^2\ . \tag{3.6}$$

The function

$$g = f(\chi_{R_3} * \varphi_\delta * \psi_{\frac{10}{T}})^2$$

satisfies the following properties, by (3.5)

$$|g| \le |f| \cdot (\chi_{R_4} * \psi_{\frac{10}{T}}) \tag{3.7}$$

$$\operatorname{supp} \hat{g} \subset 3(0,3T)\backslash 3(0,\tfrac{T}{2})\ . \tag{3.8}$$

Also

$$\|f_1 - g\|_2 \le 2\|f \cdot (\chi_{R_4} * \psi_{\frac{10}{T}})\|_2 \cdot \|\chi_{R_3} * \varphi_\delta * (\delta_0 - \psi_{\frac{10}{T}})\|_\infty$$

where δ_0 stands for Dirac measure at 0. Further

$$\|\chi_{R_3} * \varphi_\delta * (\delta_0 - \psi_{\frac{10}{T}})\|_\infty \le \|\varphi * (\delta_0 - \psi_{\frac{10}{\delta T}})\|_1 \le \frac{c}{\delta T}$$

and thus, by (3.1), (3.6),

$$\|A_\varepsilon f|_R - A_\varepsilon g|_R\|_2 \le \ell\|f_1 - g\|_2 < \frac{c}{\delta T}\|f(\chi_{R_4} * \psi_{\frac{10}{T}})\|_2 . \tag{3.9}$$

Our purpose now is to evaluate $\|A_\varepsilon g|_R\|_2$. By (3.1), we may write

$$A_\varepsilon g(x) = \int \hat{g}(\xi) e^{i\langle x,\xi\rangle} \hat{\alpha}(\varepsilon\langle v(x),\xi\rangle)\beta(T^{-1}|\xi|)d\xi \tag{3.10}$$

where β is a C^∞-function, $\beta = 1$ on $[\frac{1}{2},3]$ and vanishing outside $[\frac{1}{4},4]$.

Estimate by duality

$$\|A_\varepsilon g|_R\|_2 = \int \hat{g}(\xi)[\int h(x)e^{i\langle x,\xi\rangle}\hat{\alpha}(\varepsilon\langle v(x),\xi\rangle)dx]\beta(T^{-1}|\xi|)d\xi$$

$$\le \|g\|_2 |\iint h(x)\overline{h(x')}[\int e^{i\langle x-x',\xi\rangle}\hat{\alpha}(\varepsilon\langle v(x),\xi\rangle)\hat{\alpha}(\varepsilon\langle v(x'),\xi\rangle)\beta(T^{-1}|\xi|d\xi]dxdx']| .$$

We simply appeal to Shur's inequality to estimate the integral above. Thus by (3.7), (3.9), this gives the following bound

$$\|A_\varepsilon f|_R\|_2 \le c[\frac{1}{\delta T} + S^{1/2}]\|f(\chi_{R_4} * \psi_{\frac{10}{T}})\|_2 \tag{3.11}$$

denoting

$$S = \sup_{x\in R}\int_R |\int e^{i\langle x-x',\xi\rangle}\hat{\alpha}(\varepsilon\langle v(x),\xi\rangle)\hat{\alpha}(\varepsilon\langle v(x'),\xi\rangle)\beta(T^{-1}|\xi|)d\xi|dx' \tag{3.12}$$

and which remains to be evaluated.

Put

$x-x' = |x-x'|e^{i\phi}$, $\xi = \tau e^{i\theta}$, $v(x) = |v(x)|e^{i\nu}$, $v(x') = |v(x')|e^{i\nu'}$

and rewrite in plar coordinates

$$\int e^{i\langle x-x',\xi\rangle}\hat{\alpha}(\dots)\hat{\alpha}(\dots)\beta(T^{-1}|\xi|)d\xi$$

$$= \iint e^{i|x-x'|r\cos(\theta-\phi)}\hat{\alpha}(\varepsilon r|v(x)|\cos(\theta-\nu))$$

$$\times\, \hat{\alpha}(\varepsilon r|v(x')|\cos(\theta-\nu'))\beta(T^{-1}r)\,rdzd\theta\ . \tag{3.13}$$

Put $r = tT$, $\frac{1}{4} \leq t \leq 4$, replace θ by $\overline{\theta} = \frac{\pi}{2} + \theta$ and estimate (3.13) as

$$T^2\int\left|\int e^{i|x-x'|tT\sin(\overline{\theta}-\phi)}\hat{\alpha}(\varepsilon tT|v(x)|\sin(\overline{\theta}-\nu))\right.$$

$$\left.\times\, \hat{\alpha}(\varepsilon tT|v(x')|\sin(\overline{\theta}-\nu'))\beta(t)\,tdt\right|d\overline{\theta}\ . \tag{3.14}$$

Evaluate the t-integral in (3.14). First, by partial integration and the fact that $\hat{\alpha} \in \mathcal{S}$, there is a bound

$$C(1 + |x-x'|T|\sin(\overline{\theta}-\phi)|)^{-6}\ . \tag{3.15}$$

Secondly, using just the decay of $\hat{\alpha}$, there is also the estimate

$$C(1 + \varepsilon T|v(x)||\sin(\overline{\theta}-\nu)| + \varepsilon T|v(x')||\sin(\overline{\theta}-\nu')|)^{-6}\ . \tag{3.16}$$

Note that for $x \in R$, $|v(x) - v(x_0)| \leq B|x-x_0| \leq 2B\varepsilon|v(x_0)|$, hence $|v(x)| \sim |v(x_0)|$. Also $|x-x'| < 3\varepsilon|v(x_0)|$ for $x,x' \in R$.

Recombining the terms of (3.15), (3.16), one gets therefore

$$[1 + \varepsilon T|v(x_0)||\sin(\overline{\theta}-\nu)| + \varepsilon T|v(x_0)||\sin(\nu-\nu')| + T|x-x'||\sin(\phi-\nu)|]^{-6}$$

bounded by

$$[1 + \varepsilon T|v(x_0)||\sin(\overline{\theta}-\nu)|]^{-2}$$

$$\times\, [1 + \frac{T}{|v(x_0)|}(|\det[x-x',v(x)]| + \varepsilon|\det[v(x),v(x')]|)]^{-4}\ .$$

Integrating in $\bar{\theta}$

$$(3.14) \leq \frac{T}{\epsilon|v(x_0)|} [1 + \frac{T}{|v(x_0)|} (|\det[x-x', v(x)]| + \epsilon|\det[v(x), v(x')]|)]^{-4} . \tag{3.17}$$

We now come back to (3.12) and integrate on R. For fixed $x \in R$, write $v(x)^{\perp}$ for the normalized direction perpendicular on $v(x)$ and for $x' \in R$

$$x'-x = \alpha v(x) + \beta v(x)^{\perp} \quad (|\alpha| \leq \epsilon, \ |\beta| \leq \delta) .$$

Since

$$|\det[x-x', v(x)]| \sim |\beta||v(x_0)|$$

and

$$|\det[v(x), v(x')]| \geq |\det[v(x), v(x + \alpha v(x))]| - B|v(x)||x'-x-\alpha v(x)|,$$

integration of (3.17) gives after change of coordinates

$$\frac{T}{\epsilon} \iint_{\substack{|\alpha| < \epsilon \\ |\beta| < \delta}} [1 + T|\beta| + \frac{\epsilon T}{|v(x_0)|} |\det[v(x), v(x+\alpha v(x))]|]^{-4} \, d\alpha \, d\beta$$

$$\leq c\, \epsilon^{-1} \int_{|\alpha| < \epsilon} [1 + \epsilon T|v(x_0)|^{-1} \omega_x(\alpha)]^{-2} d\alpha . \tag{3.18}$$

Substitution of (3.18) in (3.12), (3.11) gives finally

$$\|A_\epsilon f|_R\|_2 \leq C\|f(\chi_{R'} * \psi_{\frac{1}{T}})\|_2 \cdot \{\frac{1}{\delta T} + \sup_{x \in R} \|(1 + \epsilon T|v(x_0)|^{-1}\omega_x)^{-1}\|_{L^2[-\epsilon,\epsilon]}\} \tag{3.19}$$

where R' is some multiple of R, supp $\hat{f} \subset B(0,2T)\backslash B(0,T)$ and where $L^2[-\varepsilon,\varepsilon]$ is L^2 on $[-\varepsilon,\varepsilon]$ with normalized measure.

We will specify (3.19) under hypothesis (2.2).

Observe first that (2.2) easily implies a doubling estimate

$$\sup_{|t|<2\varepsilon} \omega_x(t) \leq C \sup_{|t|<\varepsilon} \omega_x(t) . \tag{3.20}$$

We continue to use the letter B for the Lipschitz constant of v.

<u>Lemma 3.21</u>: Assume x,x' two points in Ω, which are "close" in the sense that $x' = x + \alpha v(x) + y$ where $|\alpha| < C\varepsilon$, $|y| < C\varepsilon|v(x)|^{-1} \sup_{|t|<\varepsilon} \omega_x(t)$. Then (provided $\varepsilon < \varepsilon_0$ assumed small enough),

$$\sup_{|t|<\varepsilon} \omega_x(t) \sim \sup_{|t|<\varepsilon} \omega_{x'}(t) . \tag{3.22}$$

<u>Proof</u>: We show the inequality

$$\sup_{|t|<\varepsilon} \omega_x(t) \leq C \sup_{|t|<\varepsilon} \omega_{x'}(t) \tag{3.23}$$

(the other inequality is similar, in fact simpler).

Notice first that since for $|t| < \varepsilon$

$$\omega_x(t) \leq |v(x)||v(x)-v(x + tv(\alpha))| \leq B\varepsilon|v(x)|^2 ,$$

one has

$$|v(x)-v(x')| \leq B|x-x'| \leq CB(\varepsilon|v(x)| + C\varepsilon B\varepsilon|v(x)|) \leq \frac{1}{2} |v(x)| ,$$

hence

$$|v(x)| \sim |v(x')| . \tag{3.24}$$

Thus the length is essentially preserved on the neighborhood described above.

Next estiate

$$|\det[v(x),v(x+tv(x))]| \leq C|\det[v(x'),v(x'-\alpha v(x)-y)]|$$

$$+ C|\det[v(x'),v(x'+(t-\alpha)v(x)-y)]| \qquad (3.25)$$

and the first term by

$$|v(x')|B|y| + |\det[v(x'),v(x'-\beta v(x'))]|$$

$$+ |v(x')||v(x'-\beta v(x')) - v(x'-\alpha v(x)-y)| \qquad (3.26)$$

where $\beta \sim \alpha$ will be introduced later.

First term in (3.26) is at most $CB\epsilon\|\omega_x\|_{L^\infty(-\epsilon,\epsilon)}$.

Second term in (3.26) is bounded by

$$\|\omega_{x'}\|_{L^\infty(-C\epsilon,C\epsilon)} \sim \|\omega_{x'}\|_{L^\infty(-\epsilon,\epsilon)} \quad \text{by (3.20)} .$$

Third term in (3.26) may be for an appropriate choice of β estimated by

$$B|v(x)||-\beta v(x') + \alpha v(x) - y|$$

$$\leq B|v(x)||y| + B|v(x)| \leq B|y| + B|v(x)||\alpha v(x) - \beta v(x + \alpha v(x))|$$

$$\leq CB\epsilon\|\omega_x\|_{L^\infty(-\epsilon,\epsilon)} + B\epsilon|\det[v(x),v(x + \alpha v(x)]| .$$

The second term in (3.25) is bounded similarly to the first (replacing α by $\alpha-t$). Collecting previous estimates, (3.25) is bounded by

$$C\|\omega_{x'}\|_{L^\infty(-\epsilon,\epsilon)} + BC\epsilon_0\|\omega_x\|_{L^\infty(-\epsilon,\epsilon)}$$

hence

$$\|\omega_x\|_{L^\infty(-\epsilon,\epsilon)} \leq C\|\omega_{x'}\|_{L^\infty(-\epsilon,\epsilon)} + \frac{1}{2}\|\omega_x\|_{L^\infty(-\epsilon,\epsilon)} \quad \text{implying (3.23).}$$

This proves the lemma.

We now come baca to (3.19). By previous lemma,

$\|\omega_x\|_{L^\infty(-\epsilon,\epsilon)} \sim \|\omega_{x_0}\|_{L^{-\infty}(-\epsilon,\epsilon)}$ for $x \in R$. Moreover, for fixed $x \in R$, it follows from (2.2)

$$\int_{-\epsilon}^{\epsilon} [1 + \epsilon T|v(x_0)|^{-1}\omega_x(t)]^{-2}\, dt$$

$$\leq \text{mes}\{|t| < \epsilon|\omega_x(t) < \tau\|\omega_x\|_{L^\infty(-\epsilon,\epsilon)}\}$$

$$+ \epsilon[\epsilon T|v(x_0)|^{-1}\tau\|\omega_x\|_{L^\infty(-\epsilon,\epsilon)}]^{-2}$$

$$\leq C\tau^c\epsilon + \epsilon(\tau T\delta)^{-2} \tag{3.27}$$

where $1 > \tau > 0$ may be arbitrarily chosen. Therefore, for an appropriate choice of τ in (3.27)

<u>Lemma 3.28</u>: Under hypothesis (2.2) and with R,R' and f as in (3.19),

$$\|A_\epsilon f|_R\|_2 \leq C(T\delta)^{-c}\,\|f(\chi_{R'} * \psi_{\frac{1}{T}})\|_2 \tag{3.29}$$

for some constants $0 < c, C < \infty$ depending on (2.2).

4. Geometrical Properties of Associated Rectangles

We use the notation R',R", etc. for dialtes of a given rectangle R. For $x \in \Omega$, $\epsilon < \epsilon_0$, define $R_{x,\epsilon}$ as the rectangle considered in previous section, i.e., with center x, length $\epsilon|v(x)|$ and width $\delta(R_{x,\epsilon}) = \epsilon|v(x)|^{-1}\|\omega_x\|_{L^\infty(-\epsilon,\epsilon)}$. As a consequence of Lemma 3.21 and its proof, there is the following property.

Lemma 4.1: Let $x' \in R'_{x,\epsilon}$, then

$$|v(x)| \sim |v(x')| \tag{4.2}$$

$$\|\omega_x\|_{L^\infty(-\epsilon,\epsilon)} \sim \|\omega_{x'}\|_{L^\infty(-\epsilon,\epsilon)} \tag{4.3}$$

$$\delta(R_{x,\epsilon}) \sim \delta(R_{x',\epsilon}) \tag{4.4}$$

$R_{x,\epsilon}$ is contained in a multiple of $R_{x',\epsilon}$ and vice versa . (4.5)

There is the following corollary.

Lemma 4.6: Assume $R'_{x,\epsilon} \cap R'_{x_1,\epsilon_1} \neq \phi$ and $\epsilon_1 < C\epsilon$. Then

$$R_{x_1,\epsilon_1} \subset R''_{x,\epsilon} \quad \text{and} \quad \epsilon^{-1}\delta(R_{x,\epsilon}) \geq c\epsilon_1^{-1}\delta(R_{x_1,\epsilon_1}) .$$

Proof: Take $x_2 \in R'_{x,\epsilon} \cap R'_{x_1,\epsilon_1}$. Then by Lemma 4.1, since also $x_2 \in R'_{x_1,\epsilon}$,

$$R_{x_1,\epsilon_1} \subset R'_{x_1,\epsilon} \subset R''_{x_2,\epsilon} \subset R'''_{x,\epsilon} .$$

Also, since

$$|v(x)| \sim |v(x_2)| \sim |v(x_1)|, \quad \|\omega_x\|_{L^\infty(-\epsilon,\epsilon)} \sim \|\omega_{x_2}\|_{L^\infty(-\epsilon,\epsilon)} \sim \|\omega_{x_1}\|_{L^\infty(-\epsilon,\epsilon)}$$

$$\epsilon_1^{-1}\delta(R_{x_1,\epsilon_1}) = |v(x_1)|^{-1}\|\omega_{x_1}\|_{L^\infty(-\epsilon_1,\epsilon_1)} \le C|v(x)|^{-1}\|\omega_x\|_{L^\infty(-\epsilon,\epsilon)} = C\epsilon^{-1}\delta(R_{x,\epsilon})$$

Let $\mathcal{S}$ stand for the collection of rectangles $R_{x,\epsilon}$ where $x \in \Omega$, $\epsilon > 0$.

Lemma 4.7: Let $\{R_j = R_{x_j,\epsilon_j}\}$ be a sequence in $\mathcal{S}$ and $\delta > 0$ such that

(i) $\delta(R_j) \sim \delta$

(ii) x_{j+1} does not belong to $R_1 \cup \ldots \cup R_j$

Then

$$\| \Sigma \chi_{R_j'} \|_\infty < C . \tag{4.8}$$

Proof: Suppose $R'_{x_j,\epsilon_j} \cap R'_{x_k,\epsilon_k} \neq \phi$ and $\epsilon_j \le \epsilon_k$ By (4.6) and (4.7)(i), it follows that

$$\epsilon_k^{-1}\delta > c\epsilon_j^{-1}\delta \Rightarrow \epsilon_j \sim \epsilon_k \quad \text{and} \quad R_k \subset R_j' .$$

From this observation, it is easily seen that having a fixed point in too many R_j'-rectangles must contradict (ii).

The next lemma is a Besicovitch-type covering property.

Lemma 4.9: Every subset $\mathcal{S}_0$ of $\mathcal{S}$ has a further subset $\mathcal{S}_1 \subset \mathcal{S}_0$ satisfying

(i) $\displaystyle \Big| \bigcup_{R \in \mathcal{S}_0} R' \Big| \le C \sum_{R \in \mathcal{S}_1} |R| .$

(ii) $\| \sum_{R \in \mathcal{S}_1} \chi_{R'} \|_\infty \le C$.

<u>Proof</u>: For $s > 0$, define $\mathcal{D}_s = \{R \in \mathcal{S}_0 | q^{-s-1} \le \delta(R) < q^{-s}\}$.
Construct $\overline{\mathcal{D}}_s \subset \mathcal{D}_s$ satisfying condition (4.7)(ii) and also

$$R_{x,\epsilon} \in \mathcal{D}_s \Rightarrow x \in \bigcup_{R \in \overline{\mathcal{D}}_s} R .$$

Hence we have by (4.5), (4.8),

$$\bigcup_{R \in \mathcal{D}_s} R' \subset \bigcup_{R \in \overline{\mathcal{D}}_s} R'' \tag{4.10}$$

$$\| \sum_{R \in \overline{\mathcal{D}}_s} \chi_{R'} \|_\infty \le C . \tag{4.11}$$

Construct now by induction systems $\mathcal{E}_s \subset \overline{\mathcal{D}}_s$ as follows:

$$\mathcal{E}_0 = \overline{\mathcal{D}}_0$$

$$\mathcal{E}_{s+1} = \{R \in \overline{\mathcal{D}}_{s+1} | R' \text{ does not intersect } R_0' \text{ for any } R_0 \in \mathcal{E}_1 \cup \dots \cup \mathcal{E}_s\} .$$

By construction and (4.11), denoting $\mathcal{S}_1 = \cup\, \mathcal{E}_s$,

$$\| \sum_{R \in \mathcal{S}_1} \chi_{R'} \|_\infty = \sup_s \| \sum_{R \in \mathcal{E}_s} \chi_{r'} \|_\infty \le C \text{ , i.e. (ii)}$$

Assume $R \in \overline{\mathcal{D}}_{s+1} \backslash \mathcal{E}_{s+1}$. Then R' intersects R_0' for some $R_0 \in \mathcal{E}_{s'}$, $s' \le s$.
If $R = R_{x,\epsilon}$, $R_0 = R_{x_0,\epsilon_0}$, it is easily deduced from (4.6) that $\epsilon \le C\epsilon_0$, since the other alternative leads to a contradiction. Applying again (4.6), we then get $R \subset R_0'$. Consequently

$$\bigcup_{R \in \overline{\mathcal{D}}_{s+1}} R' \subset \bigcup_{s' \le s+1} \bigcup_{R \in \mathcal{E}_{s'}} R'' .$$

Using (4.10), it follows that

$$\bigcup_{R \in \mathcal{B}_0} R' = \bigcup_s \bigcup_{R \in \mathcal{B}_s} R' \subset \bigcup_{s'} \bigcup_{R \in \mathcal{B}_{s'}} R''' = \bigcup_{R \in \mathcal{B}_1} R'''$$

$$\Big| \bigcup_{R \in \mathcal{B}_0} R' \Big| \leq C \sum_{R \in \mathcal{B}_1} |R| \text{ , i.e. (i)}$$

Lemma 4.9 has the following corollary.

Lemma 4.12: The maximal function

$$M_{\mathcal{B}} f(x) = \sup_{R \in \mathcal{B},\, x \in R} \left\{ \frac{1}{|R|} \int_R |f| \right\}$$

has a weak-type estimate and hence is L^2-bounded.

5. Estimation of the Maximal Function

Define for $0 < \epsilon < \epsilon_0$, $s > 0$,

$$\Omega_{\epsilon,s} = \{x \in \Omega \mid 2^{-s-1} \leq \epsilon |v(x)|^{-1} \|\omega_x\|_{L^\infty(-\epsilon,\epsilon)} = \delta(R_{x,\epsilon}) < 2^{-s}\} . \tag{5.1}$$

For real analytic vectorfields, either the integral curves of the vectorfield are straight lines (in which case the problem is 1-dimensional) or for each $\epsilon > 0$, the complement of the set $\bigcup_s \Omega_{\epsilon,s}$ has measure 0. It then suffices to prove the L^2-inequality on the maximal function $M_v f = \sup_{\epsilon > 0} |A_\epsilon f|$ restricted to $\bigcap_{\epsilon > 0} \bigcup_{\delta > 0} \Omega_{\epsilon,s}$. Hence, it is no restriction to assume for each $\epsilon > 0$ that

$$\Omega = \bigcup_{s > 0} \Omega_{\epsilon,s}$$

(since, a priori, only finitely many ϵ's have to be considered).

In the general case of v satisfying (2.2), it obviously suffices to prove that L^2-bound on $\Omega^* = \{x \in \Omega | v(x) \neq 0\}$. Since it suffices to prove the inequality for functions with finitely supported Fourier transform, the rectangles $R_{x,\epsilon}$, $x \in \Omega^*$, may always be redefined such that $\delta(R_{x,\epsilon}) \geq \tau\epsilon|v(x)|$ for some $\tau > 0$. In particular, the rectangles will be non-degenerated and again $\Omega^* = \bigcup_{s>0} \Omega^*_{\epsilon,s}$ for each $\epsilon > 0$.

For $\epsilon < \epsilon_0$, we only consider dyadic values 2^{-j}, $j > j_0$. Clearly, for fixed s, the $\Omega_{\epsilon,s}$ are essentially disjoint. From (4.4), if $x \in \Omega_{\epsilon,s}$ and $x' \in R'_{x,\epsilon}$, then $\delta(R_{x',\epsilon}) \sim 2^{-s}$, hence $x' \in \bigcup_{|s-s'|<C} \Omega_{\epsilon,s'}$. Hence

$$\Omega'_{\epsilon,s} \equiv \bigcup_{x \in \Omega_{\epsilon,s}} R'_{x,\epsilon} \subset \bigcup_{|s-s'|<C} \Omega_{\epsilon,s'}$$

and therefore

$$\|\sum_{\epsilon} \chi_{\Omega'_{\epsilon,s}}\|_\infty < C \, . \tag{5.2}$$

Let $f \in L^2$ and $f = \sum_{T \text{ dyadic}} f_T$ be a Littlewood-Paley decomposition where $\operatorname{supp} \hat{f} \subset B(0,2T)\backslash B(0,T)$. Thus $A_\epsilon f = \sum A_\epsilon f_T$ and for $x \in \Omega_{\epsilon,s}$ evaluated as

$$|A_\epsilon f(x)| \leq |\sum_{T \leq 2^s} A_\epsilon f_T(x)| + \sum_{T > 2^s} |A_\epsilon f_T(x)| \tag{5.3}$$

where the second term in (5.3) is bounded by

$$\sum_{s>0} \sum_{T>2^s} |A_\epsilon f_T| \chi_{\Omega_{\epsilon,s}} = \sum_{j>0} [\max_\epsilon \left(\sum_{s>0} |A_\epsilon(f_{2^{s+j}})|^2 \chi_{\Omega_{\epsilon,s}}\right)^{1/2}] \tag{5.4}$$

an expression independent of ϵ. Evaluate in L^2-norm replacing the ϵ-supremum by the square function to get

$$\|(5.4)\|_2 \le \sum_j \{ \sum_{\epsilon,s} \|A_\epsilon(f_{2^{s+j}})\chi_{\Omega_{\epsilon,s}}\|_2^2 \}^{1/2} . \tag{5.5}$$

To evaluate the inner L^2-norms, use (3.29). First cover $\Omega_{\epsilon,s}$ with rectangles $R_\ell \in \mathcal{B}$ such that condition (ii) of (4.7) holds, hence $\|\sum_\ell \chi_{R'_\ell}\|_\infty \le C$ and therefore

$$\sum_\ell \chi_{R'_\ell} \le C\,\chi_{\Omega'_{\epsilon,s}} . \tag{5.6}$$

Write, using (3.9) with $T = 2^{s+j}$, $\delta = 2^{-s}$

$$\|A_\epsilon(f_{2^{s+j}})\chi_{\Omega_{\epsilon,s}}\|_2^2 \le \sum_\ell \|A_\epsilon(f_{2^{s+j}})\chi_{R_\ell}\|_2^2 \le 2^{-cj} \sum_\ell \|f_{2^{s+j}}(\chi_{R'_\ell} * \psi_{\frac{1}{2^{s+j}}})\|_2^2 .$$

Thus, by (5.6),

$$\|A_\epsilon(f_{2^{s+j}})\chi_{\Omega_{\epsilon,3}}\|_2 \le C2^{-cj}\|f_{2^{s+j}}(\chi_{\Omega'_{\epsilon,s}} * \psi_{2^{-s-j}})\|_2 .$$

Hence, substituting in (5.5) and exploiting (5.2),

$$\|(5.4)\|_2^2 \le C \sum_j 2^{-cj} (\sum_{\epsilon,s} \|f_{2^{s+j}}(\chi_{\Omega'_{\epsilon.s}} * \psi_{2^{-s-j}})\|_2^2)$$

$$\le \sum_j 2^{-cj} \sum_s \|f_{2^{s+j}}\|_2^2 \le C\|f\|_2^2 .$$

It remains to take care of the contribution of the first term in (5.3). The idea is to replace M_V by $M_{\mathcal{B}}$.

<u>Lemma 5.7</u>: If supp $\hat{g} \subset B(0,T)$ and $|x-x'| < \frac{1}{T}$, then $|g(x)| \le Cg^*(x')$, where g^* refers to the usual Hardy-Littlewood maximal function.

Proof: Assume $x = 0$ and φ satisfying $\hat{\varphi} = 1$ on $B(0,1)$, $|\varphi(x)| < C(1 + |x|^3)^{-1}$. Since $\hat{\varphi}_{\frac{1}{T}} = 1$ on suff $\hat{f}$, we have

$$|g(0)| = |\int \hat{g}(\xi)d\xi| = |\langle g, \varphi_{\frac{1}{T}}\rangle| \leq C \sum_{k \geq 0} T^2 8^{-k} \int_{B(0,\frac{2^k}{T})} |g(x)|dx \leq Cg^*(x') .$$

Taking for g the function $\sum_{T < 2^s} f_T$ and $x \in \Omega_{\epsilon,s}$, by (5.7)

$$|A_\epsilon g(x)| = |\int g(x + \epsilon tv(x))\alpha(t)dr|$$

$$\leq c \int 4^s \{ \int_{B(x+\epsilon tv(x), 2^{-s})} g^*(y)dy \} \alpha(t)dt$$

$$\leq C \frac{2^5}{\epsilon|v(x)|} \int_{R_{x,\epsilon}} g^* \leq CM_{\mathscr{B}}(g^*)(x) \leq CM_{\mathscr{B}}(f^{**})(x)$$

since we may assume $|\sum_{T < 2^s} f_T| \leq Cf^*$, for each s.

Invoking (4.12), it follows that the L^2-contribution of the first term in (5.3) is bounded by

$$\|M_{\mathscr{B}} f^{**}\|_2 \leq C\|f^{**}\|_2 \leq C\|f\|_2$$

which completes the proof of Theorem 1.

References

[Br] J. Brianson: Weierstrass prepare a la Hironaka, Astinisque, 7-8, p. 67.

[Chr] M. Christ. Differentiation along variable curves and related singular integral operators, preprint 1985.

[G] M. de Guzman: Real variable methods in Fourier analysis, North-Holland, 1987.

[N-S-W] A. Nagel, E. M. Stein, S. Wainger: Hilbert transforms and maximal functions related to variable curves, Proc. Symp. Pure Math. XXXV, Part 1, 95-98.

[P-S] D. H. Phong, E. M. Stein: Hilbert integrals, singular integrals and Radon transforms I, Acta Math. 157, 1-2, pp. 99-157.

HANKEL OPERATORS ON H^p

Joseph Cima and David A. Stegenga

ABSTRACT. Janson, Peetre and Semmes [JPS] and Tolokonnikov [T] have recently characterized the Hankel operators mapping H^p boundedly into itself for the range $0 < p \leq 1$. We improve this result for the case $0 < p < 1$ by showing that these operators map H^p into the space of Cauchy transforms of L^1 functions. In addition, we give new proofs for the case of Hankel operators mapping H^1 into itself and H^p into H^1 for $0 < p < 1$.[1]

1. INTRODUCTION. Let $b \in H^2$ and $f \in H^\infty$, then we define $H_b f = P(b\bar{f})$ where P is the orthogonal projection mapping $L^2(\partial D)$ onto $H^2(\partial D)$. The classical result of Nehari [N] states that the Hankel operator H_b with holomorphic symbol b(z) is bounded on H^2 if and only if b is in BMOA. See chapter 1 of Power's book [P] or [T] for a modern treatment of this result. The same result holds for H^p with $1 < p < \infty$. In [JPS] and [T] it is shown that for $0 < p < 1$ the answer is the space of Lipschitz functions $\Lambda_{1/p-1}$.

Let K denote the space of Cauchy integrals of measures on ∂D, where D is the unit disk in $\mathbb{C}$. Each f in K has the representation

$$f(z) = \int_{\partial D} \frac{d\mu(\xi)}{1-z\bar{\xi}}, \qquad z \in D \tag{1}$$

for some measure μ. The K–norm of f is the infimum of the total variation norms of the measures which represent f. With this norm, K is isometrically isomorphic to the dual of disk algebra A. See chapter IV of Garnett's book [G] for the details. K is called the Cauchy–Stieltjes space. If the measure μ is absolutely continuous with respect to Lebesgue measure on ∂D, then we say that $f \in K \cap L^1$. It is obvious that $H^1 \subset K$ and it follows from a theorem of Kolmorgoroff [K] that $K \subset H^1_{weak}$ and hence that $K \subset H^p$ for all $0 < p < 1$.

---

[1] 1980 Mathematical Subject Classification (1985 Revision): 30D55, 30E20, 47B35.

THEOREM 1. *Let* $0 < p < 1$. *The Hankel operator* H_b , *with holomorphic symbol* $b(z)$, *maps* H^p *boundedly into* $K \cap L^1$ *if and only if* $b \in \Lambda_{1/p-1}$.

The Lipschitz class is the same as in [JPS] and [T] and hence the range of bounded Hankel operators on H^p , for $0 < p < 1$, is considerably smaller than all of H^p. The above result is also an improvement of the H^1_{weak} result in [BM]. Surprisingly, the proof in section 3 is quite elementary. In section 4, we use the same techniques to prove other boundedness criteria.

This research took place while the second author held a visiting position at the University of North Carolina, Chapel Hill, and he wishes to thank the Mathematics Department for their fine hospitality. Both authors would like to thank the referee for bringing the Tolokonnikov paper to our attention and for many helpful suggestions.

2. PRELIMINARIES. For $\alpha > 0$, we say that $f \in \Lambda_\alpha$ provided that f is holomorphic on D and that

$$|f^{(n+1)}(z)| \le C(1-|z|)^{\alpha-(n+1)}, \quad z \in D \tag{2}$$

where C is a constant depending only on f and $n = [\alpha]$. We norm Λ_α by taking the best constant in (2). For nonintegers α , $f \in \Lambda_\alpha$ if and only if $f, f', \ldots, f^{(n)} \in A$ and

$$|f^{(n)}(z_1) - f^{(n)}(z_2)| \le C\,|z_1-z_2|^{\alpha-n}, \quad z_1, z_2 \in D, \tag{3}$$

where again $n = [\alpha]$. For positive integers n , $f \in \Lambda_n$ if and only if $f, \ldots, f^{(n-1)} \in A$ and $f^{(n-1)}$ is in the Zygmund class. See chapter 5 of Duren's book [D] for these classical results.

In section 4, we will need a new criterion for membership in Λ_α . Let f be holomorphic on D and denote by $P_n(z,a)$ the nth Taylor polynomial of f expanded about the point $z = a$.

PROPOSITION 1. *Suppose that* $\alpha > 0$ *and that* $n = [\alpha]$. *For* $f(z)$ *holomorphic on* D :

(i) For noninteger α *,* $f \in \Lambda_\alpha$ *if and only if*

$$|f(z) - P_n(z,a)| \leq C\,|z-a|^\alpha; \qquad z,a \in D \tag{4}$$

holds for some constant C .

(ii) *For all* α *,* $f \in \Lambda_\alpha$ *if and only if*

$$\int_{|z|=1} \frac{|f(z) - P_n(z,a)|}{|z-a|^{n+2}}\,|dz| \leq \frac{C}{(1-|a|)^{n+1-\alpha}}, \qquad a \in D \tag{5}$$

PROOF. We first observe that condition (5) follows from condition (4). If we assume that either condition holds, then Cauchy's Theorem yields

$$f^{(n-1)}(a) = \frac{(n+1)!}{2\pi i} \int_{|z|=1} [f(z)-P_n(z,a)]\,\frac{dz}{(z-a)^{n+2}}$$

for $a \in D$ and hence condition (5) implies that $f \in \Lambda_\alpha$.

Suppose that $\alpha > 0$ is a noninteger and that $f \in \Lambda_\alpha$. If $n = 0$, then condition (4) follows immediately from (3). If $n > 0$, then integration by parts yields that

$$f(z)-P_n(z,a) = \frac{1}{(n-1)!} \int_\Gamma (f^n(\xi)-f^n(a))(z-\xi)^{n-1}\,d\xi \tag{6}$$

where Γ is the line segment $[a,z]$. Again, (3) and (6) imply that condition (4) holds.

Finally, assume that $\alpha = n$ is a positive integer and that $f \in \Lambda_\alpha$. For simplicity, we will assume that f is defined on the upper half plane $\Pi = \{x + iy : y > 0\}$ so that (2) becomes

$$|f^{(n+1)}(x+iy)| \leq \frac{C}{y}; \quad y > 0. \tag{7}$$

Let $a = a_1 + ia_2$ and $z = x + iy$ be points in Π. Put $\epsilon = |z-a|$. We first assume that $y \leq a_2$ and define Γ_1 to be the line segments connecting the points a, $a + i\epsilon$, $x + i(a_2+\epsilon)$ and $x + ia_2$. Also, let Γ_2 be the line segment from $x + ia_2$ to z. The classical formula

$$(8) \qquad f(z) - P_n(z,a) = \frac{1}{n!} \int_{\Gamma_1 \cup \Gamma_2} f^{(n+1)}(\xi)(z-\xi)^n d\xi$$

holds and we estimate the integrals over Γ_1 and Γ_2. An immediate consequence of (7) is that the integral over Γ_2 satisfies

$$(9) \qquad \left| \int_{\Gamma_2} f^{(n+1)}(x+it)\,(t-y)^n dt \right| \leq C_1 \epsilon^n$$

Similarly,

$$\left| \int_{\Gamma_1} f^{(n+1)}(\xi)(z-\xi)^n d\xi \right| \leq (2\epsilon)^n \int_{\Gamma_1} |f^{(n+1)}(\xi)|\,(d\xi) \leq C_2\, \epsilon^n \log\left(1 + \frac{\epsilon}{a_2}\right)$$

and hence combining (9) with the above in (8) yields that

$$(10) \qquad |f(z) - P_n(z,a)| = C_3 |z-a|^n \log\left(1 + \frac{|z-a|}{\operatorname{Im} a}\right)$$

when $\operatorname{Im} a \leq \operatorname{Im} z$. A similar argument shows that (10) holds for all $a,z \in \Pi$. Integrating (10) gives

$$\int_{-\infty}^{\infty} \frac{|f(x) - P_n(x,a)|}{|x-a|^{n+2}} dx$$
$$\leq C_3 \int_{-\infty}^{\infty} \log(1 + \frac{|x-a|}{\mathrm{Im} a}) \frac{dx}{|x-a|^2}$$
$$\leq C_3 (\mathrm{Im} a)^{-1} \int_{-\infty}^{\infty} \log(1+|x-i|) \frac{dx}{x^2+1}$$

which is the desired result. This completes the proof. □

REMARK. The above result appears to be new, although there are many closely related results in the literaure. For example, Proposition 1 is quite similar to the results on pages 80–86, Nagel and Stein [NS]. In addition, see the papers related to Campanato spaces, [C], chapter 6 in [DS], [G], and [JTW].

EXAMPLE. The inequality (10) for $f \in \Lambda_n$ is the best possible. Let $f(z) = z \log(\bar{i}z)$ for $z \in \Pi$, then $y|f''(x+iy)| = y/|x+iy| \leq 1$ and so $f \in \Lambda_1$. An easy calculation shows that $|f(i) - P_1(i,i\epsilon)|$ is comparable to the right–hand side of (10) whenever $0 < \epsilon \leq 1$.

PROPOSITION 2. (Hardy and Littlewood [HL], [DRS]). *Let* $0 < p < 1$. *There is a constant* a_p *depending only on* p *so that*

(11)
$$\iint_D |f(z)| (1-|z|^2)^{\frac{1}{p}-2} dxdy \leq a_p ||f||_{H^p}$$

holds for any $f \in H^p$.

We give a short proof based on tent spaces for this classical inequality.

PROOF. Let $f \in H^p$. In [CMS] it is shown that there exist continuous functions f_j on D, with support on the tent $\hat{I}_j$ of an interval I_j ($\hat{I}_j$ is the triangular region in D with base I_j), such that the inequality $\|f_j\|_\infty \leq |I_j|^{-\frac{1}{p}}$ holds and that

$$f(z) = \sum \lambda_j f_j(z) \ , \ z \in D \tag{12}$$

where $\sum |\lambda_j|^p \le C_1^p \|f\|_{H^p}^p$.

Integrating (12) yields that

$$\iint_D |f(z)| \ (1-|z|^2)^{\frac{1}{p}-2} \ dxdy$$

$$\le \sum_j |\lambda_j| \ |I_j|^{-\frac{1}{p}} \iint_{\hat{I}_j} (1-|z|^2)^{\frac{1}{p}-2} \ dxdy$$

$$\le C_2 \sum |\lambda_j|$$

$$\le C_2 \{\sum |\lambda_j|^p\}^{\frac{1}{p}} \le a_p \|f\|_{H^p}$$

and the proof is complete. □

Finally, we will need a representation formula for Hankel operators. Based on ideas in Bonami and Mesrar [BM] we have the following:

PROPOSITION 3. *Let* $b \in H^2$ *and* $f \in H^\infty$. *Denote by* $b_n(z) = z^n b(z)$ *for* $n=0,1,\ldots$. *For* $n=0,1,\ldots$ *we have*

$$H_b f(z) = \frac{1}{\pi n!} \iint_D \frac{\overline{f(\xi)}}{1-\bar{\xi}z} \, b_{n+1}^{(n+1)}(\xi)(1-|\xi|^2)^n \ dxdy \tag{13}$$

and

$$\frac{1}{2\pi} \int_{\partial D} H_b \bar{f} g d\theta = \frac{1}{\pi n!} \iint_D \overline{f} g(\xi) \, b_{n+1}^{(n+1)}(\xi)(1-|\xi|^2)^n dxdy \tag{14}$$

whenever $g \in H^2$.

PROOF. To prove (14) it suffices to let $b = z^{\ell}$, $f = z^k$ with $\ell \geq k$ and $g = z^m$. Then $H_b f(z) = z^{\ell-k}$. If $\ell = m + k$, it follows that

$$\frac{1}{\pi n!}\iint_D \overline{f\, g(\zeta)}\, b_{n+1}^{(n+1)}(\xi)(1-|\xi|^2)^n\, dxdy$$

$$= \frac{(n+1+\ell)!}{\ell!\, n!}\int_0^1 r^{\ell}(1-r)^n\, dr$$

$$= 1 = \frac{1}{2\pi}\int_{\partial D} H_b \bar{f} \bar{g}\, d\theta.$$

On the otherhand, if $\ell \neq m + k$, then both integrals are zero. This proves (14) and (13) follows by Cauchy's formula. □

The next proposition will be used frequently.

PROPOSITION 4. ([D, page 65]). *For* $a \in D$ *and* $p \geq 1$:

$$\frac{1}{2\pi}\int_{|z|=1}\frac{(dz)}{|z-a|^p} \doteq \begin{cases} (1-|a|^2)^{1-p}, & p > 1 \\ \log\dfrac{2}{1-|a|^2}, & p = 1 \end{cases}$$

3. HANKEL OPERATORS AND THE CAUCHY–STIELTJES SPACE. Given a finite measure μ on the open unit disk D , let μ^* denote the sweep of μ . That is, μ^* is the measure on ∂D satisfying

$$\int_D g\,d\mu = \int_{\partial D} g\,d\mu^* \tag{15}$$

whenever g is harmonic on D and continuous on the closed disk.

LEMMA 1. *Let* μ *be a finite measure defined on the open unit disk. Then* μ^* *is absolutely continuous with respect to* $d\theta$.

PROOF. Let E be a compact subset of ∂D with zero Lebesque measure. Let $g \in A$ be a peak function for E, that is, $g = 1$ on E and $|g(z)| < 1$ for all $z \in \overline{D}\setminus E$. See pages 80–81 in [H] for the construction. It follows from the Lebesque bounded convergence theorem that

$$\mu^*(E) = \lim_{n\to\infty} \int_{\partial D} g^n \, d\mu^*$$

$$= \lim_{n\to\infty} \iint_D g^n \, d\mu = 0$$

and hence $d\mu^* << d\theta$. This completes the proof. □

REMARK. The above lemma can also be proved using the Fubini theorem and the Poisson representation formula.

We now prove Theorem 1 stated in the introduction.

PROOF OF THEOREM 1. Fix $0 < p < 1$ and let $1/n+1 < p \leq 1/n$ for the integer n. Let $f \in H^\infty$, $g \in A$ and $b \in \Lambda_{1/p-1}$. By (14) of Proposition 3, (2), and Proposition 2 we have

$$\left|\frac{1}{2\pi} \int_{\partial D} H_b f \, \overline{g} \, d\theta\right| \leq \frac{1}{\pi n!} \iint_D |fg| \, |b_{n+1}^{(n+1)}| \, (1-|\xi|^2)^n \, dxdy$$

$$\leq C_1 \, \|g\|_\infty \iint_D |f(\xi)| \, (1-|\xi|^2)^{1/p-2} \, dxdy$$

$$\leq C_2 \, \|g\|_\infty \, \|f\|_{H^p} .$$

This means that $H_b f$, viewed as a linear functional on A, satisfies the norm inequality $\|H_b f\|_K \leq C_2 \|f\|_{H^p}$. Since H^∞ is dense in H^p for all $p < \infty$ we see that H_b maps H^p boundedly into K.

For any $f \in H^P$, we define a measure on D by

(16) $$d\mu(\xi) = \overline{f(\xi)}\, b_{n+1}^{(n+1)}(\xi)(1-|\xi|^2)^n \frac{dxdy}{\pi n!}.$$

As we observed above, μ is a finite measure with total variation dominated by a multiple of $\|f\|_{H^p}$. By formula (13) we have

$$H_b f(z) = \iint_D \frac{d\mu(\xi)}{1-\bar{\xi}z}$$

$$= \int_{\partial D} \frac{d\mu^*(\xi)}{1-\bar{\xi}z}$$

and it now follows from Lemma 1 that $H_b f$ is in $K \cap L^1$. Actually, a limit argument is required in the above. But it follows from Proposition 2 that

(17) $$\lim_{r\to 1} \iint_D |f(r\xi)-f(\xi)|(1-|\xi|^2)^{1/p-2}\, dxdy = 0$$

whenever $f \in H^P$.

To prove the converse, assume that $\|H_b f\|_K \le C_3\|f\|_p$ whenever $f(z) = z^{n+1}/(1-\bar{a}z)^{n+1}$ and $a \in D$. A straightforward computation with residues shows that

(18) $$H_b f(z) = \begin{cases} \dfrac{b(z)-P_n(z,a)}{(z-a)^{n+1}}, & z \ne a \\[2ex] \dfrac{b^{(n+1)}(a)}{(n+1)!}, & z = a. \end{cases}$$

It follows from (1) and Proposition 4 that

$$|H_b f(a)| \le \frac{\|H_b f\|_K}{1-|a|}$$

$$\le \frac{C_3\, \|f\|_{H^p}}{1-|a|}$$

$$\le C_4(1-|a|^2)^{1/p-1-(n+1)}$$

and hence (18) implies that $b \in \Lambda_{1/p-1}$. The proof is complete. □

The proof of Theorem 1, while fairly brief, essentially contains the sufficiency part of the theorem of Duren, Romberg and Shields [DRS] that $\Lambda_{1/p-1}$ is the dual of H^p for $0 < p < 1$. By taking $g = 1$ in the proof of Theorem 1 we see that

(19) $$\left|\frac{1}{2\pi}\int_{\partial D} f\,\bar{b}\,d\theta\right| = \left|\frac{1}{2\pi}\int_{\partial D} H_b f\,d\theta\right| \le C_2\|f\|_{H^p}$$

whenver $f \in H^2$ and by (18) the linear functional Φ_b determined by b satisfies

(20) $$\begin{aligned}\Phi_b(f) &= \lim_{r\to 1}\frac{1}{2\pi}\int_D f_r\,\bar{b}\,d\theta \\ &= \frac{1}{\pi n!}\iint_D f\,\overline{b_{n+1}^{(n+1)}}(1-|z|^2)^n\,dxdy\end{aligned}$$

for arbitrary $f \in H^p$. Here $f_r(z) = f(rz)$. Of course, the same result holds for the containing Banach space B^p, defined in [DRS] by

$$\|f\|_{B^p} = \int_0^1 M_1(f,r)(1-r)^{1/p-2}\,dr < \infty$$

where we have used the usual notation for integral means in the 1–norm. Both theorems in this section can be generalized by replacing H^p with B^p. The necessity part of Theorem 1 relies on the elementary calculation that $\|(1-\bar{a}z)^{-(n+1)}\|_{B^p}$ is comparable to $(1-|a|)^{1/p-n-1}$.

Another approach to the above theorem is to observe that H^∞ could be used in place of A in the proof of Theorem 1. For a fixed $f \in H^p$ and $b \in \Lambda_{1/p-1}$ it follows that

$$\lim_{r\to 1} \frac{1}{2\pi} \int_{\partial D} H_b (f_r)\, \bar{g}\, d\theta = \iint_D \bar{g}\, d\mu$$

whenever $g \in H^\infty$. Here $f_r(z) = f(rz)$ and $d\mu$ is defined in (16). By a theorem of Mooney [M], the above implies the existence of a function $\varphi \in L^1(\partial D)$ so that

$$\iint_D \bar{g}\, d\mu = \int_{\partial D} \bar{g}\, \varphi\, d\theta \ , g \in H^\infty .$$

But an immediate consequence of formula (15) is that the measure $d\mu^* - \varphi d\theta$ annihilates $\bar{z}^k$ for $k = 0,1,...$ and hence must be absolutely continuous by the F. and M. Riesz theorem. Hence $d\mu^*$ is also absolutely continuous.

4. HANKEL OPERATORS MAPPING H^p INTO H^1. In view of Theorem 1 one might ask whether H_b actually maps H^p $(0 < p < 1)$ to H^1 whenever $b \in \Lambda_{1/p-1}$. This turns out to be false in general by a result in [T] and also [BM] characterizing the Hankel operators mapping H^p boundedly onto H^1. Their answer is slightly stronger than $\Lambda_{1/p-1}$ and hence the L^1–functions which arise in Theorem 1 are not generally in $H^1 + \overline{H^1}$.

In the next Theorem we give a variant of this result. The proof follows the ideas in the proof of Theorem 1.

THEOREM 2. *Suppose that* $0 < p < 1$ *and that* n *is an integer satisfying* $1/n+1 < p \le 1/n$, *where* $n = 0$ *if* $p > 1$. H_b *maps* H^p *boundedly into* H^1 *if and only if*

(21) $$|b^{(n)}(z)| \le C_1 \left[(1-|z|)^{n-(1/p-1)} \log\frac{2}{1-|z|}\right]^{-1}, \quad z\in D.$$

PROOF. Assume that (21) holds. Let $g \in BMOA$, then since $|g'(z)| \le C_1\|g\|_*/(1-|z|)$ we know that $M_\infty(g,r) \le C_2\|g\|_* \log\frac{2}{1-r}$. As in the proof of Theorem 1, we have that

(22) $$\left|\frac{1}{2\pi}\int_{\partial D} H_b f\bar{g}\, d\theta\right| \le C_3 \|f\|_{H^p} \|g\|_*$$

whenever f is in H^2. By the Fefferman Duality Theorem in [FS] we know that (22) implies that $\|H_b f\|_{H^1} \le C_4 \|f\|_{H^p}$ and we are done with the sufficiency proof.

To prove that (21) is necessary, assume that $\|H_b f\|_{H^1} \le C_5 \|f\|_{H^p}$ whenever $f(z) = [z/(1-\bar{a}z)]^{n+1}$. Since $b \in \Lambda_{1/p-1}$ by Theorem 1, we see that Propositions 1 and 4 imply that

$$\frac{|b^{(n)}(a)|}{n!} \int_{|z|=1} \frac{dz}{|z-a|}$$

$$\le \int_{|z|=1} |H_b f(z)|dz| + \int_{|z|=1} \frac{|b(z)-P_{n-1}(z,a)|}{|z-a|^{n+1}} |dz|$$

$$\le C_5 \|f\|_{H^p} + C_6 (1-|a|)^{(1/p-1)-n}$$

$$\le C_7 (1-|a|)^{(1/p-1)-n}$$

and hence b satisfies (21).

5. HANKEL OPERATORS ON H^1. In [JPS] and [T] it is proved that H_b maps H^1 bounded into itself if and only if

$$\sup_I \frac{\log \frac{2}{|I|}}{|I|} \int_I |b - b_I| d\theta < \infty \tag{24}$$

where I ranges over subarcs on ∂D, $|I|$ denotes the measure of I and b_I denotes the average of b on I. Without the logarithm term, (24) is the definition of BMOA. Thus, b satisfies a condition slightly stronger than BMOA.

We present a variant of this result which follows from the techniques used in Fefferman's Duality Theorem. In particular, we need the notion of Carleson measures. A measure μ on the open disk is a Carleson measure provided

$$N(\mu) = \sup_{I} \frac{\mu(S(I))}{|I|} < \infty \tag{25}$$

where for a given arc I, $S(I)$ is the region $\{re^{i\theta} : 1-r \le |I|, e^{i\theta} \in I\}$. The importance of (25) is that it characterizes the measures which satisfy

$$\int_D |f|^p \, d\mu \le C \, \|f\|_{H^p}^p , \quad f \in H^p \tag{26}$$

for any $0 < p < \infty$. In addition, $b \in BMOA$ if and only if $|g'|^2 \log \frac{1}{|z|^2} \, dxdy$ is a Carleson measure on D. See chapter VI in [G] as a reference for this material.

THEOREM 3. *The Hankel operator* H_b *maps* H^1 *boundedly into itself if and only if*

$$\sup_{I} \frac{(\log \frac{2}{|I|})^2}{|I|} \iint_{S(I)} |b'|^2 \log \frac{1}{|z|} \, dxdy < \infty . \tag{27}$$

PROOF. Let b satisfy (27) and suppose that $f \in H^1$ and that $h \in BMOA$. Using the proof of Fefferman's Duality Theorem given on pages 246–7 in [G], we may assume that $f = g^2$ where $g \in H^2$ and $g(0) = 0$. It then follows that

$$\begin{aligned} \left|\frac{1}{2\pi} \int_{\partial D} H_b f \, \bar{h} \, d\theta\right| &= \left|\frac{1}{\pi} \iint_D \overline{(fh)'} \, b' \log \frac{1}{|z|^2} \, dxdy\right| \\ &\le \frac{1}{\pi} \iint_D |g|^2 \, |h'| \, |b'| \log \frac{1}{|z|^2} \, dxdy \\ &\quad + \frac{2}{\pi} \iint_D |gg'| \, |h| \, |b'| \log \frac{1}{|z|^2} \, dxdy \\ &= I_1 + I_2 . \end{aligned}$$

Since $|h'|^2 \log \frac{1}{|z|}\,dxdy$ and $|b'|^2 \log \frac{1}{|z|}\,dxdy$ are both Carleson measures it is easily seen that

(28)
$$I_1 \le C_1 \|g\|^2_{H^2} \|h\|_* \|b\|_*$$

$$= C_1 \|f\|_{H^1} \|h\|_* \|b\|_* .$$

On the other hand, the well known identity

(29)
$$\frac{1}{2\pi} \int_{\partial D} f\bar{g}\, d\theta = \frac{1}{\pi} \iint_D f' \overline{g'} \log \frac{1}{|z|^2}\, dxdy$$

which holds for aribtrary $f, g \in H^2$, provided that $g(0) = 0$, implies that

(30)
$$I_2^2 \le \frac{2}{\pi} \iint_D |g'|^2 \log \frac{1}{|z|^2}\, dxdy \times \frac{2}{\pi} \iint_D |g|^2\, |hb'|^2 \log \frac{1}{|z|^2}\, dxdy$$

$$\le C_2 \|g\|^4_{H^2} N(\lambda)$$

$$= C_2 \|f\|^2_{H^1} N(\lambda) .$$

and $d\lambda = |hb'|^2 \log \frac{1}{|z|^2}\, dxdy$.

The sufficiency proof will be completed if we show that $N(\lambda) \le C_3 \|h\|^2_*$. Fix an arc I , which we may assume is centered at $z = 1$ and let $|I| = 1-r$ with $0 < r < 1$. Let $\varphi(z) = (z-r)/(1-rz)$ be the Möbius transformation of D which maps r to 0 . Using the conformal invariance properties of BMOA described, for example, in section 3 of Chapter VI [G] we see that $\|f \circ \varphi^{-1}\|_* \le C_4 \|f\|_*$ holds for all $f \in$ BMOA . The change of variables argument on page 238 [G] reveals that

$$\frac{1}{|I|} \iint_{S(I)} |h(z)-h(r)|^2 \, |b'(z)|^2 \, (1-|z|^2) \, dxdy$$

$$\leq C_5 \iint_{D} |h-h(r)|^2 \, |b'|^2 \, (1-|\varphi(z)|^2) \, dxdy$$

$$= C_6 \iint_{D} |h\circ\varphi^{-1}-h\circ\varphi^{-1}(0)|^2 \, |(b\circ\varphi^{-1})'|^2 \, (1-|z|^2) \, dxdy$$

$$\leq C_7 \, \|b\circ\varphi^{-1}\|_*^2 \int_{\partial D} |h\circ\varphi^{-1}-h\circ\varphi^{-1}(0)|^2 \, |dz|$$

$$\leq C_8 \, \|b\|_h^2 \, \|h\|_h^2$$

where we have used (26) and the conformally invariant definition of BMOA (see Corollary 2.4 [G]).

Finally, we must show that

(31)
$$\frac{|h(r)|^2}{|I|^2} \iint_{S(I)} |b'|^2 \log \frac{1}{|z|^2} \, dxdy \leq C_9$$

for some constant C_9. This is an immediate consequence of (27) since $|h(r)| \leq C_7\|h\|_*$ $\log \frac{2}{1-r}$ and the sufficiency proof is complete.

We now prove the necessity of condition (27). Suppose that H_b is a bounded operator on H^1, then by the result in [JPS], condition (14) holds. By a theorem of John and Nirenberg [JN] this implies that

(32)
$$\sup_I \frac{\left[\log \frac{2}{|I|}\right]^2}{|I|} \int_I |b-b_I|^2 \, d\theta < \infty .$$

Now a standard argument for functions in BMO shows that

(33) $$\sup_{\alpha\in D}\left[\log\frac{2}{1-|a|}\right]\iint_D |b'|^2 \log\left|\frac{1-\bar{a}z}{z-a}\right| dxdy < \infty$$

which is easily seen to imply (27). The proof is complete. □

REMARK. The proof of Theorem 1 is considerably simpler than that given in either [JPS] or [BM]. The proof given in [T] is also quite simple; however, in view of Theorem 2 it is unlikely that the boundary value methods employed in that paper can be used to prove the L^1 part of the conclusion to Theorem 1. The proof of Theorem 4 is about the same level of difficulty as that given in [JPS]. The necessity proof given here is unfortunate in that it does not follow the techniques used earlier. It would be interesting to have a simple proof for the necessity of condition (27).

Bibliography

[BM] Bonami, A. and Messar, Y., "Inégalités H^p pour les opérateurs de Hankel à symbole Lipshitzien", C. R. Acad. Sc. Paris, Serie I 303 (1986), 853–856.

[C] Campanato, S., "Proprietà di una famiglia di spazi funzionali", Ann. Scuola Norm. Sup. – Pisa, 18 (1964), 137–160.

[CMS] Coifman, R. R., Meyer, Y., and Stein, E. M., "Some New Function Spaces and Their Applications to Harmonic Analysis", J. Funct'l Anal. 62 (1985), 304–335.

[D] Duren, P. L., Theory of H^p Spaces, Academic Press, New York and London, 1970.

[DRS] Duren, P. L., Romberg, B. W., and Shields, A. L., "Linear functionals on H^p spaces with $0 < p < 1$ ", J. Reine Angew. Math., 238 (1969), 32–60.

[DS] Devore, R. A. and Sharpley, R. C., Maximal Functions Measuring Smoothness, A.M.S. Memoir No. 293, 47 (1984).

[FS] Fefferman, C. and Stein, E. M., "H^p spaces in several variables", Acta Math. 129 (1972), 137–193.

[G] Garnett, J. B., Bounded Analytic Functions, Academic Press, New York, 1981.

[GR] Greenwald, H., "On the theory of homogeneous Lipschitz and Campanato spaces", Pacific J. Math., 106 (1983), 87–93.

[H] Hoffman, K., Banach Spaces of Analytic Functions, Prentice Hall, Englewood Cliffs, New Jersey, 1962.

[HL] Hardy, G. H. and Littlewood, J. E., "Some properties of fractional integrals. II", Math. Z. 34 (1932), 403–439.

[JN] John, F. and Nirenberg, L., "On functions of bounded mean oscillation", Comm. Pure Appl. Math. 14 (1961), 415–426.

[JPS] Janson, S., Peetre, J., and Semmes, S., "On the action of Hankel and Toeplitz operators on some functions spaces", Duke Math. J. 51 (1984), 937–957.

[JTW] Janson, S., Taibleson, M., and Weiss, G., "Elementary characterizations of of the Morrey–Campanato spaces", in Harmonic Analysis, Proceedings, Spring Lect. Notes in Math. No. 992, 101–114.

[K] Kolmogoroff, A. N., "Sur les fonctions harmoniques conjuqueés et les séries de Fourier", Fund. Math. 7, 24–29.

[M] Mooney, M. C., "A theorem on bounded analytic functions", Pacific J. Math. 43 (1973), 457–463.

[N] Nehari, Z., "On bounded bilinear forms", Ann. of Math. 65 (1957), 153–162.

[NS] Nagel, A. and Stein, E. M., Lectures on Pseudo–Differential Operators, Princeton University Press, Princeton, New Jersey, 1979.

[P] Powers, S., Hankel Operators on Hilbert Space, Pitman Advanced Publishing Program, London, 1982.

[T] Tolokonnikov, V. A., "Hankel and Toeplitz Operators in Hardy Spaces", J. Soviet Math. 37 (1987), 1359–64.

University of North Carolina
Chapel Hill, North Carolina 27514
and
University of Hawaii at Manoa
Honolulu, Hawaii 96822

Contractive projections on ℓ_p spaces

William J. Davis
The Ohio State University
Columbus, Ohio 43210

Per Enflo
Royal Institute of Technology
Stockholm, SWEDEN
and
The Ohio State University

Abstract: A subset of ℓ_p, with $1<p<\infty$, is the range of a contractive, (non-linear) projection if and only if it is optimal. In case the set is bounded, this is also equivalent to the fact that it is the intersection of (countably many) optimal half spaces. A set C is optimal if each point outside C can be moved to be closer to all points of C. A half space is optimal if its bounding hyperplane is norm one (linearly) complemented.

In this paper, we study non-linear contractive projections onto subspaces of ℓ_p when $1 < p < \infty$, and the relationship between contractive sets and optimal sets. A subset, C, of a Banach space, E, is said to be *optimal* [B-M] if for all $x \in E\backslash C$, there is $y \in E$, $y \neq x$, such that

$$\| x - c \| \geq \| y - c \| \text{ for all } c \in C.$$

The set is *contractive* if it is the range of a contractive projection. That is, there is $\pi : E \to C$ such that $\pi^2 = \pi$, and

$$\| x - y \| \geq \| \pi(x) - \pi(y) \| \text{ for all } x, y \in E.$$

Optimal and contractive sets have been studied by Beauzamy and Maurey in [B-M], by Beauzamy in [B], and the contractive sets in $L_1[0,1]$ have been characterized by Enflo in [E]. It is clear that contractive sets are always optimal, but not much is known in the other direction. In strictly convex spaces, optimal sets must be convex and closed. However, it is true only in Hilbert space that all closed convex sets are optimal, and, indeed, also contractive, [Ph], [B-M]. The nearest point map is the contractive projection in this case.

Beauzamy and Maurey proved many general facts about optimal sets. Some of these will be used directly here, and others will be reproved for contractive sets.

Theorem (Beauzamy): If E is a separable, strictly convex, smooth, reflexive Banach space, and if $C \subset E$ is an optimal set with interior, then C is contractive, and C is the intersection of countably many optimal half-spaces.

This is a nice result, and the techniques used in the proof are useful to us. However, in studying optimal or contractive sets in ℓ_p or L_p spaces, the result has limited applicability for the following two reasons: (1) The only optimal subset of $L_p[0,1]$ with interior is the whole space, since there are no optimal hyperplanes in the space. (2) No bounded contractive subset of ℓ_p has non-empty interior. We will use the result from [B-M] which proves these assertions, so we quote it here.

Proposition (Beauzamy-Maurey): A hyperplane in $L_p(X,\Sigma,\mu)$, $1<p<\infty$, is norm one complemented if and only if it is the zero set of a functional of the form $\chi_A + b\chi_B$, where A and B are atoms of the measure space. In particular, in ℓ_p, the optimal hyperplanes are exactly the level sets of the functionals $\delta_i+b\delta_j$.

Even though Beauzamy's result above is not directly applicable for the complete characterization of optimal sets in infinite dimensional ℓ_p spaces, it does work nicely for us in the finite dimensional case. The following proposition is a corollary of Beauzamy's result.

Proposition 1: A convex subset, $C \subset \ell_p^{(n)}$, is optimal if and only if it is the intersection of optimal half-spaces.

Proof: Since the space is finite dimensional, C contains a relative interior point. We might as well assume that the point is the origin. If we then consider the set $F = \cup nC$, it is easy to see that F is the subspace generated by C in $\ell_p^{(n)}$. By [B-M], F is an optimal subspace. Therefore, it is the range of a norm one linear projection on $\ell_p^{(n)}$. By [Pe], F is isometric to $\ell_p^{(m)}$, and is spanned by a block basic sequence of the natural basis. It is easy to see that such a subspace is the intersection of optimal hyperplanes: In fact, the argument for this is the same for the finite and infinite dimensional cases is the same, but the notation in the infinite dimensional case is easier, so we give that one. Let $\{I_k\}_{k\geq 1}$ be a sequence of disjoint subsets of the natural numbers, and assume that $\cup I_k = N$. For each k, let $x_k = \sum_{j\in I_k} \alpha_j e_j$. Then,

$[x_k] = \bigcap_{k\geq 1} (\cap\{y \mid \alpha_m^{-1}<y,\delta_m> = \alpha_n^{-1}<y,\delta_n>;\ m,n \in I_k\})$. (Here, $\{\delta_k\}$ is to denote the natural coefficient functionals of the basis $\{e_n\}$.) Each of the bracketed hyperplanes is optimal. Let $\{f_k\}$ denote the coefficient functionals of the block basis. For any $j \in I_k$, $\alpha_j^{-1}\delta_j$ in ℓ_q is an extension of f_k to all of ℓ_p. Thus, optimal functionals on $[x_k]$ extend to optimal functionals on all of ℓ_p. Now, inside $F = [x_k]$, Beauzamy's theorem applies, so that C is the intersection of optimal half-spaces in F. Thus, C is the intersection of the optimal hyperplanes which define F with the optimal half spaces we get by using the optimal extensions above to all of ℓ_p. This completes the proof.

We will need to use some of the results of [B-M] concerning optimal sets. Three of them are:

Fact1. The intersection of a countable number of optimal sets is optimal.

Fact 2. In a reflexive, strictly convex space, the closure of an increasing union of optimal sets is optimal.

Fact 3. A subspace of a smooth, reflexive Banach space is optimal if and only if it is the range of a norm 1 linear projection.

From (1) and (3) we see easily that the bounding hyperplane of an optimal half-space must be norm one complemented. Assume that $0 \in H$. If H is optimal, so is -H, so by (1), $H \cap (-H)$ (the bounding hyperplane) is optimal, and hence norm one complemented.

We are ready to begin our work. We begin by stating our main result and the lemmas we use in the proof.

Theorem: For a subset C of ℓ_p $(1<p<\infty)$, the following are equivalent:

(i) C is contractive.

(ii) C is optimal.

In case C is bounded, each of these is equivalent to

(iii) C is the intersection of a countable collection of optimal half spaces.

We call a contractive projection, $p : E \to C$, *strictly* contractive if $\| p(x) - c \| < \| x - c \|$ if $x \in E \setminus C$ and $c \in C$. In a strictly convex space, if P is a contractive projection, then the map $p(x) = P\left(\frac{x+P(x)}{2}\right)$ is a strictly contractive projection.

Lemma 1: If $\pi_1, \dots, \pi_n$ are strictly contractive projections with ranges $C_1, \dots C_n$, and if $C = C_1 \cap \dots \cap C_n$, then for all $x \notin C$, and for all $y \in C$,

$$\|\pi_n \circ \dots \circ \pi_1(x) - y\| < \|x - y\|.$$

Proof : The proof is the same for $n = 2$ as for arbitrary n, so let π_1 and π_2 be contractive projections with ranges C_1 and C_2, respectively. Let $x \notin C_1 \cap C_2$ and $y \in C_1 \cap C_2$. There are two cases: If $\pi_1(x) \in C_1 \cap C_2$, then

$$(1) \qquad ||\pi_2 \circ \pi_1(x) - y|| = ||\pi_2 \circ \pi_1(x) - \pi_2(y)|| \leq ||\pi_1(x) - y|| < ||x - y||$$

by the lemma. (π_1 had to move x for this case to occur.) In the second case, the same inequalities as in (1) are used with $<$ replacing $\leq$ and $<$ replacing $\leq$, respectively, and for the same reasons.

The lemma would be enough for our purposes if the compositions of projections were projections. They aren't in general, though. The reader can see this easily in the linear case by letting π_1 and π_2 be projections onto skewed hyperplanes in two-dimensional space. By looking at such a picture, though, one easily sees that what is needed is an iteration of the procedure. For the case of two projections, we simply start with $(\pi_2 \circ \pi_1)^n$, and proceed from there. We don't know whether or not the limiting mapping as n tends to ∞ is a projection, so there will still be some work to be done.

Lemma 2: If $\{C_k\}$ is a sequence of contractive subsets of a separable, strictly convex, reflexive space, then $(C_k$ is contractive.

Proof: Let π_k be a contractive projection onto C_k. We may as well assume thatπ_k is strictly contractive by the remark above. Let us denote by p_k the composition $\pi_k \circ ... \circ \pi_1$. Now let $\Phi_k = p_k \circ ... \circ p_1$. If $y \in \cap C_k$, and $x \notin \cap C_k$, there is m such that $||\Phi_m(x) - y|| < ||x-y||$ by the corollary. We now construct a long chain of contractive maps as follows: The Φ_m's have been defined. Let Φ_ω be the weak operator limit of the Φ_m's. That is, notice that the $\Phi_m(x)$'s are pointwise bounded, so there is a weak limit along some free ultrafilter on N. Similarly, for any limit ordinal, α, let Φ_α be a weak operator limit of the Φ_β's with $\beta < \alpha$. If α is a limit ordinal, let $\Phi_{\alpha+m} = \Phi_m \circ \Phi_\alpha$. Using the previous lemma, we see that for each $x \in E$, there must be a countable ordinal β, such that $||\Phi_\gamma(x)-y|| = ||\Phi_\beta(x)-y||$ for all $\gamma > \beta$, and all $y \in C$. Therefore, since E is separable, there is such a β which works for all x. Again appealing to the lemma, we see that this forces $\Phi_\beta(x) \in C$ for all x. Therefore, Φ_β is the desired projection.

The following lemma is a complement to Fact 2 from [B-M] quoted above, and is used to conclude the proof of the main result.

Lemma 3: If $\{C_k\}$ is an increasing sequence of contractive sets in a separable reflexive space, then $cl(\cup C_k)$ is contractive.

Proof: Let π_k be the contractive projection onto C_k. This time we only need to take a weak operator limit of the sequence π_k itself (as above), since $C_k \subset C_{k+1}$ for all k. That is, we only need to verify that such a limit is a projection onto $cl(\cup C_k)$. This follows from the density of $\cup C_k$ in the closure, and the fact that all π_k's, and hence the weak limit, are contractive.

For the rest of the lemmas, we need some notation for ℓ_p. Let $\{e_i\}$ denote the natural unit vector basis of ℓ_p, $E_n = [e_1,...,e_n]$, and $F_n=[e_{n+1},....]$. Denote by P_n the natural projection of ℓ_p onto E_n.

Lemma 4: If C is a contractive (respectively, optimal) subset of ℓ_p, then $C = cl(\cup C_k)$, with each C_k both compact and contractive (respectively, optimal).

Proof: Let $\{y_k\}$ be dense in C. Write $y_k = \Sigma a_{k,i}e_i$, and let $b_{k,i} = \max_{1\le j\le k}|a_{j,i}|$. With $H_{k,i} = \{x| \ |\langle x,e_i\rangle| \le b_{k,i}|\ \}$, we see that $y_k \in H_{m,i}$ for all i and for all $m \ge k$. Each $H_{k,i}$ is a contractive half-space, and C is contractive, so we set $C_k = C \cap (\cap H_{k,i})$. This is contractive by lemma 2, and is compact since $\cap H_{k,i}$ is. In case C is optimal, the set C_k is optimal by Fact 1 of [B-M] above. This completes the proof of the lemma.

Proposition 2: If K is a compact optimal set in ℓ_p, then $P_n(K)$ is optimal in E_n, and $P_n(K)\oplus F_n$ is optimal in ℓ_p.

We need two more lemmas in order to prove Proposition 2. The first asserts the existence of (perhaps discontinuous) projections onto optimal subsets of strictly convex, reflexive spaces. The second asserts the continuity of these projections in uniformly convex spaces.

Lemma 5: If E is separable, strictly convex and reflexive, and if C is a non-empty, optimal subset of E, there is a (not necessarily continuous) projection , P, of E onto C such that $\| P(x) - c \| < \| x - c \|$ for all $x \in E\backslash C$ and all $c \in C$.

Proof: Since $x \notin C$, there is a point $y \ne x$ with $\| y - c \| \le \| x - c \|$ for all c. The point $(x+y)/2$ is strictly closer to each $c \in C$ than is x due to the strict convexity of the space. We define a family $\{x_\alpha\}$, indexed by countable ordinals as follows: Let $x_0 = x$. If we have x_α , and $x_\alpha \notin C$, let $x_{\alpha+1}$ be any point in E with $\| x_{\alpha+1} - c \| < \| x_\alpha - c \|$ for each c in C. If $x_\alpha \in C$, stop (or just define $x_{\alpha+1} = x_\alpha$). If β is a limit ordinal, let x_β be any weak cluster of the previous x_α's . Clearly this process must end at a countable ordinal, and it only ends when $x_\alpha \in C$. This point is P(x).

Lemma 6: If C is a bounded, optimal set in a uniformly convex space, E, there is a continuous projection of E onto C with $\| P(x) - c \| < \| x - c \|$ whenever $x \in E\backslash C$ and $c \in C$.

Proof: For each $x \in E$, let $\Gamma(x) = C \cap (\cap B(c,\|x-c\|))$, where the latter intersection is over C, and where B(c,r) is the closed ball of radius r about c. By lemma 5, $\Gamma(x)$ is closed, bounded, convex and non-empty for each x. We use the selection theorem of E. Michael [M]. Thus, we need to show that for each x, each $y \in \Gamma(x)$, and each $\varepsilon > 0$, there is $\delta > 0$ such that $\|x' - x\| < \delta$ implies that $\Gamma(x') \cap B(y,\varepsilon) \neq \phi$. This is clearly the case if $x \in C$, so suppose $x \in E\backslash C$, and let $\varepsilon < d(x,C)$. Let $y \in \Gamma(x)$ and $y' \in \Gamma(x')$, with $\|x' - x\| < \delta$. Then, for each $c \in C$, $\|y - c\| < \|x' - c\| + \delta$ and $\|y' - c\| \leq \|x' - c\|$. Recall that ε is fixed, and let d denote the diameter of C. Uniform convexity of E gives us that, if $\|y' - y\| \geq \varepsilon$, $\| (y'+y)/2 - c \| \leq (\|x' - c\| + \delta)(1 - \Delta(\varepsilon/d))$, where Δ denotes the modulus of convexity of the space. Therefore, for δ sufficiently small, this midpoint is in $\Gamma(x')$. Clearly, after a finite number of iterations starting at any $y' \in \Gamma(x')$, the midpoint will arrive in $B(y,\varepsilon)$. Therefore, the selection theorem gives us a continuous map $Q : E \to C$ such that $Q(x) \in \Gamma(x)$ for each x. Q may not have the strict contractivity property, so we define the projection P by $P(x) = Q\left(\frac{x + Q(x)}{2}\right)$.

Proof of Proposition 2: Let $\Phi: \ell_p \to K$ be a continuous projection as guaranteed by lemma 6. Define a map $\Psi: \ell_p \to (I-P_n)K$ by $\Psi(x) = (I - P_n)(\Phi(x_0 + x))$, where x_0 is in E_n. Notice that the map

Ψ: $(I-P_n)K \to (I-P_n)K$. By the Schauder fixed point theorem, there is a $w \in (I-P_n)K$ fixed by Ψ. That is, $y = \Phi(x_0 + w) - w \in E_n$. We claim that y, which is clearly also in $P_n(K)$, satisfies

$$||y - c|| \leq ||x_0 - c|| \text{ for all } c \text{ in } P_n(K):$$

Let $k \in K$ such that $c = P_n(k)$. Then we have

$$||\Phi(x_0 + w) - w - (k - w)|| \leq ||x_0 + w - k||.$$

$$||\Phi(x_0 + w) - w - (k - w)||^p = (||(\Phi(x_0 + w) - w) - c||^p + ||(I-P_n)k - w||^p)$$

and $$||x_0 + w - k||^p = (||x_0 - c||^p + ||(I-P_n)k - w||^p).$$

It follows that

$$||(\Phi(x_0 + w) - w) - c|| \leq ||x_0 - c||$$

as desired. This completes the proof. Notice that we used the fact that we are working in ℓ_p strongly in this proof.

Proof of Theorem: Let $C \subset \ell_p$, $1<p<\infty$, be optimal. By lemma 4, we write $C = cl(\cup C_k)$, with each C_k compact and optimal. Then, by proposition 2, each of the sets $P_n(C_k)$ is contractive. In fact,

$P_n(C_k) = \cap H_m$, where each $H_m = [\langle x, \pm e_i + b_m e_j \rangle \leq a_m]$ where $1 \leq i,j \leq n$.

These functionals, as functionals on all of ℓ_p are still optimal, and using them as such, we see that $P_n(C_k) \oplus F_n = \cap H_m$. Now, since C_k is bounded, it is easy to verify that $C_k = \cap (P_n(C_k) \oplus F_n)$, so that C_k is the intersection of optimal half-spaces. Therefore, by lemma 2, C_k is contractive. Lemma 3, then, shows that C itself is contractive.

To see that C is the intersection of optimal half spaces when C is bounded, we simply reverse the order of the steps in the above proof: We note that $P_n(C) = cl(\cup P_n(C_k))$, so that $P_n(C)$ is contractive, and therefore the intersection of a sequence of optimal half-spaces. Now, $C = \cap (P_n(C) \oplus F_n)$, due to the boundedness of C, and the result follows.

References

[B-M] B. Beauzamy and B. Maurey, *Points minimaux et ensembles optimaux dans les espaces de Banach,* **J. Funct. Anal. 24** (1977) 107-139.

[B] B. Beauzamy, *Projections contractantes dans les espaces de Banach,* **Bull. Sc. math., 2º série, 102** (1978) 43-47.

[E] P. Enflo, *Contractive projections onto subsets of $L_1[0,1]$*, (preprint).

[M] E. Michael, *Continuous selections, I,* **Ann. Math. (2) 63** (1956) 361-382.

[Pe] A. Pelczynski, *Projections in certain Banach spaces,* **Studia Math. 19** (1960) 209-228.

[Ph] R. R. Phelps, *Convex sets and nearest points, II,* **Proc. AMS 9** (1958) 867-873.

Contractive projections onto subsets of $L^1(0,1)$

*Per Enflo**

1. Introduction.

In Davis-Enflo [1], the sets which are ranges of contractive projections in (real) ℓ_p are characterised. We call these contractive sets. In Westphal [2] similar finite-dimensional results are obtained. The main result in [1] says that the bounded contractive sets in ℓ_p, $1 < p < \infty$, $p \neq 2$, are precisely the intersections of half-spaces of the form $\{\mathbf{Y} | \mathrm{ay_i} + \mathrm{by_j} \geq \mathrm{c}\}$, $\mathbf{Y} = (\mathrm{y}_1, \mathrm{y}_2, \ldots)$. In this paper we will characterise the convex, contractive subsets of $L^1(0,1)$. In $L^1(0,1)$ a half-space cannot be contractive but as we will see, the L^1- characterisation is still analogous to the ℓ_p-characterisation. In L^1 we will use restrictions on pairs of subsets of [0,1] similar to the restrictions on pairs of coordinates in ℓ_p. This will give cones in L^1 instead of half-spaces.

The method in this paper is very different from the method in [1]. In [1] important use is made of Beauzamy's theorem which says that under general conditions on the space, the contractive subsets with interior points are intersections of contractive half-spaces.

Even if a contractive subset of ℓ_p is not assumed to have interior points, since ℓ_p has contractive half-spaces, an approximation by finite-dimensional subspaces can be used to obtain the result. We have not been able to do anything similar in L^1 (or L^p).

2. The Main Result.

To characterise the contractive convex subsets of $L^1(0,1)$ we first define "The cone $C_{f,\,T,\,x}$" and "The cone $C_{E,\,+,\,x}(C_{,E,\,-,\,x})$".

We then prove that a convex subset of $L^1(0,1)$ is the range of a contractive projection if an only if it is a countable intersection of such cones. We start by defining $C_{f,\,T,\,0}$.

Consider an $f \in L^1(0,1)$ and consider 2 disjoint subsets E and F of (0,1), both of positive measure, s.t. f has constant sign on E and constant sign on F (we assume real

*partially supported by NSF grants 716 110 and 716 464.

scalars). We assume the f and $\frac{1}{f}$ are bounded on $E \cup F$. Let T be a map from the subsets of E to subsets of F satisfying the following conditions 1.-3.:

1. $T(E_1 \cup E_2) = T(E_1) \cup T(E_2)$.
2. $m(E_1) > 0 \Rightarrow m(T(E_1)) > 0$; $m(E_1) = 0 \Rightarrow m(T(E_1)) = 0$.
3. T is left continuous, that is, if (E_n) is an increasing sequence of sets then $T(\bigcup_n E_n) = \bigcup_n T(E_n)$.

$C_{f,\,T,\,0}$ consists of all functions $g \in L^1(0,1)$ s.t.

$$\operatorname*{ess\,inf}_{t \in E_1} \frac{g(t)}{f(t)} \leq \operatorname*{ess\,inf}_{t \in T(E_1)} \frac{g(t)}{f(t)} \qquad \text{for all} \quad E_1 \subset E.$$

We put $C_{f,\,T,\,x} = x + C_{f,\,T,\,0}$ so $C_{f,\,T,\,x}$ is a translate of $C_{f,\,T,\,0}$. We let $C_{E,\,+,\,0}$ $(C_{E,\,-,\,0})$ be all functions in $L^1(0,1)$ which are non-negative (non-positive) on E, and $C_{E,\,+,\,x}$ $(C_{E,\,-,\,x})$ its translate by x.

Remarks.

1. The left continuity of T does not imply right continuity as is easily seen. If T, in addition to 1.-3. and right continuity, preserved intersections, then T could be described by a point map $h : (0,1) \to (0,1)$, by $T(E_1) = h^{-1}(E_1)$. We feel, however, that the conditions 1.-3. do not in any natural way relate T to a point map.
2. We can think of $C_{E,\,+,\,x}$ as a degenerate case of $C_{f,\,T,\,x}$ corresponding to the case when f or $\frac{1}{f}$ is infinity.
3. It is not obvious from the definition that $C_{f,\,T,\,x}$ is convex, However, we prove below that this is the case.
4. The half-space $ay_i + by_j \geq c$ is "analogous" to the cone $C_{f,\,T,\,x}$ in the following way: The pair of numbers (a,b) corresponds to f, the pairing (i,j) corresponds to the pairing of subsets of E and F given by T, and the c corresponds to x. And the degenerate case $b = 0$ corresponds to the degenerate case $C_{E,\,+,\,x}$ (or $C_{E,\,-,\,x}$.)

We can now state our main result

Theorem 1. *A convex subset K of $L^1(0,1)$ is the range of a contractive projection $L^1(0,1) \to K$, if and only if K is an intersection of a family of cones $C_{f_\alpha,\,T_\alpha,\,x_\alpha}$, $C_{E_\beta,\,+,\,x_\beta}$*

and $C_{E_\gamma,\ -,\ x_\gamma}$.

We first prove the "if"-part. One important step for doing that is to prove that every cone $C_{f,\ T,\ 0}$ is the range of a contractive projection. We do that by using the Lemmas 1-6 below. We then pass to intersections by using the Lemmas 7-11. The "only if"-part is proved in the next section 5.

3. Elementary contractive projections onto cones.

We first prove

Lemma 1. *$C_{f,\ T,\ 0}$ is a closed, convex cone.*

Proof: It is obvious that $C_{f,\ T,\ 0}$ is a cone so we need only show that it is closed and convex. To see that $C_{f,\ T,\ 0}$ is closed, let $g_n \to g$ in $L^1(0,1)$ and $g_n \in C_{f,\ T,\ 0}$. Assume now that there is an $E_1 \subset E$ s.t.

$$\operatorname*{ess\,inf}_{t\in E_1} \frac{g(t)}{f(t)} > \operatorname*{ess\,inf}_{t\in T(E_1)} \frac{g(t)}{f(t)}. \tag{1}$$

There is a sequence $E_1^{(n)} \subset E_1$ s.t. $m(E_1 \backslash E_1^{(n)}) \to 0$ and s.t.

$$\operatorname*{ess\,inf}_{t\in E_1^{(n)}} \frac{g_n(t)}{f(t)} \longrightarrow \operatorname*{ess\,inf}_{t\in E_1} \frac{g(t)}{f(t)}.$$

By the left continuity of T, $m\left(T(E_1)\backslash T(E_1^{(n)})\right) \to 0$ and so by (1),

$$\operatorname*{ess\,inf}_{t\in E_1^{(n)}} \frac{g_n(t)}{f(t)} > \operatorname*{ess\,inf}_{t\in T(E_1^{(n)})} \frac{g_n(t)}{f(t)} \quad \text{if } n \text{ is large enough.}$$

This contradicts $g_n \in C_{f,\ T,\ 0}$ and so we have proved that $C_{f,\ T,\ 0}$ is closed. Since $C_{f,\ T,\ 0}$ is a cone, in order to prove convexity we prove that $g_1 \in C_{f,\ T,\ 0}$ and $g_2 \in C_{f,\ T,\ 0}$ imply $g_1 + g_2 \in C_{f,\ T,\ 0}$. We argue again by contradiction. Assume $g_1 \in C_{f,\ T,\ 0}$ and $g_2 \in C_{f,\ T,\ 0}$ but that there is an $E_1 \subset E$ and a $\delta > 0$ s.t.

$$\operatorname*{ess\,inf}_{t\in E_1} \frac{g_1(t)+g_2(t)}{f(t)} > \operatorname*{ess\,inf}_{t\in T(E_1)} \frac{g_1(t)+g_2(t)}{f(t)} + \delta \tag{2}$$

Take a subset $F_1 \subset T(E_1)$ s. t.

$$\frac{g_1(t)+g_2(t)}{f(t)} < \operatorname*{ess\,inf}_{t\in T(E_1} \frac{g_1(t)+g_2(t)}{f(t)} + \frac{\delta}{10} \quad \text{for all } t \in F_1. \tag{3}$$

By an exhaustion argument we see that there is a subset $E_2 \subset E_1$, $E_2 \neq E_1$ s.t. $m(T(E_2) \cap F_1) = 0$, but for every $E_3 \subset E_1 \backslash E_2$, $m(E_3) > 0$, we have $m(T(E_3) \cap F_1) > 0$. E_2 is a maximal set which is mapped outside F_1 by T. Then (2) and (3) imply that for every $E_3 \subset E_1 \backslash E_2$, $m(E_3) > 0$, and we have

$$\operatorname*{ess\,inf}_{t \in E_3} \frac{g_1(t) + g_2(t)}{f(t)} > \operatorname*{ess\,inf}_{t \in T(E_3)} \frac{g_1(t) + g_2(t)}{f(t)} + \frac{9\delta}{10} \tag{4}$$

Now consider an $E_3 \subset E_1 \backslash E_2$, $m(E_3) > 0$, s.t.

$$\frac{g_2(t)}{f(t)} < \operatorname*{ess\,inf}_{t \in E_1 \backslash E_2} \frac{g_2(t)}{f(t)} + \frac{\delta}{10} \text{ for all } t \in E_3. \tag{5}$$

And consider an $E_4 \subset E_3$, $m(E_4) > 0$ s.t.

$$\frac{g_2(t)}{f(t)} < \operatorname*{ess\,inf}_{t \in E_3} \frac{g_2(t)}{f(t)} + \frac{\delta}{10} \text{ for all } t \in E_4. \tag{6}$$

So (5) and (6) give

$$\operatorname*{ess\,inf}_{t \in E_4} \frac{g_1(t) + g_2(t)}{f(t)} < \operatorname*{ess\,inf}_{t \in E_4} \frac{g_1(t)}{f(t)} + \operatorname*{ess\,inf}_{t \in E_4} \frac{g_2(t)}{f(t)} + \frac{2\delta}{10} \leq$$

$$\operatorname*{ess\,inf}_{t \in T(E_4)} \frac{g_1(t)}{f(t)} + \operatorname*{ess\,inf}_{t \in T(E_4)} \frac{g_2(t)}{f(t)} + \frac{2\delta}{10} \leq \operatorname*{ess\,inf}_{t \in T(E_4)} \frac{g_1(t) + g_2(t)}{f(t)} + \frac{2\delta}{10}.$$

But this contradicts (4) and so Lemma 1 is proved.

In order to define a projection onto $C_{f,T,0}$ we start by defining "elementary projections" (R:s).

Let E' and F' be disjoint sets of positive measure on $(0,1)$ and let $r \geq 0$. If $g \in L^1(0,1)$ we put $R(E', F', r)(g) = R' \circ S(g)$, where R' and S are the following maps:

1. $(Sg)(t) = g(t)$ if $t \notin E' \cup F'$.
2. $(Sg)(t) = \frac{\int_{E'} g(u)du}{m(E')}$ if $t \in E'$ and $(Sg)(t) = \frac{\int_{F'} g(u)du}{m(F')}$ if $t \in F'$.
3. $R'(h)(t) = h(t)$ if $t \in E' \cup F'$.

If h is constant $= b_1$ on E' and constant $= b_2$ on F' then $R'h$ is constant $= c_1$ on E' and constant $= c_2$ on F' and we have the following:

1. If $b_1 \leq rb_2$ then $c_1 = b_1$ and $c_2 = b_2$.

2. If $b_1 > rb_2$ then c_1 and c_2 are defined s.t. $c_1 = rc_2$ and

3. $c_1\, m(E') + c_2 m(F') = b_1 m(E') + b_2 m(F')$.

If $r < 0$, $R(E', F', r)(g)$ is defined by first multiplying g by -1 on F', then applying $R(E', F', |r|)$ and then multiply the rsult by -1 on F'. We now have

Lemma 2. *$R(E', F', r)$ is a contractive projection.*

Proof: It is obviously enough to consider $r > 0$. We now consider two functions g and h which are constant on E' and on F'. We can WLOG assume that g and h are positive on $E' \cup F'$. For otherwise we can consider the functions $g + d$ and $h + d$ where d is a function that takes a value c on E' and rc on F'. Obviously $R(E', F', r)(g + d) = R(E', F', r)(g) + R(E', F', r)(d)$ and $R(E', F', r)(h + d) = R(E', F', r)(h) + R(E', F', r)(d)$ for such a function d. We now consider some different cases.

Case 1. $g \geq h$ on both E' and F'. Then it is easy to see that $R(E', F', r)(g) \geq R(E', F', r)(h)$ on both E' and F' and since $F(E', F', r)$ is norm-preserving for positive functions we obviously have
$\|g - h\| = \|R(E', F', r)(g) - R(E', F, r)(h)\|$.

Case 2. $g \geq h$ on E' but $g < h$ on F', We consider first the

Case 2A. $h|_{E'} \leq rh|_{F'}$, where $h|_X$ denotes the restriction of h to X. Then $R(E', F', r)(h) = h$. If $\|g|_{E' \cup F'}\| \geq \|h|_{E' \cup F'}\|$, then applying $R(E', F', r)$ to g will diminish the distance between g and h on E'. Since $\|R(E', F', r)(g)\| = \|g\|$ the distance between g and h on F' cannot increase so this case is settled. If $\|g|_{E' \cup F'}\| < \|h|_{E' \cup F'}\|$, then applying $R(E', F', r)$ to g will diminish or keep the distance between g and h on F'. The distance between g and h on E' cannot increase so this case is settled. We finally settle the

Case 2B. $h|_{E'} > rh|_{F'}$. Since this is a special case of Case 2. we have $g|_{E'} \geq rh|_{F'} > rg|_{F'}$. Assume first $\|g|_{E' \cup F'}\| \geq \|h|_{E' \cup F'}\|$. Then the application of $R(E', F', r)$ to g and h we will diminish their norms on E' by positive amounts a_1 and a_2, $a_1 >$

a_2, and we will diminish the difference between their norms on E' by $a_1 - a_2$. The norm of g on F' will be increased by a_1 and the norm of h on F' will be increased by a_2. Since $h > g$ on F' and $R(E', F', r)(g) \geq R(E', F', r)(h)$ on F' the difference between their norms on F' has increased by a number $< a_1 - a_2$. Thus this case is settled. The case $\|g|_{E' \cup F'}\| < \|h|_{E' \cup F'}\|$ follows similarly by considering first the set F'. This completes the proof of the lemma.

To define a contractive projection P onto $C_{f,\,T,\,0}$ we approximate f by simple functions f_n and use them to define contractive projections Q_n. We then show that for each m, $\lim Q_{r+m} \circ Q_{r-1+m} \circ \ldots \circ Q_m = P_m$ exists where the limit is taken in the strong operator topology. We also show that P_m takes $L^1(0,1)$ into $C_{f,\,T,\,0}$. We finally show that for some subsequence of the $P_m : s, \lim\ P_m g = Pg$ exists in the weak topology on $L^1(0,1)$ and that P is the identity map on $C_{f,\,T,\,0}$.

Let (f_n) be a sequence of simple functions which tends to f uniformly on $E \cup F$. Let (g_m) be a sequence of bounded functions s.t.

$$\{g_m\} \text{is dense in } L^1(0,1), \text{ and } \{g_m\} \cap C_{f,\,T,\,0} \text{ is dense in } C_{f,\,T,\,0}. \tag{7}$$

Since f and $\frac{1}{f}$ are bounded on $E \cup F$ it is obvious that there is such a sequence (g_m). We assume WLOG that the union of the sets where some g_m takes a value of the form $\frac{k}{2^n}$ has measure 0 (k and n integers).

We now make a sequence of splittings of E and of F in the following way. In the nth stage E is divided into pairwise disjoint subsets $E_{n1}, E_{n2}, \ldots, E_{nN_n}$ which are the minimal sets in the algebra of subsets of E generated by the following sets:

1. The sets of constancy of f_m, $1 \leq m \leq n$, on E
2. The subsets of E where g_m, $1 \leq m \leq n$, takes values between $\frac{k}{2^n}$ and $\frac{k+1}{2^n}$, k an integer.

In the nth stage F is divided into pairwise disjoint subsets $F_{n1}, F_{n2}, \ldots, F_{nM_n}$ which are the minimal sets in the algebra of subsets of F generated by the following sets:

1. The sets of constancy of f_m, $1 \leq m \leq n$, on F
2. The subsets of F where g_m, $1 \leq m \leq n$, takes values between $\frac{k}{2^n}$ and $\frac{k+1}{2^n}$, k an

integer.

3. The sets $T(E_{ni})$, $1 \leq i \leq N_n$.

It is obvious from the definition that the sets in the $(n+1)$1st stage are subsets of those in the nth stage.

In order to define Q_n we now enumerate all pairs (E_{ni}, F_{nj}) for which

$$T(E_{ni}) \supset F_{nj}. \tag{8}$$

We now assume that f_n has the same sign as f on E and on F for all n. This can be done since it is always true if n is large enough. We define r_{nij} by $f_{n|E_{ni}} = r_{nij} f_{n|F_{nj}}$. Let (E_{ni_1}, F_{nj_1}), $(E_{ni_2}, F_{nj_2}) \ldots (E_{ni_L}, F_{nj_L})$ be the enumeration from (8). We put

$$Q_n = R(E_{ni_L}, F_{nj_L}, r_{ni_Lj_L}) \circ R(E_{ni_{L-1}}, F_{nj_{L-1}}, r_{ni_{L-1}j_{L-1}}) \circ \ldots \circ R(E_{ni_1}, F_{nj_1}, r_{ni_1j_1}).$$

It is easy to see that Q_n is a contractive projection on $L^1(0,1)$. Obviously $Q_n g$ is constant on every set E_{pi} and F_{pj} if $p \geq n$. We have the following

Lemma 3. *Lim* $Q_{m+s} \circ Q_{m+s-1} \circ \cdots \circ Q_m(g)$ *exists for every* $g \in L^1(0,1)$.

Proof: We can assume WLOG that f is positive on $E \cup F$. We observe that $Q_{m+s} \circ Q_{m+s-1} \circ \cdots \circ Q_m(g)$ decreases monotonically with s on E and increases monotonically with s on F. Since it is obviously bounded from below on E and from above on F by some function in $L^1(0,1)$ the lemma follows.

We now put

$$P_m g = \lim_{s \to \infty} \; Q_{m+s} \circ Q_{m+s-1} \circ \cdots \circ Q_m(g).$$

We have

Lemma 4. $P_m g \in C_{f,\,T,\,0}$ *for every* $g \in L^1(0,1)$.

Proof: We assume $P_m g \notin C_{f,\,T,\,0}$ for some g and argue by contradiction. If $P_m g \notin C_{f,\,T,\,0}$ then there is an $E_1 \subset E$ and a $\delta > 0$ s.t.

$$\text{ess inf } \frac{P_m g}{f}|_{E_1} > \text{ess inf } \frac{P_m g}{f}|_{T(E_1)} + \delta$$

We can now find a subset $F_1 \subset T(E_1)$ s.t. the following holds:

There is an ω s.t.

$$\frac{P_m g(t)}{f_n(t)} < \text{ess inf } \frac{P_m g}{f}|_{T(E_1)} + \frac{\delta}{10} \tag{9}$$

for all $n \geq \omega$ and for all $t \in F_1$

By removing from E_1 a maximal set E_1' s.t. $m(T(E_1') \cap F_1) = 0$ we can find a subset $E_2 \subset E_1 \backslash E_1'$ with $m(E_2) > 0$ s.t. the following holds:

$$\text{There is an } \omega \text{ s.t.} \frac{P_m g(t)}{f_n(t)} \text{ varies less than } \frac{\delta}{10} \text{ when } t \text{ runs over } E_2$$

$$\text{and } n \text{ runs over all integers } \geq \omega. \tag{10}$$

$$m(T(E_3) \cap F_1) > 0 \text{ for all } E_3 \subset E_2,\ m(E_3) > 0. \tag{11}$$

We get by (9), (10) and (11)

$$\text{ess inf}\frac{P_m g}{f}|_{E_3} > \text{ess inf}\frac{P_m g}{f}|_{T(E_3)} + \frac{\delta}{2} \text{ for all } E_3 \subset E_2. \tag{12}$$

(11) ensures that for every n there is a pair (E_{ni}, F_{nj}) in the enumeration (8) s.t. $m(E_{ni} \cap E_2) > 0$ and $m(F_{nj} \cap F_1) > 0$. For such a pair we will have for any $h \in L^1(0,1)$

$$\frac{Q_n h|_{E_{ni}}}{f_n|_{E_{ni}}} \leq \frac{Q_n h|_{F_{nj}}}{f_n|_{F_{nj}}}. \tag{13}$$

But since composing to the left by Q_{n+s} will only decrease values on E_{ni} and increase values on F_{nj} we get for every $g \in L^1(0,1)$

$$\text{ess sup } \frac{P_m g}{f_n}|_{E_{ni}} \leq \text{ess inf } \frac{P_m g}{f_n}|_{F_{nj}} \tag{14}$$

Thus

$$\begin{aligned}
&\text{ess inf}\frac{P_m g}{f}|_{E_{ni} \cap E_2} \leq \quad \text{(by (10))} \\
&\text{ess inf}\frac{P_m g}{f_n}|_{E_{ni} \cap E_2} + \frac{\delta}{10} \leq \text{ ess sup}\frac{P_m g}{f_n}|_{E_{ni} \cap E_2} \leq \quad \text{(by (14))} \\
&\text{ess inf}\frac{P_m g}{f_n}|_{F_{nj}} + \frac{\delta}{10} \leq \text{ ess inf}\frac{P_m g}{f_n}|_{F_{nj} \cap F_1} + \frac{\delta}{10} \leq \quad \text{(by (9))} \\
&\text{ess inf}\frac{P_m g}{f}|_{T(E_1)} + \frac{2\delta}{10} \leq \text{ ess inf}\frac{P_m g}{f}|_{T(E_{ni} \cap E_2)} + \frac{2\delta}{10} \text{ with } E_3 = E_{ni} \cap E_2.
\end{aligned}$$

This contradicts (12) and so the Lemma is proved.

We are almost ready to define our contractive projections of $L^1(0,1)$ onto $C_{f,T,0}$. Our next lemma will be used to show that these operators are the identity map on $C_{f,T,0}$.

Lemma 5. *Let $g_m \in C_{f,T,0}$ be in the sequence (7). Then, given $\varepsilon > 0$, there is an $h \in C_{f,T,0}$ s.t. $\|h - g_m\| < \varepsilon$ and h has the following properties:*

1. *There is an n s.t.*

$$h \text{ is constant on each set } E_{ni} \text{ and } F_{nj} \text{ in the } n\text{th partition;} \tag{15}$$

2.
$$\frac{h}{f} \leq \frac{g}{f} \text{ on } E \text{ and } \frac{h}{f} \geq \frac{g}{f} \text{ on } F. \tag{16}$$

Proof: We can assume WLOG that f is positive on $E \cup F$. Since g_m is used to define the nth subdivision, we can choose n so large that g_m is almost constant on each E_{ni} and F_{nj}. We then define h to be equal to g_m outside $E \cup F$. On each E_{ni}, $h = \text{ess inf } g_{m|_{E_{ni}}}$, and on each F_{nj}, $h = \text{ess sup } g_{m|_{F_{nj}}}$. This h obviously has the required properties, and $h \in C_{f,T,0}$ follows from 2.

We can now finally prove

Lemma 6. *Every cone $C_{f,T,0}$ is the range of a contractive projection on $L^1(0,1)$.*

Proof: Since for every bounded function g in $L^1(0,1)$, $(P_m g)$ is a sequence of uniformly bounded functions we can, by a standard diagonal procedure, extract a subsequence P_{m_γ} s.t. $P_{m_\gamma} g$ converges weakly for a dense set of g's in $L^1(0,1)$, say to P_g. Then P can obviously be extended to a contractive map on $L^1(0,1)$ and by Lemmas 1 and 4, $Pg \in C_{f,T,0}$ for every $g \in L^1(0,1)$.

Now finally assume WLOG that f is positive on $E \cup F$. Choose $g \in C_{f,T,0}$. Choose an $\varepsilon > 0$ and choose $g_m \varepsilon C_{f,T,0}$ with $\|g - g_m\| < \varepsilon$. Choose h as in Lemma 5 s.t. $\|g_m - h\| < \varepsilon$. Define h' as follows: $h' = h$ outside $E \cup F$, $h' = h - \delta$ on E and $= h + \delta$ on F where δ is chosen s.t. $\|h - h'\| < \varepsilon$. Since $f_n \to f$ uniformly on $E \cup F$ there is an ω s.t. if $n \geq \omega$ then

$Q_n h' = h'$. This implies that $P_m h' = h'$ if m is sufficiently large and so $Ph' = h'$. Since $\|g - h'\| < 4\varepsilon$, ε arbitrary, $Pg = g$, and so the P constructed as above is a contractive projection $L^1(0,1) \to C_{f,T.0}$.

4. Contractive projections onto intersections of cones.

We now consider the intersection of a family of cones. Since $L^1(0,1)$ is a Lindelöf space, it is enough to consider countable families. To project onto an intersection of cones we make compositions of projections onto the different cones, where each projection may occur many times in the composition. A difficulty that occurs is that the range of such compositions may never converge to the intersection of the ranges. We now illustrate this difficulty by a 2-diminsional example and illustrate the idea of how to overcome the difficulty.

Let ℓ_2^1 - the 2-dimensional ℓ^1 - be represented by all pairs (x,y). Let $C_1 = \{(x,y)|x \geq y\}$ and $C_2 = \{(x,y)|2y \geq x\}$. Let P_1 be the contractive projection onto C_1 obtained by moving every point outside C_1 in the direction of $(1,-1)$ until it meets the boundary of C_1. Let P_2 be the contractive projection onto C_2 obtained by moving every point outside C_2 in the direction of $(-1,1)$ until it meets the boundary of C_2. Now consider $z = (x,y)$, $x < 0$, $y < 0$. We see that the sequence $P_1 z$, $P_2 \circ P_1 z$, $P_1 \circ P_2 \circ P_1 z$, $\cdots$ moves back and forth between C_1 and C_2 and never approaches $C_1 \cap C_2$, which is a subset of the first quadrant. However if $z = (x,y)$ has $x \geq 0$, $y \geq 0$ then the same sequence will reach $C_1 \cap C_2$ after one or two steps and then stay there. So if we first project onto the first quadrant and then use compositions of P_1 and P_2 we will get a projection onto $C_1 \cap C_2$. By the following definition the first quadrant will be the box containing $C_1 \cap C_2$.

Definition. Let M be a subset of $L^1(0,1)$. The box containing M, denoted by $B(M)$, is defined as follows: $h \in B(M)$ if $h \in L^1(0,1)$ and for every set $E' \subset (0,1)$ we have

$$\inf_{g\in M} \operatorname*{ess\,inf}_{t\in E'} g(t) \leq \operatorname*{ess\,inf}_{t\in E'} h(t) \leq \operatorname*{ess\,sup}_{t\in E'} h(t) \leq \sup_{g\in M} \operatorname*{ess\,sup}_{t\in E'} g(t).$$

We have the simple

Lemma 7. *For every M, $B(M)$ is the range of a contractive projection.*

Proof: We observe that $\inf_{g\in M} \operatorname{ess\,inf}_{t\in E^1} g(t)$ and $\sup_{g\in M} \operatorname{ess\,sup}_{t\in E'} g(t)$ define two measurable functions – possibly taking the values $+\infty$ and $-\infty$ – and that truncation at these functions defines a contractive projection $L^1(0,1) \to B(M)$.

We now let K be a countable intersection of cones defined as by Theorem 1. We denote by $P^{(n)}$ the contractive projection onto C_{f_n,T_n,x_n} defined as by the Lemmas 1-6. We denote by $Q_i^{(n)}$ the projections used to form $P^{(n)}$. We now let (g_i) denote a sequence which is dense in K and $B(K)$, and in $L^1(0,1)$. We donote by $(g_i^{(n)})$ a sequence which is dense in C_{f_n,T_n,x_n} and $L^1(0,1)$ s.t. for every i, $g_i^{(n)} - x_n$ is a bounded function.

The sequence $g_i^{(n)}$ is used to construct $P^{(n)}$ as above. We do not assume that the sequence (g_i) consists of bounded functions. We now have

Lemma 8. *Let $M_S = \{g_{i_1}, g_{i_2}, \cdots, g_{i_r}\} \subset K$ be a finite set of g^1,s. Then for every $h \in B(M_S)$, and for every n, we have $P^{(n)}h \in B(M_S)$.*

Proof: By approximation, it is obviously enough to prove that for every set $M_S^{(n)} = \{g_{i_1}^{(n)}, g_{i_2}^{(n)}, \cdots, g_{i_r}^{(n)}\} \subset C_{f_n,T_n,x_n}$ and every $h \in B(M_S^{(n)})$ we have $P^{(n)}h \in B(M_S^{(n)})$. As in the construction of h' in the proof of Lemma 5 we can also assume that there is an ω s.t. $Q_i^{(n)} g_{i_s}^{(n)} = g_{i_s}^{(n)}$ for all $i \geq \omega$, $1 \leq s \leq r$. Since $h \in B(M_S^{(n)})$ we can – by approximation – assume that $h - x_n$ is constant on every subset $E_{il}^{(n)}$ and $F_{im}^{(n)}$ for $i \geq \omega$. We can also WLOG assume $x_n = 0$.

We observe also that it is enough to prove that for each of the elementary projections $R(E_{il}^{(n)}, F_{im}^{(n)}, r_{ilm}^{(n)})$ in $Q_i^{(n)}$ we have $h \in B(M_S^{(n)})$

$$R(E_{il}^{(n)}, F_{im}^{(n)}, r_{ilm}^{(n)})h \in B(M_S^{(n)}). \tag{17}$$

To prove (17) we put $u_1 = \max_s(g_{i_s}^{(n)}|_{E_{il}^{(n)}})$ $u_2 = \max_s(g_{i_s}^{(n)}|_{F_{im}^{(n)}})$, $\ell_1 = \min_s(g_{i_s}^{(n)}|_{E_{il}^{(n)}})$ and $\ell_2 = \min_s(g_{i_s}^{(n)}|_{F_{im}^{(n)}})$. We have

$$u_1 \leq r_{ilm}^{(n)} u_2 \text{ and } \ell_1 \leq r_{ilm}^{(n)} \ell_2. \tag{18}$$

Since $h \in B(M_S^{(n)})$ implies $\ell_1 \le h|_{E_{il}^{(n)}} \le u_1$ and $\ell_2 \le h|_{F_{im}^{(n)}} \le u_2$ we get by (18) and the definition of $R(E_{il}^{(n)}, F_{im}^{(n)}, r_{ilm}^{(n)})$ that

$$\ell_1 \le R(E_{il}^{(n)}, F_{im}^{(n)}, r_{ilm}^{(n)})h|_{E_{il}^{(n)}} \le u_1$$

and

$$\ell_2 \le R(E_{il}^{(n)}, F_{im}^{(n)}, r_{ilm}^{(n)})h|_{F_{im}^{(n)}} \le u_2.$$

This proves (17) and so the lemma is proved.

We can now easily prove

Lemma 9. *If $h \in B(K)$, then $P^{(n)}h \in B(K)$ for every n.*

Proof: Given $\varepsilon > 0$, there is an h' with $\|h' - h\| < \varepsilon$ and a finite set of $g_1's$ in K s.t. $h' \in B(M_S)$. This gives the lemma as a consequence of Lemma 8. We can also easily prove

Lemma 10. *For every $h \in B(K)$ and every sequence $(j_1, j_2, \cdots, j_n, \cdots)$ of positive integers, the sequence $h, P^{(j_1)}h, P^{(j_2)}P^{(j_1)}h, \cdots, P^{(j_n)} \dots P^{(j_2)}P^{(j_1)}h, \cdots$ has a weakly convergent subsequence.*

Proof: Given $\varepsilon > 0$ we consider an h' s.t. $\|h' - h\| \le \varepsilon$ and $h' \in B(M_S)$ where M_S is a finite set $\{g_{i_1}, g_{i_2}, \cdots, g_{i_r}\} \subset K$. By Lemma 9,

$$\min\{g_{i_1}, g_{i_2}, \cdots, g_{i_r}\} \le P^{(j_n)} \dots P^{(j_2)}P^{(j_1)}h' \le \max\{g_{i_1}, g_{i_2}, \cdots, g_{i_r}\}$$

for every n. Since $\min\{g_{i_1}, g_{i_2}, \cdots, g_{i_r}\}$ and $\max\{g_{i_1}, g_{i_2}, \cdots, g_{i_r}\}$ are in $L^1(0,1)$ the lemma follows.

In order to complete the construction of a contractive projection onto K, we would like the weakly convergent subsequence in Lemma 10 to converge to an element in K. But it is enough – the simple argument is given below – that it decreases the distance between h and some element in K if $h \in K$.

And this follows from

Lemma 11. *If $h \in B(K)$ but $h \notin C_{f_n,T_n,x_n}$, then there is a $g_j \in K$ s.t. $\|P^{(n)}h - g_j\| < \|h - g_j\|$.*

Proof: We can assume WLOG that $f_n > 0$ on $E^{(n)} \cup F^{(n)}$ and that $x_n = 0$. We can now find a subset $E_1^{(n)}$ of $E^{(n)}$ where we have the following:

1. There is a $\delta_1 > 0$ s.t. $P^{(n)}h < h - \delta_1$ on $E_1^{(n)}$;

2. There is a $B > 0$ s.t. $|h| \leq B$ and $P^{(n)}h \leq B$ and for all i and m $|Q_{i+m}^{(n)} \circ \cdots \circ Q_{i+1}^{(n)} \circ Q_i^{(n)}h| \leq B$ on $E_1^{(n)}$;

3. There is a $\delta_2 > 0$ and a function $g_j \in K$ s.t. $P^{(n)}h \leq g_j \leq h - \delta_2$ on $E_1^{(n)}$.

It is obvious how to find a subset s.t. 1. and 2. hold and since by Lemma 1.8 both h and $P^{(n)}h$ are in $B(K)$. We get 3. by passing to a subset. The g_j in 3. will be the g_j of the lemma. To see that we will approximate h and g_j. We have the following:
Given $\delta_3 > 0$ there exist h' and g'_j and $\omega > 0$ s.t. the following holds:

1A. h' and g'_j are constant on every subset of the ith subdivision of $E^{(n)}$ and $F^{(n)}$ for $i \geq \omega$.

2A. $\|h' - h\| \leq \delta_3$ and $\|g'_j - g_j\| \leq \delta_3$.

3A. $Q_i^{(n)}g'_j = g'_j$ for all $i \geq \omega$.

4A. $P^{(n)}h' \leq g'_j \leq h' - \frac{\delta}{2}$ on a set $E_2^{(n)}$ s.t. $m(E_2^{(n)} \cap E_1^{(n)}) \geq \frac{9}{10}m(E_1^{(n)})$.

We now prove that there is an $\varepsilon > 0$ s.t.

$$\|P^{(n)}h' - g'_j\| < \|h' - g'_j\| - \varepsilon \text{ where } \varepsilon \text{ does not depend on } \delta_3. \tag{19}$$

Since δ_3 is arbitrary > 0, (19) gives the lemma. To prove (19) it is obviously enough to prove that there is an $\omega > 0$ s.t. for $i \geq \omega$ and m large enough depending on i we have

$$\|Q_{i+m}^{(n)} \circ \cdots \circ Q_{i+1}^{(n)} \circ Q_i^{(n)}h' - g'_j\| < \|h' - g'_j\| - \varepsilon \tag{20}$$

To prove (20) we consider the elementary projections in $Q_{i+m}^{(n)} \circ \cdots \circ Q_{i+1}^{(n)} \circ Q_i^{(n)}$. Let $R(E', F', r')$ be such a projection. Then by 3A. $R(E', F', r')g'_j = g'_j$. Now if h'_1 is a function that is constant on E' and $h'_1 \geq g'_j$ on E' then we have

$$(R(E',F',r')h'_1 - g'_j)|_{E'} \leq (h'_1 - g'_j)|_{E'} - 2\,m(E')\min\{(h'_1 - g'_j)|_{E'}, (h'_1 - R(E',F',r')h'_1)|_{E'}\} \tag{21}$$

And if i is large enough and then m is chosen large enough then we have $Q^{(n)}_{i+m} \circ \cdots \circ Q^{(n)}_{i+1} \circ Q^{(n)}_i h' \leq g_j + \dfrac{\delta}{10}$ on a subset of $E^{(n)}_2 \cap E^{(n)}_1$ of measure

$$m(E^{(n)}_1) \cdot \varepsilon(\delta_2, B), \text{ where } \varepsilon(\delta_2, B) \text{ depend just on } \delta_2 \text{ and } B. \tag{22}$$

(22) holds since $P^{(n)} = \lim_{\gamma\to\infty}(\lim_{m\to\infty} Q^{(n)}_{j_\gamma+m} \circ \cdots \circ Q^{(n)}_{j+\gamma+1} \circ Q^{(n)}_{j+\gamma})$ where the outer limit is taken in the weak topology and $P^{(n)}h' \leq g'_j$ on all of $E^{(n)}_1$. Now (20) follows from (21) and (22) and since $h' - g'_j \geq \frac{\delta}{2}$ on $E^{(n)}_2 \cap E^{(n)}_1$ we get (20) with $\varepsilon = 2(\frac{\delta}{2} - \frac{\delta}{10})m(E^{(n)}_1) \cdot \varepsilon(\delta_2, B)$. This proves the lemma.

With the Lemmas 7-11 we can now complete the construction of a projection onto K where K is a countable intersection of cones $C_{f_n,T_n,x_n}, C_{E_\beta,+,x_\beta}, C_{E_\gamma,-,x_\gamma}$. We have

Proposition 1. Every countable intersection of cones of the form $C_{f_\alpha,T_\alpha,x_\alpha}, C_{E_\beta,+,x_\beta}, C_{E_\gamma,-,x_\gamma}$ is the range of a contractive projection on $L^1(0,1)$.

Proof: We first observe that $B(K) \subset (\cap_\beta C_{E_\beta+,x_\beta}) \cap (\cap_\gamma C_{E_\gamma,-,x_\gamma})$, and so $K = B(K) \cap (\cap_\gamma C_{f_\gamma,T_\gamma,x_\gamma})$. We first project onto $B(K)$ by Lemma 7. and so it is enough to find a projection P from $B(K)$ onto K. We consider the sequence (g_j) as above. We then consider only those j for which $g_j \in B(K)$, and by dropping one index we denote this subsequence (g_j). Consider the first j s.t. $g_j \notin K$. We move g_j into K by the following procedure. Consider the first i s.t. $g_j \notin C_{f_i,T_i,x_i}$. Put $T_1 = P^i$. If T_α is defined for a countable ordinal α then we define $T_{\alpha+1}$ as follows: Consider the first i s.t. $T_\alpha g_j \notin C_{f_i,T_i,x_i}$. Then $T_{\alpha+1} = P^{(i)} \circ T$. If α is a countable limit ordinal we define T_α by taking a sequence $\alpha_n \nearrow \alpha$ s.t. $T_{\alpha_n} g_r$ converges weakly for every r. This can be done by Lemma 10. By the Lemmas 9-11 we get $T_\beta g_j \in K$ for some countable ordinal β. Put $T_\beta = T^{(1)}$. Now we consider the smallest $m > j$ s.t. $T^{(1)} g_m \notin K$. We apply the same procedure to $T^{(1)} g_m$ as was applied to g_j. We get a $T^{(2)}$ s.t. $T^{(2)} \circ T^{(1)} g_m \varepsilon K$. By continuing this procedure we get a sequence of contractions s.t. $\lim_{p\to\infty} T^{(p)} \circ T^{(p-1)} \circ \cdots \circ T^{(2)} \circ T^{(1)} g_r = P g_r$ exists for every r. In fact, for every r, the sequence eventually becomes constant. P is obviously a contractive projection from $B(K)$ onto K. Thus the "if"-part of Theorem 1 is proved.

5. The proof of Theorem 1.

Proposition 1 proves the "if"-part of Theorem 1. In this section we prove the "only if"-part of Theorem 1. We let K be a convex set onto which there is a contractive projection. Let x be a support point of K, that is a point in K where there is a supporting hyperplane. It is well-known that K cannot have interior points without being all of $L^1(0,1)$. We exclude this trivial case and so it is well-known that the set of support points of K is dense in K. We let K_x be the closed cone generated by K at a support point x, so

$$K_x = \{x + g \mid \text{ there is an } \varepsilon > 0 \text{ s.t. } x + \varepsilon g \in K\}.$$

It is well-known and easy to see that K is the intersection of all the cones K_x at support points x. So we only need to prove that every cone K_x is of the form given by Theorem 1. In the Lemma 13 below we prove that K_x has a property slightly weaker than that of being the range of a contractive projection and then we use Lemmas 14-17 to prove that that property implies that K_x is of the form given by Theorem 1. We say that a subset M of $L^1(0,1)$ is contractive if there is a contractive projection from $L^1(0,1)$ onto M. We now start with the simple

Lemma 12. *If A and B are closed convex sets in a Banach space, $A \subset B$ and $A \neq B$, then there is a support point x of A and a unit vector y and an $\varepsilon > 0$ s.t. $x + ty \in B$ for all sufficiently small positive t, but $d(x + ty, A) \geq \varepsilon t$ for all t, $0 \leq t \leq 1$.*

We omit the simple proof.

If g_1 and g_2 are in $L^1(0,1)$ we let M_{g_1,g_2} denote the union of metric lines between g_1 and g_2 so M_{g_1,g_2} consists of all functions h for which $\|g_1 - g_2\| = \|g_1 - h\| + \|g_2 - h\|$. We observe that M_{g_1,g_2} is weakly compact. We now have

Lemma 13. *Let K be a contractive, convex subset of $L^1(0,1)$. Let K_x be the cone generated*

by K at a support point x. Let $x+f$ be a support point of K_x. Then, for every $\delta > 0$, there is a contractive projection from

$$K_x \cup M_{x+f,x+(1+\delta)f} \quad \text{onto} \quad K_x.$$

Proof: For simplicity of notation we assume WLOG that $x = 0$. If f is a support point of K then there is a sequence $f_n \to f$ and a sequence of positive real numbers α_n s.t. $\alpha_n f_n$ and $(1+\delta)\alpha_n f_n \in K$. Let P be the contractive projection $L^1(0,1) \to K$. Let (h_i) be a sequence dense in $M_{f,(1+\delta)f}$. For every i let $h_{ni} \to h_i$, where $h_{ni} \in M_{f_n,(1+\delta)f_n}$. Since P is contractive and $P(\alpha_n f_n) = \alpha_n f_n$ and $P((1+\delta)\alpha_n f_n) = (1+\delta)\alpha_n f_n$ for every n we get $P(\alpha_n h_{ni}) \in M_{\alpha_n f_n,(1+\delta)\alpha_n f_n}$ for every n and i. And so $\dfrac{1}{\alpha_n} P(\alpha_n h_{ni}) \in M_{f_n,(1+\alpha)f_n}$ for every n and i. We can now by a diagonal procedure pick out a subsequence s.t. $\dfrac{1}{\alpha_{n_\gamma}} P(\alpha_{n_\gamma} h_{n_\gamma i})$ converges weakly for every i. It is now easy to see that $h_i \to \lim\limits_{\gamma\to\infty} \dfrac{1}{\alpha_{n_\gamma}} P(\alpha_{n_\gamma} h_{n_\gamma i})$ defines a contractive projection from $K_0 \cup M_{f,(1+\delta)f}$ onto K_0. This proves the lemma.

We now consider a cone K_0 as above and we let f be a support point of K_0. By an easy exhaustion argument, either Case 1. or Case 2. below occurs.

Case 1. There is a set $E \subset (0,1)$, $m(E) > 0$, and an $\varepsilon > 0$ s.t.

A. f and $\dfrac{1}{f}$ are bounded on E and f has constant sign on E

B. Either $d(f + t\chi_{E'}, K_0) \geq \varepsilon t\; m(E')$ for every $t \geq 0$ and every $E' \subset E$,
or $d(f - t\chi_{E'}, K_0) \geq \varepsilon t\; m(E')$ for every $t \geq 0$ and every $E' \subset E$,
where $\chi_{E'}$ is the characteristic function of E'. (23)

Case 2. There is a set $E \subset (0,1)$, $m(E) > 0$, and an $\varepsilon > 0$ s.t.

A. $f \equiv 0$ on E;

B. Either $d(f + t\chi_{E'}, K_0) \geq \varepsilon t\; m(E')$ for every $t \geq 0$ and every $E' \subset E$,
or $d(f - t\chi_{E'}, K_0) \geq \varepsilon t\; m(E')$ for every $t \geq 0$ and every $E' \subset E$. (24)

We now have

Proposition 2. Assume that Case 1 (23) occurs. Then there is a function T so that $K_0 \subset C_{f,T,0}$. The proof of this proposition will require the lemmas below. We can assume WLOG that f is positive on E and by multiplying f by a positive constant and possible passing to a subset we can assume $1 \leq f \leq 1 + \frac{\varepsilon}{100}$ on E. We first assume a plus-sign in B. Let P be the ontractive projection from $K_0 \cup M_{f,2f}$ onto K_0 given by Lemma 13. We consider a subset E' of E, $m(E') > 0$, and a δ, $0 \leq \delta \leq \frac{1}{2}$. We say that we apply the P-construction to $f + \delta\chi_{E'}f$ by forming the sequences (g_n) and $\tilde{T}_{\delta n}(E')$ which we are now going to define. Put $P(f + \delta\chi_{E'}f) = \tilde{T}_{\delta 1}(E')$. Let g_1 be the function which is $(1+\delta)f$ on E' and $\tilde{T}_{\delta 1}(E')$ on the rest of (0,1). Inductively, let $P(g_n) = \tilde{T}_{\delta(n+1)}(E')$ and let g_{n+1} be $(1+\delta)f$ on E' and $\tilde{T}_{\delta(n+1)}(E')$ on the rest of (0,1). By this definition we get that g_n is on a metric line between $\tilde{T}_{\delta_n}(E')$ and $(1+\delta)f$, that is, $\|\tilde{T}_{\delta n}(E') - (1+\delta)f\| = \|\tilde{T}_{\delta_n}(E') - g_n\| + \|g_n - (1+\delta)f\|$. Thus g_n and $\tilde{T}_{\delta_n}(E')$ converge in $L^1(0,1)$, and it is easy to see that they converge to the same function which we call $g = \tilde{T}_\delta(E')$. We observe that $\delta_1 \leq \delta_2$ implies $\tilde{T}_{\delta_1}(E') \leq \tilde{T}_{\delta_2}(E')$ a.e. since in every step of the P-construction applied to $f + \delta_1\chi_{E'}f$ and to $f + \delta_2\chi_{E'}f$ we obviously have

$$\tilde{T}_{\delta_1 n}(E') \leq \tilde{T}_{\delta_2 n}(E') \text{ a.e.} \tag{25}$$

Put supp $\left(\tilde{T}_\delta(E) - f\right) = E \bigcup F_{(\delta)}$ with $F_{(\delta)} \bigcap E = \phi$. Here we define supp g to be the set where $g \neq 0$ so supp g is not closed in general. By (25) the sets $F_{(\delta)}$ decrease with δ and so $\lim_{\delta\to 0} F_{(\delta)}$ exists, say $\lim_{\delta\to 0} F_{(\delta)} = F$. Now let $E' \subset E$. Put $T_\delta(E') = \text{supp}\left(\tilde{T}_\delta(E') - f\right) \bigcap F$. By (25) $\lim_{\delta\to 0} T_\delta(E')$ exists and we put $T(E') = \lim_{\delta\to 0} T_\delta(E')$. In the following lemmas we prove that T has the properties 1. – 3. given in the definition of $C_{f,T,0}$.

Lemma 14. $T(E_1 \bigcup E_2) = T(E_1) \bigcup T(E_2)$.

Proof: Since, by (25), $T(E')$ increases with E' we have $T(E_1) \bigcup T(E_2) \subset T(E_1 \bigcup E_2)$. Now assume that there is a subset F' of $T(E_1 \bigcup E_2)$, $m(F') > 0$, s.t. F' is disjoint from $T(F_1) \bigcup T(E_2)$. Then there is a $\delta > 0$ and $F'' \subset F'$, $m(F'') > 0$, s.t. F'' is disjoint from

$T_\delta(E_1)$ and $T_\delta(E_2)$. Consider the function $\frac{1}{2}\left(\tilde{T}_\delta(E_1)+\tilde{T}_\delta(E_2)\right)$. This function is in K_0 and it is $\geq (1+\frac{\delta}{2})f$ on $E_1 \bigcup E_2$. This implies that all the functions g_n in the P-construction applied to $f+\frac{\delta}{2}\chi_{(E_1 \bigcup E_2)}f$ will be on a metric line between f and $\frac{1}{2}\left(\tilde{T}_\delta(E_1)+\tilde{T}_\delta(E_2)\right)$. But the P-construction applied to $f+\frac{\delta}{2}\chi_{(E_1 \bigcup E_2)}f$ will by assumption give functions which are $xx > f$ on F'' and by assumption $\frac{1}{2}\left(\tilde{T}_\delta(E_1)+\tilde{T}_\delta(E_2)\right) = f$ on F''. This is a contradiction and so the lemma is proved.

Lemma 15. *$m(E_1) > 0$ implies $m(T(E_1)) > 0$.*

Proof: We have

$$\int_{F_{(\delta)}} [P(f+\delta\chi_{E'}f) - f]\, dt > \frac{\epsilon}{2}\delta\, m(E') \tag{26}$$

where ϵ is the same as in (23). We get (26) in the following way: The contractivity of P implies that $P(f+\delta\chi_{E'}f)$ is on a metric line between f and $(1+\delta)f$, and that $\int_0^1 /P(f+\delta\chi_{E'}f)-f)dt = \int_0^1((f+\delta\chi_{E'}f)-f)dt$. Since $\|P(f+\delta\chi_{E'}f)-(f+\delta\chi_{E'}f)\| \geq \delta\epsilon\, m(E')$ we get (26). Now there is an N independent of δ s.t.

$$\|(\delta f - \delta f)_{\delta N}\| \leq \frac{1}{10}\delta\epsilon\, m(E'), \tag{27}$$

where $(\delta f)_{\delta N} = \delta f$ if $\delta f \leq \delta N$, and δN otherwise. (26) and (27) now give

$$\frac{\epsilon}{2}\delta\, m(E') \leq \int_{F(\delta)} (P(f+\delta\chi_{E'}f) - f)\, dt =$$

$$= \int_{\substack{F(\delta)\\ \delta \geq N}} (P(f+\delta\chi_{E'}f) - f)\, dt + \int_{\substack{F(\delta)\\ f<N}} (P(f+\delta\chi_{E'}f) - f)\, dt. \tag{28}$$

Since the integrands of both terms of (28) are between 0 and δf, (27) gives that the first term is $\leq \frac{\delta}{10}\,\epsilon\, m(E')$. Thus (28) gives

$$\int_{\substack{F(f)\\ f<N}} (P(f+\delta\chi_{E'}f) - f)\, dt \geq (\frac{1}{2} - \frac{1}{10})\delta\epsilon\, m(E').$$

Thus $m(F_{(\delta)}) > (\frac{1}{2} - \frac{1}{10})\frac{\epsilon}{N} \cdot m(E')$, and since this is independent of δ the lemma is proved.

We now prove

Lemma 16. *T is left continuous, that is for every $E_1 \subset E$, there is a function $\delta_{E_1(\epsilon)}$, $\delta_{E_1}(\epsilon) \to 0$ as $\epsilon \to 0$, s.t. if $E_1' \subset E_1$ and $m(E_1' > m(E_1) - \epsilon$ then $m(T(E_1')) > m(T(E_1)) - \delta_{E_1}(\epsilon)$.*

Proof: We first observe that the definition of left continuity given in this lemma is equivalent to the one in 3. in the definition of $C_{f,T,0}$. We now consider $T(E_1)$. It is near to supp $\left(\tilde{T}_\delta(E_1) - f\right) \bigcap F$ if δ is small enough. And for fixed δ, supp $\left(\tilde{T}_\delta(E_1) - f\right)$ is near to supp $\left(\tilde{T}_{\delta n}(E_1) - f\right)$ if n is large enough. And for $\tilde{T}_{\delta n}(E_1) - f$ we can obviously argue by left continuity. This proves the lemma.

After the next lemma we will find a T_1 such that $K_0 \subset C_{f,T_1,0}$. To do that it remains to prove that $h \in K_0$ implies $\operatorname{ess\,inf}_{t \in E_1} \frac{h(t)}{f(t)} \leq \operatorname{ess\,inf}_{t \in T(E_1)} \frac{h(t)}{f(t)}$ for all $E_1 \subset E$. That is obviously equivalent to the following

Lemma 17. *If $h \in K_0$ and $h(t) \geq f(t)$ a.e. on E_1, then $h(t) \geq f(t)$ a.e. on $T(E_1)$. This holds for all $E_1 \subset E$.*

Proof: By the left continuity of T, it is enough to prove the following:
For every $\delta > 0$, if $h \in K_0$ and $h \geq (1+\delta)f$ on E_1, then $h \geq f$ on $T(E_1)$. To prove that we consider a $\delta > 0$ and consider the P-construction on $f + \delta\chi_{E_1} f$. We have $f \leq P(f + \delta\chi_{E_1} f) \leq (1+\delta)f$ on $(0,1)$ and in particular on E_1, since $P(f + \delta\chi_{E_1} f)$ is on a metric line between f and $(1+\delta)f$. We now put $\int_{E_1} ((1+\delta)f - P(f + \delta\chi_{E_1} f))\, dt = A$. the contractivity of P gives

$$\int_{(0,1)\setminus E_1} (P(f + \delta\chi_{E_1} f) - f)\, dt = A \tag{29}$$

If $h \geq (1+\delta)f$ on E_1, $h \in K_0$, then

$$\int_{E_1} (h - P(f + \delta\chi_{E_1} f))\, dt = \int_{E_1} (h - (f + \delta\chi_{E_1} f))\, dt + A.$$

Since P is contractive and $f + \delta\chi_{E_1} f = f$ on $(0,1)\setminus E_1$ this gives

$$\begin{aligned} \int_{(0,1)\setminus E_1} |h - P(f + \delta_{\chi_{E_1}} f)|dt &\leq \int_{(0,1)\setminus E_1} |h - (f + \delta_{\chi_{E_1}} f)|dt - A \\ &= \int_{(0,1)\setminus E_1} |h - f|dt - A \end{aligned} \tag{30}$$

By (29) this can only happen if

$$h \geq P(f + \delta\chi_{E_1} f) \geq f \text{ on supp } (P(f + \delta\chi_{E_1} f) - f) \bigcap ((0,1)\backslash E_1).$$

So

$$h \geq f \text{ on supp } (P(f + \delta\chi_{E_1} f) - f),$$

$$\text{and in particular on supp } \left(\tilde{T}_{\delta 1}(E_1) - f\right) \bigcap F. \tag{31}$$

Now g_1, formed as above in the P-construction applied to $f + \delta\chi_{E_1} f$, is on a metric line between $\tilde{T}_{\delta 1}(E_1)$ and $(1+\delta)f$ and so is $P(g_1)$. If $g_1 \in K_0$, then $T_\delta(E_1) = \text{supp}(\tilde{T}_{\delta 1}(E_1) - f) \bigcap F$, and so the lemma is proved. If $g_1 \notin K_0$ put $\int_{E_1} (g_1 - P(g_1))dt = A_1 > 0$. That gives, as above,

$$\int_{(0,1)\backslash E_1} (P(g_1) - \tilde{T}_{\delta 1}(E_1))dt = A_1.$$

If $h \geq (1+\delta)f$ on E_1, $h \in K_0$, then as above

$$\int_{E_1} (h - P(g_1)|dt = \int_{E_1} (h - g_1)dt + A_1.$$

And as above we get

$$\int_{(0,1)\backslash E_1} |h - P(g_1|dt \leq \int_{(0,1)\backslash E_1} |h - g_1|dt - A_1 = \int_{(0,1)\backslash E_1} |h - \tilde{T}_{\delta 1}(E_1)| - A_1.$$

This can happen only if $h \geq P(g_1) \geq f$ on $\text{supp}(P(g_1) - \tilde{T}_{\delta 1}(E_1)) \bigcap (0,1)\backslash E_1$. So $h \geq f$ on $\text{supp}(\tilde{T}_{\delta 2}(E_1) - \tilde{T}_{\delta 1}(E_1)) \bigcup \text{supp}(\tilde{T}_{\delta 1}(E_1) - f)$. By repeating this argument we get that if $h \geq (1+\delta)f$ on E_1, then $h \geq f$ on $\bigcup_n \text{supp}(\tilde{T}_{\delta n}(E_1) - T_{\delta(n-1)}(E_1))$, that is, $h \geq f$ on $T_\delta(E_1)$. By letting $\delta \to 0$ we get the lemma. So we have proved that $K_0 \subset C_{f,T,0}$, provided f and $\frac{1}{f}$ are bounded on $E \bigcup F$. If not, we construct T as follows. Assume f and $\frac{1}{f}$ are bounded on $F' \subset F$. Consider a maximal subset E' of E s.t. $T(E') \subset F\backslash F'$. Define T_1 on subsets of $E\backslash E'$ by $T_1(E_1) = T(E_1) \bigcap F'$. We now obviously get

$$K_0 \subset C_{f,T_1,0}. \tag{32}$$

If Case 1. (23) occurs with a minus-sign in B., we reduce the case to a plus-sign in B. as follows. Consider $f - \delta\chi_{E_1} f$ and $P(f - \delta\chi_{E_1} f)$ which are both on metric lines between f

and $(1-\delta)f$. By copying the discussion in the P-construction (or, in fact, just using the first step of a P-construction) we can define a T s.t. if $h \in K_0$ and $h \le f$ on E_1, then $h \le f$ on $T(E_1)$. And it is easy to see then that there is an $\epsilon > 0$ and an $F' \subset F$, $m(F') > 0$, s.t. if F'' is any subset of F' then $d(f + t\,\chi_{F''}, K_0) \ge \epsilon t\; m(F'')$ for $t \ge 0$. This completes the proof of Proposition 2.

Remark: If, in the definition of the map T, we had just used the first step of the P-construction, we would easily get the lemmas 15-17. However, we would not have gotten the preservation of union given by Lemma 14. This can be seen by the following example.

Example: Let K be the cone of non-negative, increasing functions in $L^1(0,1)$. A contractive projection P from $L^1(0,1)$ onto K is obtained by mapping every $h \in (L^1(0,1)$ onto the increasing rearrangement of h. It is easy to see that a support point of K is an increasing function f s.t. f is constant on some interval $[t_0, t_1] \subset [0,1]$. Suppose now that we have such an f, which is in addition continuous and strictly increasing on both $[0, t_0]$ and $[t_1, 1]$. Let the set E be some interval $[t_0, t_0 + \gamma]$ where γ is much smaller than $t_1 - t_0$. If we had used just the first step of the P-construction to define T, then the set F would be the interval $[t_1 - \gamma, t_1]$ and a subset E_1 of E would be mapped onto the interval $[t_1 - m(E_1), t_1]$. So obviously T would not preserve unions. When the full P-construction is used to define T the set F will be the interval $[t_0 + \gamma, t_1]$ and every subset of E will be mapped onto all of F by T. And $C_{f,T,0}$ will consist of all those functions g s.t. for every subset E_1 of $[t_0, t_0 + \gamma]$ we have

$$\operatorname*{ess\,inf}_{t \in E_1} g(t) \le \operatorname*{ess\,inf}_{t \in [t_0 + \gamma t_1]} g(t),$$

which is equivalent to

$$\operatorname*{ess\,sup}_{t \in [t_0,\, t_0 + \gamma]} g(t) \le \operatorname*{ess\,inf}_{t \in t_0 + \gamma_1 t_1]} g(t).$$

Completion of the proof of Theorem 1:

As remarked above, by a simple exhaustion argument, either the assumptions of Proposition 2 are fulfilled or Case 2 (24) occurs. We now assume that Case 2 (24) occurs. Then either there is a set $E' \subset E$ s.t. for all $h \in K_0$ we have $h \ge 0$ (or $h \le 0$) on E', or for every

$E' \subset E$ there is an $E'' \subset E'$, $m(E'') > 0$, and an $h \in K_0$ s.t. $h > 0$ on E''. In the first case we get

$$K_0 \in C_{E',+,0}(\text{ or } C_{E',-,0}). \tag{33}$$

In the second case we can construct a new support point f_1 of K_0, for which Case 1. (23) applies, by the following procedure. Assume B. occurs with a plus-sign in (24). Put $f + h = h_0$. Now if $h_m \in K_0$ is constructed we form $h_{m+1} \in K_0$ by letting $h_{m+1} = h_m + h'$, where h' satisfies

$$\|h' - \frac{\|h'\|}{m(E'')}\chi_{E''}\| \leq \frac{\epsilon}{10}\|h'\|.$$

This process obviously stops at some countable ordinal α and we put $f_1 = h_\alpha$. We observe that f_1 can be assumed to be arbitrarily close to f. We now see, by an obvious exhaustion argument, that there is a subset $E''' \subset E''$, $m(E''') > 0$, for which Case 1. (23) occurs and so we get

$$K_0 \subset C_{f_1,T_1,0}. \tag{34}$$

Finally we prove that K_0 is equal to the intersection of all cones formed in (32), (33) and (34). To do that we consider an intersection C of such cones s.t. $K_0 \subset C$ and $K_0 \neq C$. By lemma 2.1 we find a support point f of K_0, an $\epsilon > 0$ and an h s.t.

$$f + h \in C \text{ but } d(f + th,\ K_0) \geq \epsilon t \text{ for all } t \geq 0. \tag{35}$$

We construct a possibly new support point f_1 of K_0 by the following procedure. Let $h_0 = f$ and if h_m is defined for an ordinal number m let h_{m+1} be an element in K_0 for which $\|h_{m+1} - (h + f)\| \leq \|h_m - (h + f)\| - (1 - \frac{\epsilon}{2})\|h_{m+1} - h_m\|$. If possible $h_{m+1} \neq h_m$. Obviously h_m becomes constant from some countable ordinal m on and if we put

$$h_m = f_1 \text{ we obviously have a support point } f_1 \text{ of } K_0. \tag{36}$$

We can now WLOG assume that $h \geq 0$ on $(0, 1)$. Then we obviously have $f + h > f_1 \geq f$ on some subset E of $(0, 1)$. If $f > 0$ on $E_1 \subset E$, $m(E_1) > 0$, then by applying the P-construction to $f_1 + \delta\chi_{E_1 f_1}$ we get by (35) and (36) a cone $C_{f_1,T_1,0}$ s.t. $K_0 \subset C_{f_1,T_1,0}$ but $f + h \notin C_{f_1,T_1,0}$. If $f_1 \equiv 0$ on E_1 we find as above either an $E_2 \subset E_1$, s.t. $K_0 \subset C_{E_2,+,0}$ $(C_{E_2,-,0})$ but $f + h \notin C_{E_2,+,0}$ $(C_{E_2,-,0})$, or we find a new support point $f_2 \in K_0$, s.t.

$K_0 \subset C_{f_2,T_1,0}$ but $f + h \notin C_{f_2,T_1,0}$. Thus K_0 is equal to the intersection of all cones formed by (32), (33) and (34). And this completes, the proof that every contractive convex set K is an intersection of a family of cones $C_{f_\alpha,T_\alpha,x_\alpha}$, $C_{E_\beta,+,y_\beta}$, $C_{E_\delta,-,z_\delta}$. And so the proof of Theorem 1 is complete.

References

1. W. Davis and P. Enflo, Contractive projections on ℓ_p spaces, (to appear in Illinois J. of Math.).
2. U. Westphal, Cosuns in $\ell^p(n)$, (To appear in J. of Approximation Theory).

Some Banach space properties of translation invariant subspaces of L^p

KATHRYN E. HARE NICOLE TOMCZAK-JAEGERMANN*

Abstract. We study Banach space properties of translation invariant subspaces of L^p and L^∞ on a compact abelian group. Using methods from the operator ideal theory we give a characterization of $\Lambda(p)$ sets and Sidon sets in terms of local unconditional structure of related invariant subspaces.

In this note we apply concepts and techniques from Banach space theory to harmonic analysis, in particular, to the study of translation invariant subspaces of L^p and L^∞ on a compact abelian group. The main theorem gives characterizations of $\Lambda(p)$ sets and Sidon sets in terms of an unconditional structure of related invariant subspaces. The theorem is a generalization of some results of [3] and [9]. Similarly as in these papers, the proof here depends on powerful methods of the theory of operator ideals. To make this note accessible to non-specialists in this theory we try to present crucial points with all needed details.

Let G denote a compact abelian group and $\widehat{G}$ its discrete dual group. If X is a space of functions on G and $F \subset \widehat{G}$ let X_F denote the space of functions f in X whose Fourier transform $\hat{f}$ is supported on F.

Let m denote Haar measure on G. For $p > 0$ we denote the quasi-norm on L^p by

$$\|f\|_p \equiv \left(\int_G |f|^p \, dm \right)^{1/p}.$$

A subset F of $\widehat{G}$ is called a $\Lambda(p)$ *set*, $0 < p < \infty$, if there is $0 < r < p$ and a constant $\Lambda_{p,r}$ so that whenever $f \in \mathrm{Trig}_F(G) \equiv \{\text{polynomials } f : G \to \mathbb{C} \text{ with supp } \hat{f} \subseteq F\}$,

$$\|f\|_p \le \Lambda_{p,r} \|f\|_r. \tag{1}$$

1980 *Mathematics Subject Classification* (1985 *Revision*): 43A46, 46B99.
*Research partially supported by NSERC Grant A8854.

If $p > 2$ a duality argument shows that F is a $\Lambda(p)$ set if and only if

$$\Big(\sum_{\chi\in F} |\hat{f}(\chi)|^2\Big)^{1/2} \leq A_{p,2}\|f\|_{p'},$$

whenever $f \in$ Trig G, where $\frac{1}{p'} + \frac{1}{p} = 1$.

Let us also recall that a subset F of $\hat{G}$ is called a *Sidon set* if there exists a constant c such that

$$\sum_{\chi\in F} |\hat{f}(\chi)| \leq c\|f\|_\infty,$$

whenever $f \in C_F(G)$. For standard results on $\Lambda(p)$ sets and Sidon sets see [11] and [5].

For a standard notation in the Banach space theory we refer the reader to [4]. A detailed discussion of notions from the local theory of Banach spaces, such as type and cotype, as well as all notions from the theory of operator ideals used in this note, can be found e.g. in [8] and [12].

A basis $\{x_\alpha\}$ of a Banach space X is unconditional if there exists a constant c such that $\|\sum \varepsilon_\alpha a_\alpha x_\alpha\| \leq c\|\sum a_\alpha x_\alpha\|$ for all choices of signs $\varepsilon_\alpha = \pm 1$ and any sequence of scalars $\{a_\alpha\}$ with only finitely many non-zero terms. The least constant c is called the unconditional basis constant of $\{x_\alpha\}$ and is denoted by $ubc\{x_\alpha\}$. It is well known that for $p \neq 2$, F is a $\Lambda(q)$ set with $q = \max(p, 2)$ if and only if $\{\chi : \chi \in F\}$ forms an unconditional basis of L^p_F. Indeed, if $\|\sum_{\chi\in F} \varepsilon_\chi \hat{f}(\chi)\chi\|_p \sim \|f\|_p$ for all $f \in L^p_F$ and $\varepsilon_\chi = \pm 1$ for $\chi \in F$, then integrating over all ε and using Khintchine's inequality we get $\Big(\sum_{\chi\in F} |\hat{f}(\chi)|^2\Big)^{1/2} \sim \|f\|_p$ for all $f \in L^p_F$ and this means that F is a $\Lambda(q)$ set.

In the theory of Banach spaces the notion of unconditional basis leads to more general notions of local unconditional structure. In the context of translation invariant spaces we introduce the following definition which is a natural modification of the concept of local unconditional structure (l.u.st) considered in [1] (cf. also [8]).

Definition. We will say that a Banach space of functions X_F has *invariant l.u.st.* if there exist a sequence of finite increasing sets F_n with $\cup F_n = F$ and bases $\{x_{F_n}(\alpha)\}$ in

X_{F_n} such that

$$\sup_n (\text{ubc}\{x_{F_n}(\alpha)\}) < \infty.$$

Actually, we will work with a weaker property, a variant of the Gordon–Lewis property introduced in [2] (*cf.* also [9]). To describe it we need to recall the fundamental notion of p-absolutely summing operators. Let $0 < p < \infty$. An operator $T : X \to Y$ is called *p-absolutely summing* if there is a constant c so that

$$\sum \|Tx_n\|^p \le c^p \sup\{\sum |x^*(x_n)|^p : x^* \in X^*,\ \|x^*\| \le 1\}$$

for all finite sequences $\{x_n\}_0$ in X. The least such c is denoted $\pi_p(T)$. For standard results on p-absolutely summing operators we refer the reader to [8] and [12].

Definition. A Banach space of functions X_F defined on G has the *invariant Gordon-Lewis property* (invariant GL) if every translation invariant 1-absolutely summing operator from X_F to $L^2(G)$ factors through L^1. For an operator $T : X_F \to L^2(G)$, $\gamma_1(T)$ will denote the norm of the factorization of T through L^1, i.e., $\gamma_1(T) = \inf \|A\|\ \|B\|$ where the infimum is taken over all measure spaces (Ω, μ) and all bounded operators
$A : X_F \to L^1(\mu)$ and $B : L^1(\mu) \to L^2(G)$ such that $T = BA$. $GL(X_F)$ will denote the least constant so that $\gamma_1(T) \le GL(X_F)\pi_1(T)$ for all translation invariant 1-absolutely summing operators $T : X_F \to L^2(G)$.

A space X_F has the *invariant local* GL property if $F = \cup F_n$ for F_n finite increasing sets, and $\sup_n GL(X_{F_n}) < \infty$.

It follows from Banach space theorems that if a space X_F has an unconditional basis then it has GL (*cf.* e.g. [2], [8], [12]). Clearly this shows that invariant l.u.st. implies invariant local GL. Thus if F is a $\Lambda(p)$ set, $p > 2$, then L^p_F has invariant local GL.

As another non-trivial example consider the Hardy spaces H^p, $1 < p < \infty$. For G the circle group, set $F = \mathbb{Z}^+ \subset \widehat{G}$, so that L^p_F is the Hardy space H^p. Let $F_n = \{0, 1, \ldots, n\}$, so that $L^p_{F_n} = H^p_n = \text{span}\{1, z, \ldots, z^n\}$. A result due to Marcinkiewicz states that H^p_n

is isomorphic to ℓ_p^{n+1}, with constants of isomorphism independent of n (cf. [13, X.7.10]). Therefore H_n^p has an unconditional basis with the constant $c = c_p$ independent of n, and so H^p has invariant l.u.st. and yet $\mathbb{Z}^+$ is not a $\Lambda(q)$ set for any $q > 0$.

The main result of this note states.

THEOREM 1. *Let E be a subset of $\hat{G}$.*

(i) Let $p > 2$ and $\frac{1}{p} + \frac{1}{p'} = 1$. If L_E^p has the invariant local GL property and E is a $\Lambda(p')$ set, then E is a $\Lambda(p)$ set.

(ii) If $C_E(G)$ has the invariant local GL property and E is a $\Lambda(1)$ set, then E is a Sidon set.

In [3] and [9] similar results were proved under the assumption that E is a $\Lambda(2)$ set. The problem whether every $\Lambda(p)$ set is a $\Lambda(2)$ set, with $0 < p < 2$, is still open. It should be also noted that while part (i) of the theorem follows formally from the result of [3] (by Proposition 2 below), part (ii) requires an approach different then previously used.

The proof of the theorem requires an additional notation. For $F \subset \widehat{G}$ and $q < p$ put

$$\Lambda_{p,q}(F) = \sup\left\{\frac{\|f\|_p}{\|f\|_q} : f \in \mathrm{Trig}_F(G)\right\}.$$

We start the proof of the theorem with an easy extrapolation fact.

PROPOSITION 2. *Let $p > 2$ and let $F \subset \widehat{G}$ be a finite subset. If for some $q > p'$,*

$$\Lambda_{p,2}(F) \le A\,\Lambda_{2,q}(F),$$

then

$$\Lambda_{p,2}(F) \le A^{\theta/(2\theta-1)},$$

where $1/2 = (1-\theta)/p + \theta/q$.

Proof. Set $B = A\Lambda_{2,q}(F)$. By Hölder's inequality we have

$$\|f\|_p \le B\|f\|_2 \le B\|f\|_p^{1-\theta}\|f\|_q^{\theta},$$

for all $f \in \mathrm{Trig}_F(G)$. Thus $\|f\|_p \le B^{1/\theta}\|f\|_q$, for all $f \in \mathrm{Trig}_F(G)$. Therefore,

$$\|f\|_2 \le \|f\|_p^{1-\theta}\|f\|_q^{\theta} \le B^{(1-\theta)/\theta}\|f\|_q \qquad \text{for } f \in \mathrm{Trig}_F(G).$$

Hence

$$\Lambda_{2,q}(F) \le B^{(1-\theta)/\theta} = A^{(1-\theta)/\theta}\left(\Lambda_{2,q}(F)\right)^{(1-\theta)/\theta}.$$

The hypotheses ensure that $\theta > 1/2$ thus

$$\Lambda_{2,q}(F) \le A^{(1-\theta)/(2\theta-1)},$$

and

$$\Lambda_{p,2}(F) \le A^{\theta/(2\theta-1)}. \quad \blacksquare$$

We also require the following quantiative result.

PROPOSITION 3. *Let $p > 2$ and let F be a subset of $\widehat{G}$. Then the $\Lambda(p)$-constant of F satisfies*

$$\Lambda_{p,2}(F) \le c\sqrt{p}\Lambda_{2,1}(F)GL(X), \tag{2}$$

where c is a universal constant and $X = L_F^p$ or $X = L_F^\infty$.

The case $X = L_F^p$ was proved (although not stated) in [3]. A similar approach works for $X = L_F^\infty$. For sake of completeness we outline the proof in both cases.

The argument is based on properties of nuclear translation invariant operators. Let us recall that an operator $T : X \to Y$ is said to be *nuclear* if there are $\{x_j^*\} \subset X^*$ and $\{y_j\} \subset Y$ so that $T(x) = \sum x_j^*(x)y_j$ for all $x \in X$. The nuclear norm of T is defined by

$$\nu(T) = \inf\{\sum \|x_j^*\|_{X^*}\|y_j\|_Y : T(x) = \sum x_j^*(x)y_j\}.$$

We shall use the following formula (well-known in the theory of operator ideals cf. e.g. [12.I.1.2]) valid for all operators $T : X \to Y$ when at least one of X or Y is a finite dimensional Banach space. We have

$$\nu(T) = \sup\{\nu(TS) : S : Y \to X, \quad \|S\| \leq 1\}.$$

For $f \in L^p$ let T_f denote the operator given by convolution with f.

LEMMA 4. *Let $F \subset E \subset \widehat{G}$. Let $T : L^\infty_E \to L^p_F$ be a nuclear translation invariant operator. For every $\varepsilon > 0$ there exists $f \in L^p_F$ such that $T = T_f$ and $\|f\|_p \leq \nu(T)(1+\varepsilon)$.*

Proof. Fix $\varepsilon > 0$. There exist $x_j^* \in (L^\infty_E)^*$ and $h_j \in L^p_F$ such that $Tg = \sum_j x_j^*(g)h_j$ and $\sum \|x_j^*\| \; \|h_j\|_p \leq (1+\varepsilon)\nu(T)$. Let $\mu_j \in M(G)$ be a Hahn–Banach extension of x_j^*, with $\|\mu_j\|_{M(G)} = \|x_j^*\|$, for $j = 1, 2, \ldots$. It is not difficult to check that for $g \in \mathrm{Trig}_F(G)$,

$$Tg = \sum_j \mu_j(g)h_j = \sum_j \mu_j * h_j * g.$$

Set $f = \sum_j \mu_j * h_j$. Then

$$\begin{aligned} \|f\|_p &\leq \sum_j \|\mu_j * h\|_p \leq \sum_j \|\mu_j\|_{M(G)} \; \|h_j\|_p \\ &\leq \sum_j \|x_j^*\| \; \|h_j\|_p \leq (1+\varepsilon)\nu(T) \end{aligned}$$

and clearly $Tg = T_f g$. ∎

Proof of Proposition 3. Fix $f \in \mathrm{Trig}\,(G)$ with $\|f\|_{p'} = 1$ and consider the following diagram.

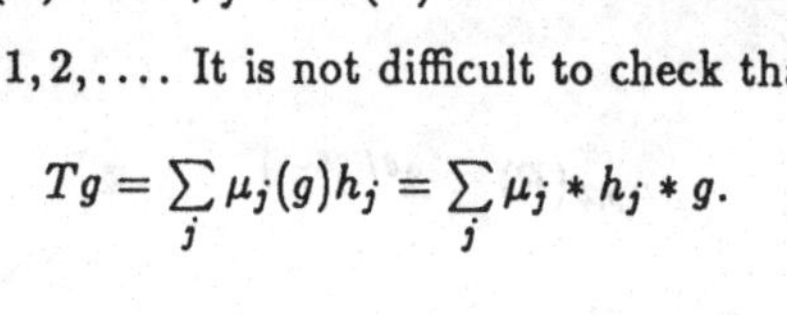

Observe that $\|T_f : L^p_F \to L^\infty_F\| \leq \|f\|_{p'} = 1$. By the definition of the GL constant there exists a measure space (Ω, μ) and operators

$$u_1 : L^p_F \to L^1(\Omega, \mu) \quad \text{and} \quad v : L^1(\Omega, \mu) \to L^2_F$$

such that $T_f = vu_1$, $\|u_1\| \leq 1$ and $\|v\| \leq \Lambda_{2,1}(F)GL(X)$. For $X = L^p_F$ this follows directly from the fact that

$$\pi_1(T_f : L^p_F \to L^2_F) \leq \|T_f : L^p_F \to L^\infty_F\| \pi_1(id : L^\infty_F \to L^1_F)\|id : L^1_F \to L^2_F\|$$
$$\leq \|f\|_{p'} \Lambda_{2,1}(F) = \Lambda_{2,1}(F).$$

For $X = L^\infty_F$, observe that there exist $u' : L^\infty_F \to L^1$ and $v : L^1 \to L^2_F$ such that $id = vu'$, $\|u'\| = 1$ and

$$\|v\| \leq \pi_1(id : L^\infty_F \to L^2_F)GL(L^\infty_F) \leq \Lambda_{2,1}(F)GL(L^\infty_F).$$

Then set $u_1 = u'T_f$.

It is well-known and follows directly from Khintchine's inequality that the type 2 constant of L^p is at most $c\sqrt{p}$ and that the cotype 2 constant of L^1 is $\sqrt{2}$. Thus by Maurey's extension theorem ([6], cf. also [12.II.6.13]) an operator $u_1 : L^p_F \to L^1$ admits a Hilbertian extension. That is, there exist $w : L^p \to H$ and $u : H \to L^1$ such that $u_1 = uw|_{L^p_F}$ and $\|u\|\ \|w\| \leq T_2(L^p)C_2(L^1)\|u_1\| \leq c\sqrt{p}\ \|u_1\| \leq c\sqrt{p}$.

Let $F_1 = F \cap \operatorname{supp} \hat{f}$. For every operator $S : L^2_{F_1} \to L^\infty_F$, $\|S\| = 1$ consider the following diagram where P is the orthogonal projection from L^2_F onto $L^2_{F_1}$.

$$\begin{array}{ccccccccc}
L^2_{F_1} & \xrightarrow{S} & L^\infty_F & \xrightarrow{id} & L^p_F & \xrightarrow{T_f} & L^2_F & \xrightarrow{P} & L^2_{F_1} \\
 & & \downarrow & & \downarrow & & \nearrow vu & & \\
 & & L^\infty & \xrightarrow[id]{} & L^p & \xrightarrow[w]{} H & & &
\end{array}$$

Recall two facts which go back to Grothendieck (*cf.* [8, 5.4 and 5.10] and [12, II 3.8 and II 3.10]). Every operator $w : L^\infty \to H$ or $w : L^1 \to H$ satisfies $\pi_2(w) \leq \sqrt{\pi/2}\|w\|$. Thus

$$\pi_2(wS : L^2_{F_1} \to H) \leq \|S : L^2_{F_1} \to L^\infty\| \pi_2(w : L^\infty \to H)$$
$$\leq \|S : L^2_{F_1} \to L^\infty\| \sqrt{\pi/2}\|w\| = \sqrt{\pi/2}\|w\|,$$
$$\pi_2(Pvu : H \to L^2_{F_1}) \leq \|u\|\pi_2(v) \leq \|u\|\sqrt{\pi/2}\|v\|.$$

Hence

$$\begin{aligned}\nu(T_f S : L^2_{F_1} \to L^2_{F_1}) &\le \pi_2(wS : L^2_{F_1} \to H)\pi_2(Pvu : H \to L^2_{F_1})\\ &\le (\pi/2)\|w\|\,\|u\|\,\|v\|\\ &\le c'\sqrt{p}\Lambda_{2,1}(F)GL(X) = A.\end{aligned}$$

Since the estimate holds for every $S : L^2_{F_1} \to L^\infty_F$ with $\|S\| = 1$, it follows that

$$\nu(T_f : L^\infty_F \to L^2_{F_1}) \le A.$$

By Lemma 4 this implies that for every $\varepsilon > 0$

$$\Big(\sum_{\chi\in F} |\hat f(\chi)|^2\Big)^{1/2} = \Big(\sum_{\chi\in F_1} |\hat f(\chi)|^2\Big)^{1/2} \le A(1+\varepsilon)\|f\|_{p'}.$$

Thus $\Lambda_{p,2} \le A$. ∎

Proof of Theorem 1. (i) Since L^p_E has the invariant local GL property, $E = \cup F_n$ where F_n are finite, increasing sets and $\sup_n GL(L^p_{F_n}) < \infty$. It clearly suffices to show $\sup_n \Lambda_{p,2}(F_n) < \infty$ in order to prove that E is a $\Lambda(p)$ set. By Rosenthal's theorem [10] E is a $\Lambda(q)$ set for some $q > p'$.

Since $\Lambda_{2,1}(F_n) \le \Lambda_{2,q}(F_n)\Lambda_{q,1}(E)$, by Proposition 3 we have

$$\begin{aligned}\Lambda_{p,2}(F_n) &\le c\sqrt{p}\Lambda_{2,q}(F_n)\Lambda_{q,1}(E)GL(L^p_{F_n})\\ &\le c\sqrt{p}\Lambda_{2,q}(F_n)\Lambda_{q,1}(E)\sup_m GL(L^p_{F_m}).\end{aligned}$$

By Proposition 2

$$\Lambda_{p,2}(F_n) \le \big(c\sqrt{p}\Lambda_{q,1}(E)\sup_m GL(L^p_{F_m})\big)^{\frac{\theta}{2\theta-1}},$$

where $\frac{1}{2} = \frac{1-\theta}{p} + \frac{\theta}{q}$ and this bound is independent of n.

The proof of (ii) requires an additional notation. Set $\psi(t) = \exp t^2 - 1$ for $t > 0$. For $f \in \mathrm{Trig}(G)$ define

$$\|f\|_\psi = \inf\{c > 0 : \int \psi(|f|/c) \le 1\}.$$

It is well known that $\| \ \|_\psi$ is a norm. Pisier [7] has shown that $E \subset \widehat{G}$ is a Sidon set if and only if there exists a constant c so that for some sequence of finite subsets F_n of E satisfying $\bigcup_{n=1}^{\infty} F_n = E$,

$$\|f\|_\psi \leq c\|f\|_2, \tag{3}$$

whenever $f \in \mathrm{Trig}_{F_n}(G)$, $n = 1, 2, \ldots$.

Assume $E = \cup F_n$, F_n finite increasing sets satisfying $\sup_n GL(L^\infty_{F_n}) < \infty$. By Proposition 3

$$\Lambda_{p,2}(F_n) \leq c\sqrt{p}\Lambda_{2,1}(F_n) \sup_m GL(L^\infty_{F_m}).$$

Therefore for $f \in L^2_{F_n}$,

$$\|f\|_p \leq \sqrt{p}\, D\, \|f\|_2, \tag{4}$$

where $D = c\Lambda_{2,1}(F_n) \sup_m GL(L^\infty_{F_m})$.

An elementary calculation shows that there exists $a > 0$ and $b > 0$ such that

$$\frac{1}{a}\|f\|_\psi \leq \sup_{p>2} \frac{\|f\|_p}{\sqrt{p}} \leq b\|f\|_\psi$$

for $f \in \mathrm{Trig}(G)$. Therefore for $f \in L^2_{F_n}$,

$$\|f\|_\psi \leq a\ D\ \|f\|_2. \tag{5}$$

By Rosenthal's theorem [10] there exists $q > 1$ such that E is a $\Lambda(q)$ set. We are now in a situation similar to Proposition 2. Fix $p > q'$ so that if $\frac{1}{2} = \frac{1-\theta}{p} + \frac{\theta}{q}$ then $\theta > \frac{1}{2}$. By (4) we have, as in the proof of Proposition 2,

$$\|f\|_p \leq \left(\sqrt{p}\ D\right)^{1/\theta} \|f\|_q,$$

for $f \in \mathrm{Trig}_{F_n}(G)$. Thus for $f \in \mathrm{Trig}_{F_n}(G)$,

$$\|f\|_2 \leq \left(\sqrt{p}\ D\right)^{\frac{1-\theta}{\theta}} \|f\|_q.$$

Since $\Lambda_{2,1}(F_n) \leq \Lambda_{2,q}(F_n)\Lambda_{q,1}(E)$ this implies that

$$\Lambda_{2,q}(F_n) \leq \left(c\sqrt{p}\,\Lambda_{q,1}(E)\,\sup_m GL(L^{\infty}_{F_n})\right)^{\frac{1-\theta}{\theta}} \left(\Lambda_{2,q}(F_n)\right)^{\frac{1-\theta}{\theta}}.$$

So

$$\Lambda_{2,q}(F_n) \leq \left(c\sqrt{p}\,\Lambda_{q,1}(E)\,\sup_m GL(L^{\infty}_{F_m})\right)^{\frac{1-\theta}{2\theta-1}}.$$

Thus, by (5),

$$\begin{aligned} \|f\|_{\psi} &\leq aD\|f\|_2 \leq ac\Lambda_{2,q}(F_n)\Lambda_{q,1}(E)\sup_m GL(L^{\infty}_{F_m})\|f\|_2 \\ &\leq c' p^{\frac{1-\theta}{2(2\theta-1)}}\Lambda_{q,1}(E)^{\frac{\theta}{2\theta-1}} \sup_m GL(L^{\infty}_{F_m})^{\frac{\theta}{2\theta-1}}\|f\|_2, \end{aligned}$$

for $f \in \mathrm{Trig}_{F_n}(G)$. This shows (3) and completes the proof of (ii). ∎

COROLLARY 4.

(i) Let $p > 2$ and suppose that $E \subset \widehat{G}$ is a $\Lambda(p')$-set. Then L^p_E has invariant l.u.st. if and only if E is a $\Lambda(p)$-set.

(ii) Suppose $E \subset \widehat{G}$ is a $\Lambda(1)$ set. Then $C_E(G)$ has invariant l.u.st. if and only if E is a Sidon set.

REFERENCES

1. E. Dubinsky, A. Pelczynski, and H.P. Rosenthal, *On Banach spaces X for which $\Pi_2(\ell_\infty, X) = B(\ell_\infty, X)$*, Studia Math. **44** (1972), 617–648.

2. Y. Gordon, and D. Lewis, *Absolutely summing operators and local unconditional structures*, Acta Math. **133** (1974), 27–48.

3. S. Kwapien and A. Pelczynski, *Absolutely summing operators and translation invariant spaces of functions on compact abelian groups*, Math. Nachr. 94 (1980), 303–340.

4. J. Lindenstrauss and L. Tzafriri, Classical Banach Spaces, Springer Verlag, **I** (1977), **II** (1979).

5. J. Lopez and K. Ross, Sidon Sets, Lecture notes in Pure and Applied Mathematics, **13**, Marcel Dekker, Inc. New York (1975).

6. B. Maurey, *Un théoreme de prolongement*, C.R. Acad. Sci. Paris, Sér. A **279** (1974), 329–332.

7. G. Pisier, *De nouvelles caracterisations des ensembles de Sidon*, Advances in Math, Suppl. St. **7** (1981), 685–726.

8. ———, Factorization of linear operators and geometry of Banach spaces, "CBMS AMS," (1985).

9. ———, *Some results on Banach spaces without local unconditional structures*, Compositio Math. **37** (1978), 3–19.

10. H.P. Rosenthal, *On subspaces of L^p*, Ann. of Math. **97** (1973), 344–373.

11. W. Rudin, *Trigonometric series with gaps*, J. Math. Mech. **9** (1960), 203–227.

12. N. Tomczak-Jaegermann, Finite-dimensional operator ideals and Banach-Mazur distances; to be published by Pitman.

13. A. Zygmund, Trigonometric Series, Cambridge University Press (1959).

Department of Mathematics
University of Alberta
Edmonton, Alberta T6G 2G1

Random multiplications, random coverings, multiplicative chaos

by Jean-Pierre Kahane
University of Paris-Sud, Orsay

First Lecture

Introduction

I shall consider three apparently very different questions, and then show how they can be approached in the same way, through multiplication of independent random weights.

First question: random coverings.

There is a beautiful recent paper on random coverings in several dimensions, by Svante Janson (Acta Mathematica, 1986). The general problem is how to cover a big body by small random bodies. In Janson's approach all small bodies are independent and have the same distribution, namely, the distribution of $aX + T$, where $a > 0$, X is a random compact and convex set in $\mathbb{R}^n$, and T is a random vector in $\mathbb{R}^n$, independent of X and equally distributed on an open and bounded set G. His paper is about $N_a(K)$, the (random) number of such sets needed in order to cover a given compact subset K of G; more precisely, his paper treats the asymptotic behaviour of the law of $N_a(K)$ as $a \to 0$.

This paper provides an opportunity to revisit another covering problem considered by A. Dvoretzky in 1956 (references can be found in my book [SRSF]= "Some random series of functions," either first or second edition). We are given a sequence (ℓ_n), $0 < \ell_n < 1$, and, usually, $\ell_n \downarrow 0$. We consider random arcs on the circle $\mathbb{T}$ $(= \mathbb{R}/\mathbb{Z})$, that is, arcs $I_n =]\omega_n, \omega_n + \ell_n[$, where the ω_n are independent random variables equally distributed on $\mathbb{T}$.

A. Dvoretzky asks whether or not $\mathbb{T} = \cup I_n$ a.s. (that is, the I_n cover $\mathbb{T}$) or what is the same, $\mathbb{T} = \overline{\lim}\, I_n$ a.s. (that is, the I_n cover $\mathbb{T}$ infinitely often). Certainly $\sum \ell_n = \infty$ is a necessary condition, since $\sum \ell_n < \infty$ implies meas $\overline{\lim}\, I_n = 0$. Actually $\sum \ell_n = \infty$ implies (Borel–Cantelli) that each given point belongs to $\overline{\lim}\, I_n$ a.s.; therefore meas $\overline{\lim}\, I_n = 1$ a.s., or meas $E = 0$ a.s., where "meas" denotes normalized Haar measure, and $E = \mathbb{T} \setminus \overline{\lim}\, I_n$.

In 1972, L. Shepp gave a necessary and sufficient condition for covering (meaning $E = \phi$ a.s.), specifically

$$\sum \frac{1}{n^2} \exp(\ell_1 + \cdots + \ell_n) = \infty$$

(here it is essential to suppose $\ell_n \downarrow 0$), or, equivalently,

$$\int_0^{1/2} \exp \sum \Delta_n(t)dt = \infty,$$

where $\Delta_n(t) = (\ell_n - |t|) \vee 0$ on $[-1,1]$. In particular, covering occurs when $\ell_n = \frac{a}{n}$ $(n > n_0)$ and $a \geq 1$, and does not occurs when $\ell_n = \frac{a}{n}$ with $a < 1$. In this case it can be proved that $\dim E = 1 - a$ a.s., where dim is the Hausdorff dimension. The necessary part of Shepp's theorem and the calculation of dim E are already in [SRSF, 1968]. We shall see in the course of these lectures how Janson's ideas apply to afford a new and powerful proof of Shepp's Theorem.

There are corresponding problems and results in several dimensions (say, with $\mathbb{T}^d$ or $\mathcal{S}^d$ instead of $\mathbb{T} (= \mathcal{S}^1)$, and random cubes or random caps instead of the I_n). Again dim E can be computed, and the analogue of Shepp's condition is necessary here too (see [SRSF, 1985]). However the "sufficient" part is not proved. There may be some hope that Janson's ideas provide a solution in this setting.

For simplicity (and sometimes necessity) we shall consider only the one-dimensional case in the sequel.

Second question: self-similar cascades.

Here is a model introduced by Benoit Mandelbrot as a possible model of turbulence (J. Fluid Mech., 1974 and Comptes Rendus Ac. Sc. Paris, 1974). A given body is decomposed into c cells, each cell of this decomposition is further decomposed into c subcells, and so on. A probability distribution is given on $\mathbb{R}^+$, and we denote by W a positive random variable with this distribution. We assume $EW = 1$. Writing C for a cell at any step of the decomposition (thus there are c^n cells C introduced at the n-th step), we consider a sequence (W_C) of independent random variables having the given distribution, and the sequence is indexed by the set of all cells.

From now on it is convenient to replace the body by $\{0,1,\dots,c-1\}^{\mathbb{N}}$ or, equivalently, by $I = [0,1[$, with the correspondence

$$(i_0, i_1, i_n, \dots) \longrightarrow \sum_{n=0}^{\infty} i_n c^{-n-1}.$$

Thus, the cells C can be considered either as cylinders in $\{0,1,\dots,c-1\}^{\mathbb{N}}$, or as c-adic intervals (closed to the left, open to the right). Starting with the Lebesgue (or Haar) measure μ_0, we define μ_n as a random measure whose density on a cell C of the n-th step is prescribed by

$$\text{density } \mu_n \text{ on } C = W_C \times \text{density } \mu_{n-1} \text{ on } C \qquad (n = 1,2,\dots)$$

In other words, if we consider a decreasing sequence of cells, $C_0, C_1, \dots, C_n$. (with $|C_j| = c^{-j}$, $|\ |$ denoting the μ_0-measure), the density of μ_n on C_n is $W_{C_0} W_{C_1} \cdots W_{C_n}$.

For each Borel set E in I the sequence $\mu_n(E)$ is a positive martingale, which converges a.s. to a limit $\mu(E)$, where μ is a random measure (these notions will be detailed in the sequel), and we write $Z = \mu(I)$. Here are a few results which were partially guessed by B. Mandelbrot, and then proved by J. Peyrière and me (Adv. Math. 1976)

1. $EZ > 0 \Leftrightarrow EZ = 1 \Leftrightarrow E(W \log W) < \log c$
2. $(h > 1)$ $0 < EZ^h < \infty \Leftrightarrow EW^h < c^{h-1}$
3. Suppose $D = 1 - \dfrac{E(W \log W)}{\log c} > 0$. Then μ is a.s. concentrated on a (random) Borel set of Hausdorff dimension D, and no (random) Borel set of dimension $< D$ carries any part of μ.

In the Kahane-Peyrière paper, 3 is not stated in full generality. We shall see later how it follows from 1 rather easily.

Third question: exponentiation of gaussian processes.

Let us consider a gaussian stationary process of $\mathbb{T}$, i.e.

$$X(t) = c_0\xi_0 + \sum_1^{\infty} c_m(\xi_m \cos 2\pi mt + \xi'_m \sin 2\pi mt)$$

where $\xi_0, \xi_1, \xi'_1, \dots$ denote independent normal r.v. $N(0,1)$ (that is, $Ee^{u\xi} = e^{(u^2)/2}$) and $\sum_1^{\infty} c_m^2 < \infty$. Then the series converges in $L^2(\Omega)$ (Ω is the probability space) for each given t, and converges a.s. for almost every t. The normalized exponential of $X(t)$ is

$$P(t) = e^{X(t)-\frac{1}{2}EX^2(t)}$$

(it is normalized in such a way that $EP(t) = 1$). Question: does the normalized exponential have any meaning when $\sum c_m^2 = \infty$? Can we define limit log-normal processes as B. Mandelbrot suggests (cf. The fractal geometry of nature, 1982) as a correct version of the "log normal hypothesis" stated by A. Kolmogorov in 1961 in a rather sketchy random model of turbulence? (For an overview of the history, see my paper in Ann. Sci. Math. Quebec, 1985.)

We can think of an analogous question. Random trigonometric series have much in common with lacunary trigonometric series, say

$$x(t) \sim \sum_1^\infty a_n \cos 3^n t,$$

with $\sum_1^\infty a_n^2 < \infty$. The exponential $e^{x(t)}$ and the product

$$\prod_1^\infty (1 + a_n \cos 3^n t)$$

(supposing $-1 < a_n < 1$) look very much the same: their quotient is a positive continuous function bounded away from zero. However the infinite product (called "Riesz product") has a meaning even when $\sum a_n^2 = \infty$. Actually, whenever $-1 < a_n < 1$ for each n, the partial products are positive and they converge weakly to a positive measure. If $\sum a_n^2 < \infty$, this measure is absolutely continuous. If $\sum a_n^2 = \infty$, it is singular with respect to Lebesgue measure (A. Zygmund). Generally, if we consider

$$\mu' \sim \prod_1^\infty (1 + a'_n \cos 3^n t) \qquad (-1 < a'_n < 1)$$

$$\mu'' \sim \prod_1^\infty (1 + a''_n \cos 3^n t) \qquad (-1 < a''_n < 1)$$

μ' and μ'' are equivalent (same null-sets) if $\sum_1^\infty (a'_n - a''_n)^2 < \infty$, and mutually singular (concentrated on disjoint Borel sets) if $\sum_1^\infty (a'_n - a''_n)^2 = \infty$ (J. Peyrière). Here is the proof of the last point. Consider

$$\varphi_n(t) = e^{i3^n t} - \frac{a_n}{2} \qquad (n = 1,2,\dots)$$

$$\mu \sim \prod_1^\infty (1 + a_n \cos 3^n t) \qquad (-1 < a_n < 1).$$

One checks easily that the φ_n are orthogonal and bounded in $L^2(\mu)$. Therefore, whenever $\sum_1^\infty c_n^2 < \infty$, $\sum_1^\infty c_n\varphi_n$ converges in $L^2(\mu)$. Apply this to μ' and μ''. Then, for some sequence n_j,

$$\sum_1^{n_j} c_n\varphi'_n \text{ converges } \mu'\text{-almost everywhere,}$$

$$\sum_1^{n_j} c_n\varphi''_n \text{ converges } \mu''\text{-almost everywhere,}$$

therefore, if μ' and μ'' are not mutually singular, the differences $\sum_{1}^{n_j} c_n(\varphi'_n - \varphi''_n)$ converges somewhere, which implies $\sum (a'_n - a''_n)^2 < \infty$.

For example, the measures μ_a corresponding to $a_n \equiv a$ are mutually singular. We may guess that random analogues of the μ_a will be obtained from series

$$\sum_{1}^{\infty} \frac{b}{\sqrt{m}} (\xi_m \cos 2\pi mt + \xi'_m \sin 2\pi mt)$$

instead of

$$\sum_{1}^{\infty} a \cos 3^n t$$

because the average energy of the first in the strip of frequencies $[3^n, 3^{n+1} - 1]$ is

$$b^2 \sum_{3^n}^{3^{n+1}} \frac{1}{m} \approx b^2 \log 3$$

while it is $\frac{a^2}{2}$ for the second series.

The general situation is as follows. We are given independent centered gaussian processes $X_n(t)$ indexed by $t \in T$ (not necessarily $\mathbb{T}$), and we consider the corresponding random weights

$$P_n(t) = \exp(X_n(t) - \frac{1}{2} E\, X_n^2(t)).$$

Can we define $\prod_1^\infty P_n(t)$ in some way?

To arrive at a common approach of the three questions, let us consider a more general situation. We are given independent random weights $P_n(t)$, with the condition $EP_n(t) = 1$ for each t. Then $P_1P_2\cdots P_n(t)$ is a positive martingale for each t. The same is true when we integrate with respect to a given measure $d\sigma(t)$. We look at $\lim_{n\to\infty} \int P_1P_2\cdots P_n d\sigma$ (which exists a.s.). The interesting case is when it is not zero. Then $\lim_{n\to\infty} P_1P_2\cdots P_n d\sigma$ is a random measure and we can try to study its properties.

In the first question, writing χ_n for the indicator function of $[0,\ell_n]$, we consider

$$P_n(t) = \frac{1 - \chi_n(t - \omega_n)}{1 - \ell_n} .$$

Then $P_1\cdots P_n$ is carried by $\mathbb{T}\backslash(I_1 \cup I_2 \cup\cdots\cup I_n)$. If the martingale $\int_{\mathbb{T}} P_1\cdots P_n dt$ converges to a non-zero r.v., the random measure $\lim P_1\cdots P_n dt$ is carried by $\mathbb{T}\backslash\bigcup_1^\infty I_n$, therefore $\mathbb{T}$ is not covered a.s. and moreover we know somehow that $\mathbb{T}\backslash\bigcup_1^\infty I_n$ is large enough to carry $\lim P_1\cdots P_n dt$. Surprisingly this proves to be the necessary and sufficient condition for non-covering. Moreover it suffices to consider the L^2-theory (equivalent in this case to any L^p-theory, $1 \le p \le 2$). Another idea is needed in order to prove that covering occurs when the martingale does not converge in L^2.

In the second question, we take

$$P_n(t) = \sum_C W_C 1_C(t),$$

the sum being extended to all c^n cells of the n-th step. Then

$$\mu_n = \mu_0 P_1 \cdots P_n,$$

and statements 1, 2, 3 answer the following questions:

- when does $\mu_n(I)$ converge in $L^1(\Omega)$?
- when does $\mu_n(I)$ converge in $L^p(\Omega)$?
- when does there exist a random Borel set of dimension d on which the random measure $\lim \mu_n$ is concentrated?

Finally in the third question we already defined the P_n. The mapping $\sigma \to \lim P_1 \cdots P_n \sigma$ (fixed measure $\to$ random measure) is what we call multiplicative chaos.

Second Lecture

Positive martingales and Q-operators

We shall review a few properties of positive martingales. Considering positive martingales $Q_n(t)$ indexed by $t \in T$ (T a locally compact metric space), we shall define the corresponding Q-operator, which maps $\sigma \in M^+(T)$ (space of positive Radon measures on T) into $Q\sigma$, a random Radon measure. Then we shall study the case when $Q_n(t) = P_1(t)\cdots P_n(t)$, a product of independent random weights, and show how it applies in the theory of self-similar cascades.

A short review on martingales.

We are given $(\Omega,\mathcal{A},P)$, a probability space; $(\mathcal{A}_n)_{n\in\mathbb{N}}$, an increasing sequence of sub-σ-fields of $\mathcal{A}$; $(Z_n)_{n\in\mathbb{N}}$, a sequence of real random variables, adapted to $(\mathcal{A}_n)_{n\in\mathbb{N}}$ (i.e. each Z_n is $\mathcal{A}_n$-measurable). The martingale condition is

$$Z_n = E(Z_{n+1}|\mathcal{A}_n) \qquad (n \in \mathbb{N}).$$

Replacing $=$ by $\leq$ (resp., $\geq$) we get submartingales (resp., supermartingales). When $(Z_n,\mathcal{A}_n)$ is a martingale, and $\mathcal{C}_n$ is the σ-field generated by $Z_0,Z_1,\ldots,Z_n$ ($n \in \mathbb{N}$), then $(Z_n,\mathcal{C}_n)$ is also a martingale. This remark allows to say that (Z_n) is a martingale, without mentioning $(\mathcal{A}_n)$. The same for sub- or supermartingales.

In order to write the martingale condition we have to suppose $Z_n \in L^1(\Omega)$. The martingale condition gives $EZ_n = EZ_0$, and

$$Z_n = E(Z_{n+p}|\mathscr{A}_n) \qquad (n \in \mathbb{N},\ p \in \mathbb{N}).$$

This suggests the most important example of martingale:

$$Z_n = E(Z|\mathscr{A}_n)$$

where $Z \in L^1(\Omega)$. Not all martingales can be written in this way. Here is the necessary and sufficient condition.

The L^1-convergence condition. The following are equivalent (TFAE)

(a) $Z_n = E(Z|\mathscr{A}_n)$ for some $Z \in L^1(\Omega)$;

(b) Z_n converges in $L^1(\Omega)$ (actually, to Z);

(c) Z_n is equi-integrable, meaning $\sup E\,\varphi(|Z_n|) < \infty$ for some function $\varphi : \mathbb{R}^+ \longrightarrow \mathbb{R}^+$ such that $\lim\limits_{t\to\infty} \frac{\varphi(t)}{t} = \infty$.

Thus $E|Z_n| = O(1)$ is necessary, but not sufficient. The situation is simpler for L^p $(1 < p < \infty)$.

The L^p-convergence condition. TFAE (given $1 < p < \infty$):

(a) $Z_n = E(Z|\mathscr{A}_n)$ for some $Z \in L^p(\Omega)$;

(b) Z_n converges in $L^p(\Omega)$ (to Z);

(c) (Z_n) is bounded in $L^p(\Omega)$.

What about almost sure convergence?

An a.s.-convergence condition. Suppose (Z_n) bounded in $L^1(\Omega)$ (true if (Z_n) is a positive martingale). Then Z_n converges a.s.

Moreover, when (Z_n) is a positive martingale, L^1-convergence holds if and only if $E \lim Z_n = EZ_0$. The case $\lim Z_n = 0$ a.s. is called degenerate (example: $\Omega = [0,1[,\ Z_n = 2^n 1_{[0,2^{-n}[}$).

A.s. convergence holds also for positive supermartingales.

Positive martingales indexed by T.

Now we are given (T,d), a locally compact metric space, together with $(\Omega,\mathscr{A},P)$. $(\mathscr{A}_n)_{n\in\mathbb{N}}$ is as before. We consider a sequence of random functions $Q_n(t,\omega)$ $(n \in \mathbb{N}, t \in T, \omega \in \Omega)$ such that for each t $(Q_n(t,\cdot))_{n\in\mathbb{N}}$ is a positive martingale adapted to $(\mathscr{A}_n)_{n\in\mathbb{N}}$ (positive means ≥ 0) and for almost all ω the $(Q_n(\cdot,\omega)$ are positive Borel functions on T. For brevity, we write $(Q_n)_{n\in\mathbb{N}}$, and we call the latter a positive T-martingale. We also write $Q_n(t)$ for $Q_n(t,\cdot)$, and

$$q(t) = EQ_n(t). \tag{1}$$

Given $\sigma \in M^+(T)$ (Radon measure on T) we shall look at the random measures $Q_n\sigma$. Here is the first result.

<u>Proposition 1</u>. Suppose $q \in L^1(\sigma)$. Then $Q_n\sigma$ converges weakly a.s. to a random measure S: $Q_n\sigma \xrightarrow{*} S$ a.s.

<u>Proof</u>. Let Φ be a countable family of bounded Borel functions on T. When $\varphi \in \Phi$ the sequence $\int \varphi\, Q_n d\sigma$ is a positive martingale (we use $q \in L^1(\sigma)$), and therefore it converges a.s. to a limit, $S(\varphi)$. Choose $\Phi = \Phi_0$, a dense subset of $C(T)$ (resp., $C_0(T)$) if T is compact (resp., if T is not compact). Clearly a.s. the $Q_n\sigma$ are norm bounded on T (resp., on every compact subset of T); since they converge on Φ_0 they converge also on the whole of $C(T)$ (resp. $C_0(T)$), and the limit S satisfies

$$S(\varphi) = \int \varphi dS \qquad (\varphi \in \Phi). \tag{2}$$

Remark. It is not completely obvious that the last formula holds when we choose $\Phi \supset \Phi_0$ instead of $\Phi = \Phi_0$. However it is true, as we shall check in a moment.

Now we consider the operator $Q : \sigma \to S$, where $\sigma \in M^+(T)$ and $\int q d\sigma < \infty$. We say that Q dies on σ if $S = 0$ (the zero measure on T). We say that Q lives on σ if

$$(3) \qquad E \int \varphi dS = \int \varphi q d\sigma$$

for each $\varphi \in C(T)$ resp. $C_0(T)$. This is the same as

$$(4) \qquad E \int dS = \int q d\sigma.$$

For, considering $\varphi_n \in C(T)$, resp., $C_0(T)$, such that $0 \leq \varphi_n \leq 1$ and $\sum \varphi_n = 1$ (n ∈ finite, resp., countable, set of integers), we see that (3) implies (4) (Beppo-Levi), and (4) together with

$$E \int \varphi_n dS \leq \int \varphi_n q d\sigma$$

(Fatou) implies (3).

Proposition 2. Given $(Q_n)_{n \in \mathbb{N}}$ (a positive T-martingale) and $\sigma \in M^+(T)$ such that $\int q d\sigma < \infty$ there is a unique decomposition of (Q_n) as a sum of two positive T-martingales (Q'_n) and (Q''_n) such that Q'' dies on σ and Q' lives on σ.

Proof Let us write

(5) $$E(S|\mathcal{A}_n) = S'_n,$$

meaning that for each $\varphi \in C(T)$ (let us assume T compact for simplicity)

(6) $$E(\int \varphi dS|\mathcal{A}_n) = \int \varphi dS'_n$$

(we begin as above with $\varphi \in \Phi_0$, then define S'_n as a random measure, then prove (6) when $\varphi \in C(T)$). Clearly $S'_n \le Q_n\sigma$ (Fatou for conditional expectations). Therefore

(7) $$S'_n = Q'_n\sigma, \quad 0 \le Q'_n \le Q_n$$

and $(Q'_n)_{n\in\mathbb{N}}$ is a positive T-martingale. Now (5) implies that $\int \varphi dS'_n$ converges to $\int \varphi dS$ in $L^1(\Omega)$; since $\int \varphi dS'_n$ converges a.s. to $\int \varphi dS'$ $(S' = Q'\sigma)$, we obtain $S = S'$, that is

$$(Q - Q')\sigma = 0.$$

Moreover, writing

(8) $$q'(t) = EQ'_n(t),$$

we get from the L^1-convergence

$$E \lim \int Q'_n d\sigma = \int q' d\sigma.$$

That is, Q' lives on σ.

The remark after proposition 1 is now easy to prove. Suppose $0 \lesssim \varphi \lesssim 1$ (and T compact), and we want to prove (2). Since $S = S'$ we may assume that Q lives on σ. Then we have (3) (same proof), and therefore $\int \varphi Q_n d\sigma$ converges to $\int \varphi dS$ in $L^1(\Omega)$. Since we suppose that $\int \varphi Q_n d\sigma$ converges to $S(\varphi)$ a.s., we have the desired conclusion (2).

Multiplication of random independent weights.

Now suppose that we are given independent normalized random weights $P_n(t,\omega)$ ($t \in t, \omega \in \Omega, n \in \mathbb{N}$). That is

(1) for almost all ω, the $P_n(\cdot,\omega)$ are positive Borel functions

(2) for all t, the $P_n(t,\cdot)$ are positive r.v., with expectation one

(3) the σ-fields generated by the $(P_n(t,\cdot))_{t\in T}$ are independent.

Writing

$$Q_n(t,\omega) = \prod_{m=0}^{n} P_m(t,\omega)$$

we have a T-martingale and $q(t) \equiv 1$. Again we write (P_n) and $P_n(t)$. Here we can improve Proposition 2.

<u>Proposition 3</u>. Given $(P_n)_{n\in\mathbb{N}}$ and $\sigma \in M^+(T)$, there exists a Borel set B on T such that, writing

(9) $\sigma' = 1_B\sigma, \qquad \sigma'' = (1 - 1_B)\sigma$

Q lives on σ' and dies on σ''. Moreover

$$(10) \qquad \sigma' = EQ\sigma,$$

and the operator EQ (which maps σ on σ') is a projection.

Proof. Given n, the T-martingale

$$Q_m^{(n)} = P_{n+1}P_{n+2}\cdots P_{n+m} \qquad (m = 1,2,\dots)$$

defines the operator $Q^{(n)}$, and clearly

$$\begin{aligned} E(Q\sigma|\mathcal{A}_n) &= E(P_1\cdots P_n Q^{(n)}\sigma|\mathcal{A}_n) \\ &= Q_n E(Q^{(n)}\sigma). \end{aligned}$$

By taking (5), (7), (8) into account, we see that the expectation of the first member is $q'\sigma$. Therefore $E(Q^{(n)}\sigma) = q'\sigma$, and

$$E(Q\sigma|\mathcal{A}_n) = Q_n q'\sigma,$$

that is, $Q_n' = q'Q_n$ and $Q_n'' = (1 - q')Q_n$ in the notation of Proposition 2. For every Borel set A in T

$$(11) \qquad \begin{cases} \lim_{n\to\infty} \int_A q'Q_n d\sigma = \int_A dS & \text{(a.s. and in } L^1(\Omega)) \\ \lim_{n\to\infty} \int_A (1 - q')Q_n d\sigma = 0 & \text{(a.s.)} \end{cases}$$

It follows that the t-set where $q'(t)(1 - q'(t)) > 0$ has zero σ-measure, therefore $q' = 1_B$ for some Borel set B. Writing (9) as a definition of σ' and σ'', (11) means that Q lives on σ' and dies on σ''. Moreover we already observed that $q'\sigma = EQ\sigma$, and clearly $EQ\sigma' = \sigma'$; therefore EQ is a projection.

A particular case: the Q_α's.

Let us fix c (integer ≥ 2) and consider the self-similar cascade attached to $W = W_\alpha$ with the following two-valued distribution:

$$\begin{cases} P(W = c^\alpha) = c^{-\alpha} & (\alpha > 0); \\ P(W = 0) = 1 - c^{-\alpha}. \end{cases}$$

At each step we kill a cell C when $W_C = 0$ and we keep it when $W_C = c^\alpha$. The population of living cells at time n is a birth and death process. The limit case is $\alpha = 1$, and the population disappears a.s. whenever $\alpha \geq 1$, and lives forever with a positive probability when $\alpha < 1$. This is also a "covering" problem (if we want to cover the whole space with dead cells).

Let us write Q_α for the operator associated with c and W_α with $0 < \alpha < 1$. We claim that Q_α lives on measures σ which are regular enough (we call them α-regular) and dies on measures which are singular in a strong sense (we call them α-singular). Then EQ_α is an α-regularizing operator in the sense that it keeps the α-regular part of a measure and kills its α-singular part. It is possible to see, in many ways, that EQ_α is a decreasing family of projections when α increases from 0 to 1. In other words, the set S_α of α-singular measures increases and the set R_α of α-regular measures decreases.

One way to see that S_α increases is to use the formula

$$Q_{\alpha+\beta} = Q_\alpha Q_\beta \qquad\qquad (0 < \alpha + \beta < 1)$$

meaning that you can obtain $Q_{\alpha+\beta}$ (that is, a random operator with the same law) by considering two independent operators, Q_α and Q_β, letting Q_β act first and then Q_α. This is due to the fact that

$$W_{\alpha+\beta} = W_\alpha W_\beta$$

(equality in law, when W_α and W_β are independent). As a consequence $Q_{\alpha+\beta}$ dies whenever Q_β dies, that is, $S_{\alpha+\beta} \supset S_\beta$.

Another way is just to investigate what $\sigma \in S_\alpha$ means. Here is a simple result. Let us write $\text{meas}_\alpha \sigma < \infty$ when σ is concentrated on a Borel set B such that $\text{meas}_\alpha B < \infty$ ($\text{meas}_\alpha B$ is the Hausdorff measure in dimension α). Then

$$\text{meas}_\alpha \sigma < \infty \Rightarrow \sigma \in S_\alpha \Rightarrow \text{meas}_{\alpha+\epsilon} \sigma < \infty$$

for each $\epsilon > 0$. The proof is not very long, but nevertheless I prefer to skip it. The reference is "multiplications aleatoires et dimensions de Hausdorff", to appear in Annales de l'Institut Henri Poincaré.

An application to self-similar cascades.

Let us show how statement 3 in Lecture 1 (about Hausdorff dimensions of Borel sets which carry the random measure μ) can be derived from statement 1 (that is $EZ > 0 \Leftrightarrow EZ = 1 \Leftrightarrow E(W \log W) < \log c$).

For given W (a general positive random variable with expectation 1), and $\sigma = \mu_0$ (the Haar measure on the space $(\mathbb{Z}/c\mathbb{Z})^{\mathbb{N}}$), statement 1 provides that Q dies on σ if $E(W \log W) \geq c$ and lives on σ if $E(W \log W) < c$.

Now let us consider Q_α as above, independent of Q. The operator $Q_\alpha Q$ is associated with the weight $W_\alpha W$, and a simple calculation shows

$$\begin{aligned} E(W_\alpha W)\log(W_\alpha W)) &= E(W \log W) + E(W_\alpha \log W_\alpha) \\ &= E(W \log W) + \alpha \log c \end{aligned}$$

$Q\sigma$ is α-singular if $Q_\alpha Q\sigma = 0$ a.s., that is

$$E(W \log W) + \alpha \log c \geq \log c.$$

$Q\sigma$ is α-regular when the reverse is true. Writing $D = 1 - \dfrac{E(W \log W)}{\log c}$, we have that $Q\sigma$ is concentrated on Borel sets of finite $(D + \epsilon)$-measure whenever $\epsilon > 0$ (therefore, on a Borel set of dimension D), and it is not concentrated on any Borel set of dimension $< D$.

Third Lecture

Covering the circle

We consider the following covering problem (see question 1 in the first lecture). We are given a sequence (ℓ_n) $(0 < \ell_n < \frac{1}{2})$ and a closed subset of $\mathbb{T}$, F. We try to decide whether or not $F \subset \overline{\lim}\, I_n$ a.s. (the I_n have indicator functions $\chi_n(t - \omega_n)$ and $\int \chi_n(t)dt = \ell_n$), and when F is not covered we shall investigate the uncovered set $F \setminus \cup I_n$ or the finitely covered set $F \setminus \overline{\lim}\, I_n$.

The answer involves the following kernel

$$k(t) = \exp \sum (\ell_n - |t|)^+ \qquad (|t| < \tfrac{1}{2});$$

we also consider k as defined on $\mathbb{T}$. We shall prove that the necessary and sufficient condition for $T = \overline{\lim}\, I_n$ a.s. is $k \notin L^1(\mathbb{T})$. For $F \subset \overline{\lim}\, I_n$ a.s. the corresponding necessary and sufficient condition is $\mathrm{Cap}_k F = 0$, meaning that F carries no probability measure of finite energy with respect to k. Finally, when $\mathrm{Cap}_k F > 0$ and we write $F \setminus \cup I_n = F'$,

$$P(\mathrm{Cap}_{k'} F' > 0) > 0 \Longleftrightarrow \mathrm{Cap}_{kk'} F > 0.$$

There are three ideas: L^2-martingales, stopping time for Poisson coverings, and independent covering families. Only the second one is new (it is used in a recent paper of Svante Janson, 1983[1]); it proves to be the right thing in order to achieve the theory. Therefore, the corresponding multi-dimensional results may be much more difficult to prove.

---

[1]Random coverings of the circle with arcs of random lengths. Probability and Mathematical Statistics Essays in honour of Carl-Gustav Esseen, 1983, pp. 62-73.

L^2-martingales. A sufficient condition for non-covering.

Suppose $\mathrm{Cap}_k F > 0$, that is,

$$(1) \qquad \iint_{F^2} k(t-s)\,d\sigma(t)\,d\sigma(s) < \infty$$

for some $\sigma \in M_1^+(F)$ (probability measure carried by F). Defining

$$P_n(t) = \frac{1 - \chi_n(t-\omega_n)}{1 - \ell_n}$$

we get a random operator Q. Let $S = Q\sigma = \lim Q_n\sigma$ a.s. Let us prove that L^2-convergence holds. It is enough to show

$$(2) \qquad E\left(\int Q_n\,d\sigma\right)^2 = O(1).$$

Now

$$\begin{aligned} E\left(\int Q_n\,d\sigma\right)^2 &= E\iint Q_n(t)Q_n(s)\,d\sigma(t)\,d\sigma(s) \\ &= \iint E\,Q_n(t)Q_n(s)\,d\sigma(t)\,d\sigma(s) \\ &= \iint \prod_0^n E\,P_m(t)P_m(s)\,d\sigma(t)\,d\sigma(s) \end{aligned}$$

$$EP_m(t)P_m(s) = (1-\ell_m)^{-2}(1 - 2\ell_m + \Delta_m(t-s))$$

where $\Delta_n = \chi_n * \tilde{\chi}_n$, that is $\Delta_n(t) = (\ell_n - |t|)^+$ when $|t| < \frac{1}{2}$. A simple calculation shows

$$EP_m(t)P_m(s) = \exp \Delta_m(t-s) + O(\ell_m^2).$$

Therefore (1) together with $\sum \ell_m^2 < \infty$ provides (2). Now (1) implies $k \in L^1(\mathbb{T})$ (which corresponds to $d\sigma(t) = dt$) and $k \in L^1(\mathbb{T})$ implies $\log k \in L^1(\mathbb{T})$, that is $\sum \ell_m^2 < \infty$. Therefore (1) implies (2). Hence $E(S(F))^2 > 0$, which proves

$$P(F \not\subset \cup I_n) > 0;$$
$$P(F \not\subset \overline{\lim}\, I_n) = 1.$$

Moreover, given another kernel k' (associated with another sequence (ℓ'_n)) and assuming

$$(3) \qquad \iint (kk')(t - s)d\sigma(t)d\sigma(s) < \infty,$$

we obtain

$$(4) \qquad \iint k'(t - s)dS(t)dS(s) < \infty \quad \text{a.s.}$$

(actually, the first member is in $L^1(\Omega)$). Therefore the uncovered set F' carries a non-zero measure of finite energy with respect to k', with a positive probability, that is

$$(5) \qquad \mathrm{Cap}_{kk'} F > 0 \Rightarrow P(\mathrm{Cap}_{k'} F' > 0) > 0.$$

If we start with F with a positive Lebesgue measure, we can choose $d\sigma(t) = dt$. Then condition (1) reads

$$(6) \qquad k \in L^1(\mathbb{T}) \Rightarrow P(\mathbb{T} \not\subset \cup I_n) > 0$$

and (5) reads

(7) $kk' \in L^1(\mathbb{T}) \Rightarrow P(\mathrm{Cap}_{k'}(\mathbb{T} \setminus \bigcup I_n) > 0) > 0.$

Now we intend to prove the converse, that is

(8) $k \notin L^1(\mathbb{T}) \quad \Rightarrow P(\mathbb{T} \subset \bigcup I_n) = 1$

(9) $kk' \notin L^1(\mathbb{T}) \Rightarrow P(\mathrm{Cap}_{k'}(\mathbb{T} \setminus \bigcup I_n) = 0) = 1$

(10) $\mathrm{Cap}_k F = 0 \quad \Rightarrow P(F \subset \bigcup I_n) = 1$

(11) $\mathrm{Cap}_{kk'} F = 0 \Rightarrow P(\mathrm{Cap}_{k'}(F \setminus \bigcup I_n) = 0) = 1.$

There will be a number of steps. Before going to the hard work, let us point out how (9) and (11) derive from (8) and (10). Let us suppose that (8) is proved, and consider the random family of intervals consisting of (I_n), associated with (ℓ_n), and (I'_m) associated with (ℓ'_m), (I_n) and (I'_m) being independent families; let us call it (I_n, I'_m). Assuming $kk' \notin L^1(\mathbb{T})$, we know that $\mathbb{T}$ is covered a.s. by (I_n, I'_m); therefore $\mathbb{T} \setminus \bigcup I_n$ is covered by (I'_m) a.s. The necessary condition for covering applies, and gives $\mathrm{Cap}_{k'}(\mathbb{T} \setminus \bigcup I_n) = 0$ a.s., proving (9). In the same way (11) is proved using (10) and the auxiliary family (I'_m).

We then have to prove (8) and (10). For a technical reason we shall assume

(12) $\forall p > 1, \ \sum \ell_n^{\,p} < \infty.$

This is not a restriction because it is known, and easy to prove, that covering holds when (12) fails (SRSF 1985 p. 145).

First step. Reduction to a Poisson covering of the line.

Here is what we call a Poisson covering of the line. We are given μ, a positive locally bounded measure on $]0,\infty[$, and we equip the open half plane $\mathbb{R} \times]0,\infty[$ with the measure $\nu = dt \otimes \mu$. Let $(P_n) = (X_n, Y_n)$ be the Poisson point process associated with ν, that is, a random discrete set in $\mathbb{R} \times]0,\infty[$ (the ordering does not matter) such that the number of points P_n in a fixed Borel set B is a Poisson random variable with parameter $\nu(B)$, and the numbers of points in B and B' are independent if $B \cap B' = \phi$. We consider the random intervals

$$J_n =]X_n, X_n + Y_n[.$$

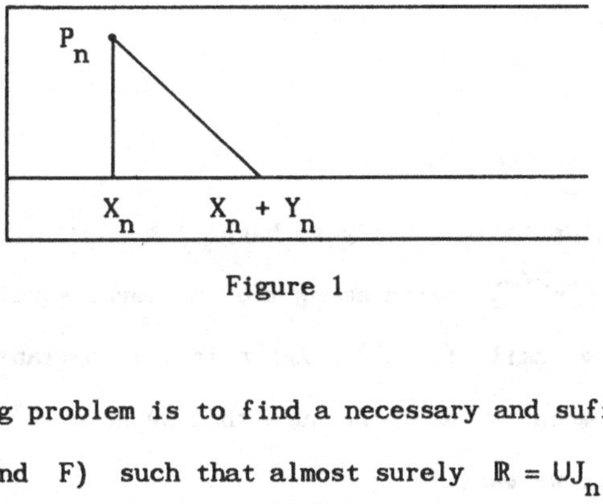

Figure 1

The Poisson covering problem is to find a necessary and sufficient condition on μ (resp. μ and F) such that almost surely $\mathbb{R} = \cup J_n$ (resp. $F \subset \cup J_n$). Surprisingly such a condition is not difficult to write, and we shall return to the question later. For the time being, a sufficient condition for covering will be enough, as will be clear in a moment.

We are given (ℓ_n) such that $k \notin L^1(\mathbb{T})$ (resp. $\overline{\text{Cap}}_k F = 0$), and we want to prove covering (the second members of (8) resp. (10)). Certainly covering occurs for (ℓ_n) if it occurs for (ℓ'_n) with $\ell'_n \le \ell_n$ $(n = 1,2,\dots)$. Let us assume $\ell_1 > \ell_2 > \cdots$ and define (ℓ'_n) in the following manner:

$$\ell'_1 = \ell_1 = \lambda_1$$
$$\ell'_2 = \ell'_3 = \ell_3 = \lambda_2$$
$$\ell'_4 = \ell'_5 = \ell'_6 = \ell_6 = \lambda_3$$

....

$$\ell'_{\frac{m(m-1)}{2}+1} = \cdots = \ell'_{\frac{m(m+1)}{2}} = \ell_{\frac{m(m+1)}{2}} = \lambda_m$$

....

Thus the sequence (ℓ'_n) takes the value λ_m m times. It is important to check that $\sum(\ell_n - \ell'_n) < \infty$, so that $k' \notin L^1(\mathbb{T})$ (resp $\mathrm{Cap}_{k'} F = 0$). Actually

$$\sum(\ell_n - \ell'_n) \le \sum m(\lambda_{m-1} - \lambda_m) = \sum \lambda_m.$$

and (12) imples

(13) $$\forall\, p > 1, \ \sum m\,\lambda_m^p < \infty.$$

Therefore, using Hölder's inequality we have $\sum \lambda_m < \infty$.

Now we remove $[m^{2/3}]$ terms among the m terms equal to λ_m. We obtain a new sequence and we call it (ℓ''_n). Again it is important to check that we did not remove too much, so as to be sure that we have $k'' \notin L^1(\mathbb{T})$ (resp. $\mathrm{Cap}_{k''}(F) = 0$). What we want is

$$\sum m^{2/3} \lambda_m < \infty,$$

and again we get it from (13) and Hölder's inequality in the form

$$\sum m^{(1/p)-(2/q)} \lambda_m \le (\sum m\lambda_m^p)^{1/p}(\sum m^{-2})^{1/q}.$$

Let us define $\mu = \sum \delta_{\ell''_n}$.

I say that the a.s. covering of the line by the corresponding J_n (resp. the covering of F by the J_n) implies the a.s. covering of $\mathbb{T}$ (resp. F) by the I'_n (corresponding to (ℓ'_n)), therefore by the I_n (corresponding to (ℓ_n)).

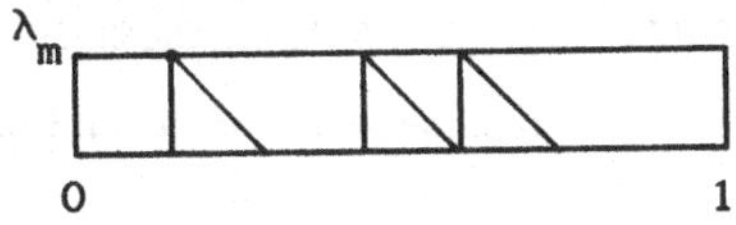

Figure 2

Suppose $F \subset]0,1[$. Assuming $F \subset \cup J_n$ a.s., F is covered a.s. by small J_n (say, $|J_n| < \varepsilon$ = distance from F to the 2-point set $\{0,1\}$). Consider the J_n with length λ_m whose origins X_n lie in $]0,1[$; their number is a Poisson variable $\mathcal{P}_m$, and the X_n can be considered (as soon as $\mathcal{P}_m$ is fixed) as $\mathcal{P}_m$ independent random variables equally distributed on $]0,1[$ (that is actually a way to define a Poisson process). Moreover the assemblage of J_n with length λ_m and all other such assemblages are independent. Let us observe that $E\mathcal{P}_m = m - [m^{2/3}]$.

Now (using the central limit theorem, for example) we have:

$$\sum P(\mathcal{P}_m > m) < \infty.$$

Therefore $\mathcal{P}_m \le m$ for large m, a.s. Therefore we can construct the I'_n (for each m, m random intervals of length λ_m) when m is large enough (so that $\mathcal{P}_m < m$) by choosing first the J_n (for each m, $\mathcal{P}_m$ random intervals of length λ_m), then, independently, $m - \mathcal{P}_m$ random intervals of length λ_m. Therefore $P(F \subset \cup J_n) = 1$ implies $P(F \subset \cup I'_n) = 1$. This holds in particular when F is a closed interval, and we finally obtain

$$P(F \subset \cup J_n) = 1 \Rightarrow P(F \subset \cup I_n) = 1;$$

$$P(\mathbb{R} = \cup J_n) = 1 \Rightarrow P(\mathbb{T} = \cup I_n) = 1.$$

These implications were the purpose of the reduction. The first idea of such a reduction is due to Benoit Mandelbrot (Z. Wahrsch. 22, 1972, 158-160).

Second step. A sufficient condition for Poisson covering (case of $\mathbb{R}$).

We consider a general measure μ, and we ask for a sufficient condition to have

$$\mathbb{R} = \cup J_n \quad \text{a.s.}$$

We use here Svante Janson's idea. Let us write

$$G_\varepsilon = \cup J_n \quad (|J_n| \geq \varepsilon)$$

$$I_\varepsilon = E \int_0^\infty e^{-t} 1_{t \notin G_\varepsilon} \, dt.$$

We shall compute I_ε in two ways.

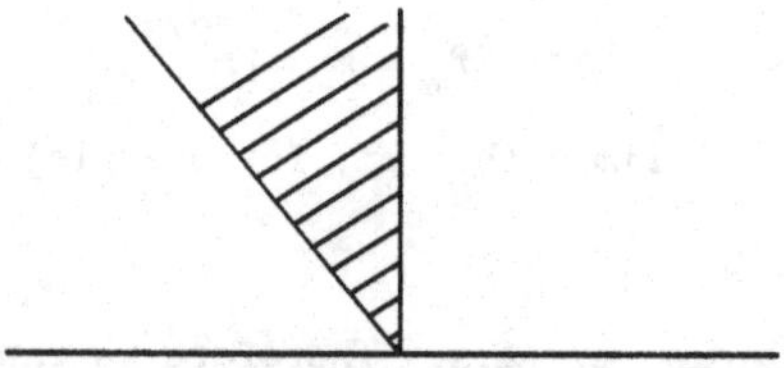

Figure 3

First way: $P(t \in G_\varepsilon)$ is the probability to have no point (X_n, Y_n) with $Y_n \geq \varepsilon$ in the angle represented in the figure. It does not depend on t, and its value is

$$\exp(-\nu_\varepsilon(\text{angle} ◥)),$$

where $\nu_\varepsilon = dt \otimes \mu_\varepsilon$ and $\mu_\varepsilon = 1_{[\varepsilon,\infty[}\mu$. Moreover,

$$\nu_\varepsilon\,(\text{angle} ◥) = \int_0^\infty y\,d\mu_\varepsilon(y) \quad (= \int_\varepsilon^\infty \mu(y,\infty)dy).$$

Therefore

$$(14) \qquad I_\varepsilon = \exp(-\int_0^\infty y\,d\mu_\varepsilon(y)).$$

Second way. We introduce

$$\tau_\varepsilon = \inf(t \geq 0, t \notin G_\varepsilon).$$

It is a stopping time, when we define the σ-field of the past of t as generated by the Poisson process associated with $\nu_{(t)}$, the restriction of ν to $]-\infty,t[\times]0,\infty[$. We write

$$I_\varepsilon = Ee^{-\tau_\varepsilon} \int_0^\infty e^{-s} 1_{\tau_\varepsilon + s \notin G_\varepsilon}\, ds$$

$$= E(e^{-\tau_\varepsilon}\, E(\int_0^\infty \,|\tau_\varepsilon)).$$

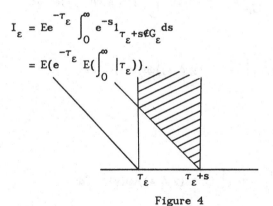

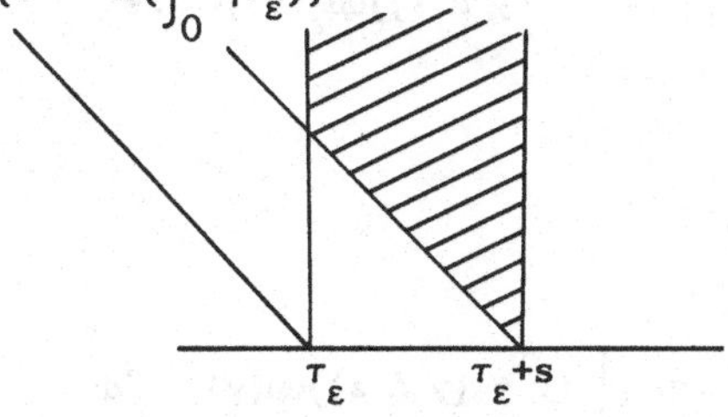

Figure 4

Actually the conditional expectation does not depend on τ_ε, because

$$P(\tau_\varepsilon + s \notin G_\varepsilon | \tau_\varepsilon)$$

is the probability to have no (X_n, Y_n) with $Y_n \geq \varepsilon$ in the domain represented on Figure 4, i.e.

$$\exp(-\nu_\varepsilon($$  $$)$$

and

$$\nu_\varepsilon($$  $$) = \int_0^\infty (y \wedge s) d\mu_\varepsilon(y);$$

hence

(15) $$I_\varepsilon = E(e^{-\tau_\varepsilon}) \int_0^\infty \exp(-\int_0^\infty (y \wedge s) d\mu_\varepsilon(y)) e^{-s} ds.$$

Now we use (14) and (15). By eliminating I_ε we obtain

$$\frac{1}{E(e^{-\tau_\varepsilon})} = \int_0^\infty \exp(\int_0^\infty y d\mu_\varepsilon(y)) \exp(-\int_0^\infty (y \wedge s) d\mu_\varepsilon(y)) e^{-s} ds$$

$$= \int_0^\infty \exp\left(\int_0^\infty (y - (y \wedge s)) d\mu_\varepsilon(y)\right) e^{-s} ds;$$

therefore

$$\lim_{\varepsilon \to 0} \frac{1}{E(e^{-\tau_\varepsilon})} = \int_0^\infty \exp\left(\int_0^\infty (y - (y \wedge s)) d\mu(y)\right) e^{-s} ds.$$

The interior integral can be written as $\int_s^\infty \mu(y,\infty)dy$. In order to have $\mathbb{R} = \cup J_n$ a.s., it is enough to have $\lim_{\varepsilon\to 0} E(e^{-\tau_\varepsilon}) = 0$. Therefore a sufficient condition is

$$\text{(16)} \qquad \int_0^\infty \exp\left(\int_s^\infty \mu(y,\infty)dy\right)e^{-s}ds = \infty.$$

Actually (16) is Shepp's necessary and sufficient condition for a.s. covering of $\mathbb{R}$. When $\mu(1,\infty) = 0$ it reads

$$\text{(17)} \qquad \int_0^1 \exp\left(\int_s^1 \mu(y,1)dy\right)ds = \infty.$$

In particular, if we choose

$$\mu = \sum \delta_{\ell_n},$$

then (17) reads

$$\int_0^1 \exp \sum(\ell_n - s)^+ ds = \infty,$$

which is exactly $k \notin L^1(\mathbb{T})$. If, as we should do after the first step, we choose $\mu = \sum \delta_{\ell''_n}$, (17) reads $k'' \notin L^1(\mathbb{T})$, a consequence of $k \notin L^1(\mathbb{T})$.

This finishes the story for $P(\mathbb{T} \subset \cup I_n)$.

Third step. A sufficient condition for the Poisson covering (case of F).

Here F will denote a closed subset of the circle $\mathbb{T}$, and $\bar{F}$ the corresponding 1-periodic set on $\mathbb{R}$. In the same way, σ_ε will be a positive measure carried by F, and $\bar{\sigma}_\varepsilon$ the corresponding 1-periodic measure on $\mathbb{R}$. Sometimes $\mathbb{T}$ is identified with a subset of $[0.,1[$ (then $F \subset [0,1[$ and $\sigma \in M^+[0,1[$).

Let us suppose $\mu(\frac{1}{2},\infty) = 0$. For $|t| \le \frac{1}{2}$, we write

$$k_\varepsilon(t) = \exp\Big(\int_{|t|}^{1/2} \mu_\varepsilon(y,\tfrac{1}{2})dy\Big).$$

This is a positive, continuous, even function, convex on $[0,\frac{1}{2}]$. Moreover, as $\varepsilon \downarrow 0$, $k_\varepsilon(t)$ increases to

$$k(t) = \exp\Big(\int_{|t|}^{1/2} \mu(y,\tfrac{1}{2})dy\Big)$$

(the notation is consistent with our preceding notation when $\mu = \sum \delta_{\ell_n}$).

Let us assume $\mathrm{Cap}_k F = 0$. Then $\mathrm{Cap}_{k_\varepsilon} F \downarrow 0$ as $\varepsilon \downarrow 0$. From potential theory on the line (see for example Kahane and Salem, Ensembles parfaits et series trigonométriques, Chap. III), there exists an equilibrium measure on F with respect to $k_\varepsilon^* = k_\varepsilon - 1$, that is, a probability measure $\sigma_\varepsilon \in M_1^+(F)$ such that (considering k_ε^* as now defined on $\mathbb{T}$)

$$\forall t \in F, \quad \int k_\varepsilon^*(t-s)d\sigma_\varepsilon(s) = (\mathrm{Cap}_{k_\varepsilon^*} F)^{-1} (= A_\varepsilon).$$

It will be convenient to introduce

(18) $$\bar{F}_\varepsilon^+ = \{t \in \bar{F} : \int_0^{1/2} k_\varepsilon^*(s)d\bar{\sigma}_\varepsilon(t+s) \geq \frac{1}{2} A_\varepsilon\};$$

$$\bar{F}_\varepsilon^- = \{t \in \bar{F} : \int_0^{1/2} k_\varepsilon^*(s)d\bar{\sigma}_\varepsilon(t-s) \geq \frac{1}{2} A_\varepsilon\};$$

$$G_\varepsilon = \cup\,]X_n, X_n + Y_n[\qquad (Y_n > \varepsilon).$$

then $\bar{F} = \bar{F}_\varepsilon^+ \cup \bar{F}_\varepsilon^-$. In order to prove $P(\bar{F} \subset \cup J_n) = 1$, it is enough to show that, for each $\ell > 0$, $P(\bar{F}_\varepsilon^+ \cap [0,\ell] \not\subset G_\varepsilon)$ tends to 0 as $\varepsilon \to 0$ (because the same is true for $\bar{F}_\varepsilon^-$, by reversing the orientation of the line).

Now we introduce our stopping time

(19) $$\tau_\varepsilon = \inf(t \geq 0,\ t \in \bar{F}_\varepsilon^+, t \notin G_\varepsilon),$$

and the new integral

$$I_\varepsilon = E\int_0^\infty e^{-t} 1_{t \notin G_\varepsilon} d\bar{\sigma}_\varepsilon(t).$$

The first way of computing I_ε gives

(20) $$I_\varepsilon = \exp(-\int_0^\infty y d\mu_\varepsilon(y)) \int_0^\infty e^{-t} d\bar{\sigma}_\varepsilon(t).$$

The second way gives

$$I_\varepsilon = E(e^{-\tau_\varepsilon} \int_0^\infty e^{-s} P(\tau_\varepsilon + s \notin G_\varepsilon | \tau_\varepsilon) d\bar{\sigma}_\varepsilon(\tau_\varepsilon + s))\ ;$$

$$(21) \qquad I_\varepsilon = E\left(e^{-\tau_\varepsilon}\int_0^\infty e^{-s}\exp\left(-\int_0^\infty (y \wedge s)\,d\mu_\varepsilon(y)\right)d\bar{\sigma}_\varepsilon(\tau_\varepsilon + s)\right).$$

We remark that

$$(e-1)^{-1} \le \int_0^\infty e^{-t}\,d\bar{\sigma}_\varepsilon(t) \le e(e-1)^{-1} = \gamma\ .$$

After this is taken into account, elimination of I_ε between (20) and (21) gives

$$E\left(e^{-\tau_\varepsilon}\int_0^\infty e^{-s}\exp\left(\int_0^\infty (y - y \wedge s)\,d\mu_\varepsilon(y)\right)d\bar{\sigma}_\varepsilon(\tau_\varepsilon + s)\right) \le \gamma.$$

The interior integral has already been computed. It is

$$\int_s^\infty \mu_\varepsilon(y,\infty)\,dy;$$

therefore its exponential is $k_\varepsilon(s)$. Taking (18) into account together with $\tau_\varepsilon \in \bar{F}^+_\varepsilon$ (see (19)), we obtain

$$E(e^{-\tau_\varepsilon}) \times \frac{1}{2} A_\varepsilon \le \gamma.$$

Therefore $\lim_{\varepsilon\to 0} E(e^{-\tau_\varepsilon}) = 0$. Since

$$P(\bar{F}^+_\varepsilon \cap [0,\ell] \not\subset G_\varepsilon) = P(\tau_\varepsilon < \ell),$$

and we have just proved that this tends to 0 as $\varepsilon \to 0$, we have finished the proof of

$$\mathrm{Cap}_k F = 0 \Rightarrow P(\bar{F} \subset \cup J_n) = 1.$$

We supposed $\mu(\frac{1}{2}, \infty) = 0$ and wrote

$$k(t) = \int_{|t|}^{\infty} \mu(y,\infty)dy = \int_{|t|}^{1/2} \mu(y,\tfrac{1}{2})dy \quad \text{for} \quad |t| \le \tfrac{1}{2}.$$

If now

$$\mu = \sum \delta_{\ell''_n}$$

as at the end of the first step, we should write k'' instead of k, in order to be consistent with the previous notation. We obtain

$$\mathrm{Cap}_{k''} F = 0 \Rightarrow P(F \subset \cup I'_n) = 1 \Rightarrow P(F \subset \cup I_n) = 1,$$

and we already observed that

$$\mathrm{Cap}_k F = 0 \Rightarrow \mathrm{Cap}_{k''} F = 0.$$

Therefore

$$\mathrm{Cap}_k F = 0 \Rightarrow P(F \subset \cup I_n) = 1,$$

which is what we had to prove.

Let us summarize the results we obtained for random coverings of the circle or part of the circle by random arcs I_n of given lengths ℓ_n. The kernel associated with the sequence (ℓ_n) is defined at the beginning of the lecture. The following theorem explicitly expresses the reversibility of the implications (8) to (11).

Theorem. Let F be a closed subset of $\mathbb{T}$. Let $k(t)$ be the kernel associated with the sequence (ℓ_n), and let $k'(t)$ be another kernel, associated with another sequence. Then

$$k \in L^1(\mathbb{T}) \Leftrightarrow P(\mathbb{T} \subset \cup I_n) = 1;$$

$$kk' \in L^1(\mathbb{T}) \Leftrightarrow P(\mathrm{Cap}_{k'}(\mathbb{T} \setminus \cup I_n) = 0) = 1;$$

$$\mathrm{Cap}_k F = 0 \Leftrightarrow P(F \subset \cup I_n) = 1;$$

$$\mathrm{Cap}_{kk'} F = 0 \Leftrightarrow P(\mathrm{Cap}_{k'}(F \setminus \cup I_n) = 0) = 1.$$

Fourth Lecture

More on random coverings.

Introduction to the case of log-normal weights

The first part of the lecture deals with random coverings and decompositions of measures into regular and singular parts.

The second part of the lecture gives a few formulas and the L^2-theory for multiplicative chaos arising from log-normal weights.

I. Decomposition of measures and random coverings.

Suppose that $k(t)$ is given as in lecture 3, that is,

$$k(t) = \exp \int_{|t|}^{\infty} \mu(y,\infty)dy,$$

where μ is a locally bounded positive measure on $]0,\infty[$, and, for practical purposes (in order to have $k(t) = 1$ when $|t| \geq \frac{1}{2}$) we shall assume $\mu(\frac{1}{2},\infty) = 0$. Lecture 3 considers the particular case

$$k(t) = \exp \sum(\ell_n - |t|)^+$$

corresponding to $\mu = \sum \delta_{\ell_n}$. In the general situation, k is nothing but the exponential of an even function, convex and decreasing to zero between 0^+ and $\frac{1}{2}$. This is known be to be a good potential kernel on the line or the circle (on the circle, we consider the kernel $k(t)1_{[-½,½]}(t)$).

Let us consider a bounded positive measure σ on $\mathbb{T}$ or $\mathbb{R}$. Let us say that it is k-regular (written $\sigma \in R_k$) if it is a countable union of positive measures of finite k-energy, and let us call it k-singular if it is orthogonal to R_k (written $\sigma \in S_k$). Each $\sigma \in M^+$ (class of bounded positive measures) can be decomposed in a unique way as

$$\sigma = \sigma' + \sigma'', \quad \sigma' \in R_k, \quad \sigma'' \in S_k.$$

Actually random covering will provide a proof. The direct way is to consider the largest measure $\leq \sigma$ and contained in R_k ; this is σ', and clearly $\sigma'' \in S_k$. Let us remark that

$$\sigma \in S_k \Leftrightarrow \sigma\{t : \int k(t - s)d\sigma(s) < \infty\} = 0.$$

The implication $\Rightarrow$ is obvious (for σ restricted to the t-set in question is in R_k). In the opposite direction, let us suppose that σ is concentrated on

$$E = \{t : \int k(t - s)d\sigma(s) = \infty\}.$$

If we had $\text{Cap}_k E > 0$, there would exist a compact $F \subset E$, $\text{Cap}_k F > 0$, and, by definition of capacity, a $\rho \in M^+(F)$ with $\int d\rho > 0$ and

$$\int k(t - s)d\rho(t) \leq 1.$$

Hence

$$\infty = \iint k(t - s)d\sigma(s)d\rho(t) \leq \int d\sigma,$$

a contradiction. Therefore $\text{Cap}_k E = 0$, and consequently $\sigma \in S_k$.

Let us remark also that σ belongs to S_k if and only if σ is concentrated on a Borel set of zero k-capacity.

Of course this can be done in a more general context, with a locally compact space T and a positive and symmetric kernel k(t,s).

Given a sequence (ℓ_n) we introduced the weights

$$P_n(t) = \frac{1 - \chi_n(t - \omega_n)}{1 - \ell_n}$$

and the corresponding operator Q. The L^2-theory says that Q lives on measures of finite k-energy, and therefore on k-regular measures. The detour through Poisson intervals and the use of Svante Janson's idea gave almost sure covering for closed sets of zero k-capacity. Therefore Q dies on k-singular measures. The operator EQ (from $M^+(T)$ to $M^+(T)$) is "k-regularising": it keeps the k-regular part of a measure and kills its k singular part. In particular, if $\ell_n = \frac{a}{n}$, we can write $Q = Q_a$. In this setting, Q_a is not the same as in the second lecture. Here the decomposition

$$\sigma = EQ_a\sigma + (\sigma - EQ_a\sigma) = \sigma' + \sigma''$$

is more attractive: σ', the a-regular part of σ, is a countable union of measures of finite a-energy (energy with respect to the kernel $|t|^{-a}$), and σ'', the a-singular part of σ, is concentrated on a Borel set of vanishing a-capacity.

We can do the same and even a little more with the Poisson intervals associated with the measure μ. First we have to define a martingale.

Though we shall not use it later on, let us introduce a very natural family of martingales depending on a parameter z, $0 \le z < 1$. We consider the case $0 < z < 1$ first.

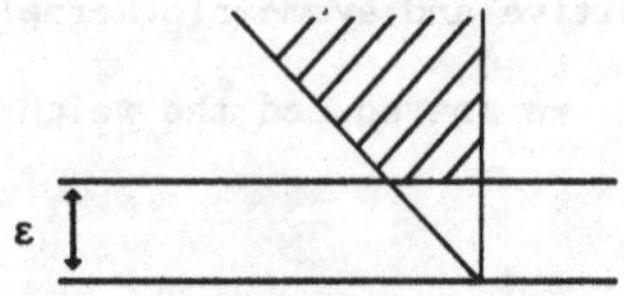

Figure 1

Let $N(t,\varepsilon)$ be the (random) number of points (X_n,Y_n) in the domain represented on figure 1, which we denote by $D(t,\varepsilon)$. It is a Poisson variable with parameter $\nu(D(t,\varepsilon))$ $(\nu = dt \otimes \mu)$. Therefore

$$Ez^{N(t,\varepsilon)} = e^{(z-1)\nu(D(t,\varepsilon))}.$$

We consider

$$Q(t,\varepsilon) = z^{N(t,\varepsilon)}\, e^{(1-z)\nu(D(t,\varepsilon))}.$$

Since $N(t,\varepsilon)$ has independent increments as we decrease ε, $Q(t,\varepsilon)$ is a martingale (we may consider a sequence $\varepsilon \downarrow 0$ and get a discrete martingale or consider ε^{-1} as a time and deal with the continuous martingale). The same is true for

$$\int Q(t,\varepsilon)d\sigma(t)$$

where $\sigma \in M_1^+(\mathbb{R})$ (probability measure on $\mathbb{R}$). This martingale converges in $L^2(\Omega)$ when

$$\iint E(Q(t,\varepsilon)Q(s,\varepsilon))d\sigma(t)d\sigma(s) = O(1).$$

Since $E(Q(t,\varepsilon)Q(s,\varepsilon))$ increases as $\varepsilon \downarrow 0$ let us pass to the limit and compute $E(Q(t,0)Q(s,0))$ in formal way, dropping ε whenever it appears, and t when it does not matter. We get

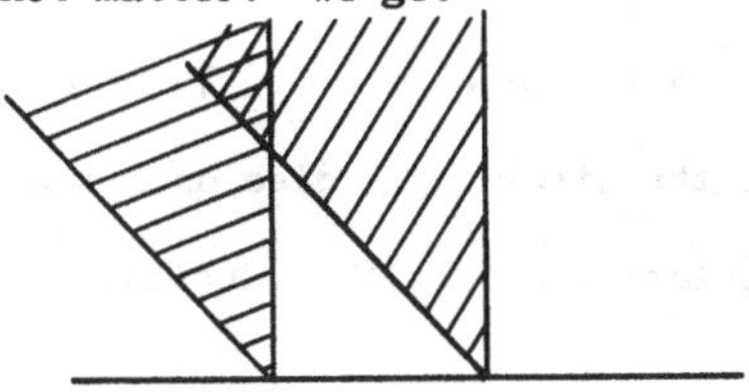

Figure 2

$$E(Q(t)Q(s)) = e^{2(1-z)\nu(D)}Ez^{N(t)+N(s)},$$

$$Ez^{N(t)+N(s)} = e^{(z-1)\nu(D(t)\ \Delta\ D(s))+(z^2-1)\nu(D(t)\cap D(s))}.$$

Therefore

$$\begin{aligned} E(Q(t)Q(s)) &= e^{(z^2-1+2(1-z))\nu(D(t)\cap D(s))} \\ &= e^{-(1-z)^2\nu(D(t)\cap D(s))} \\ &= (k(t-s))^{(1-z)^2}. \end{aligned}$$

The L^2-convergence condition reads

$$\iint (k(t-s))^{(1-z)^2} d\sigma(t)d\sigma(s) < \infty.$$

In the case $z = 0$ (using the convenient convention $0^N = 1_{N=0}$ if we want to squeeze any further calculation), the T-martingale is defined as

$$Q(t,\varepsilon) = e^{\nu(D(t,\varepsilon))}1_{N(t,\varepsilon)=0},$$

and the L^2-convergence condition for the martingale $\int Q(t,\varepsilon)d\sigma(t)$ is

$$\iint k(t - s)d\sigma(t)d\sigma(s) < \infty,$$

that is, σ has finite k-energy.

From now on, we stick to the case $z = 0$. We have the same conclusions as for random intervals on the circle: Q dies on k-singular measures, and the operator EQ is k-regularising. In the particular case

$$\mu = \sum \delta_{a/n} \qquad (0 < a < 1)$$

we have the same conclusion regarding the a-regular and a-singular part of σ.

However it is still more attractive to start with

$$\mu_1(dy) = 1_{[0,1]}(y)\,\frac{dy}{y^2}$$

and write $\mu_a = a\mu_1$. Then again $k_a(t) \approx |t|^{-a}$ and all conclusions are the same. Now, since μ_a increases with a, we may construct all Poisson processes $(0 < a < 1)$ at once. We consider the measure

$$\bar{\omega}(d\theta dt dy) = d\theta dt\,\frac{dy}{y^2}$$

in the 3-dimensional strip $0 < \theta < 1$, $-\infty < t < \infty$, $0 < y < 1$, and the corresponding Poisson process (θ_n, T_n, Y_n). For each a we consider the Poisson intervals $]T_n, T_n + Y_n[$ associated with the (θ_n, T_n, Y_n) belonging to the strip $0 < \theta < a$. Let G_a be their union and Q_a be the corresponding operator.

Then (G_a) $(0 < a < 1)$ is a random increasing open set on $\mathbb{R}$. Given $\sigma \in M^+(R)$ multiplication by 1_{G_a} kills a.s. the a-singular part of σ when a is given (therefore for almost all $a \in [0,1]$). It would require more work to see if the event "1_{G_a} kills the a-singular part of σ for all $a \in [0,1]$" has probability one.

The operators Q_a look the same as in lecture 2. We can write

$$Q'_a Q_b = Q_{a+b} \;,$$

meaning that we obtain Q_{a+b} when we apply first Q_b, then an operator independent of Q_b and having the same law as Q_a; actually

$$Q'_a = Q_{b,a+b}$$

with obvious notations ($Q_{b,a+b}$ associated with the strip $b < \theta < a + b$ as Q_a is with the strip $0 < \theta < a$).

Again, $(EQ_a)_{0<a<1}$ is a decreasing family of projections. The kernels consist of a-singular measures, the images of a-regular measures.

II. <u>The case of log-normal weights.</u>

Here again T is a locally compact metric space, usually $\mathbb{R}^d$. A gaussian process indexed by T (we mean always, centered) is a continuous mapping from T to a Hilbert space consisting of gaussian random variables (a closed subspace of $L^2(\Omega)$). We write $X(t)$ instead of $X(t,\omega)$. The law of a gaussian process $X(t)$ $(t \in T)$ is well-defined by the covariance kernel

$$p(t,s) = E(X(t)X(s)) \quad (t,s \in T).$$

The corresponding random weight is

$$P(t) = e^{X(t)-\frac{1}{2}(p(t,t))}.$$

Let us remark that $EP(t) = 1$ for all t and

$$E(P(t)P(s)) = e^{p(t,s)}.$$

Now we consider a series of independent processes (that is, processes with values in orthogonal gaussian Hilbert spaces) $X_n(t)$, the corresponding log-normal weights $P_n(t)$, and the products

$$Q_n(t) = P_1(t)P_2(t)\cdots P_n(t) \qquad (n = 1,2,\ldots).$$

We get a T-martingale, and our purpose is to investigate the corresponding operator Q.

What can we say about Q-singular measures (meaning $Q\sigma = 0$ a.s.) and Q-regular measures (meaning $EQ\sigma = \sigma$)? Given a probability measure σ, what can we say about the moments $E(\int Q\sigma)^h$ $(h \geq 1)$?

We begin with the L^2-theory. Let $\sigma \in M_1^+(T)$. We shall give a necessary and sufficient condition for

$$(1) \qquad \begin{cases} EQ\sigma = \sigma, \\ E(\int Q\sigma)^2 < \infty. \end{cases}$$

As we know, we simply have to express

$$E\left(\int Q_n d\sigma\right)^2 = O(1).$$

The first member can be written as

$$\iint EQ_n(t)Q_n(s)d\sigma(t)d\sigma(s) = \iint k_n(t,s)d\sigma(t)d\sigma(s)$$

with $k_n(t,s) = \exp(p_1 + \cdots + p_n)(t,s)$. These integrals are an increasing sequence, and the boundedness of this sequence is the necessary and sufficient condition we are looking for.

From now on we shall restrict ourselves (mainly for technical reasons) to the case

$$p_n(t,s) \geq 0 \qquad (n = 1,2,\ldots; \quad t,s \in T).$$

After we write

$$q(t,s) = \sum_1^\infty p_n(t,s) ,$$

the necessary and sufficient condition above is expressed by

$$(2) \qquad \iint \exp q(t,s)d\sigma(t)d\sigma(s) < \infty.$$

Moreover, given any positive kernel $k(t,s)$, we have

$$E \iint k(t,s)dS(t)dS(s) < \infty \qquad (S = Q\sigma)$$

if and only if

$$\iint k(t,s)\exp q(t,s)d\sigma(t)d\sigma(s) < \infty.$$

Here is a particular case. Let us assume

$$q(t,s) = u \log^{+} \frac{1}{|t - s|} + O(1) ,$$

where $|\ |$ stands for the euclidean norm. Then (2) means that σ has finite u-energy.

Therefore, if $u \geq d$, (1) fails whatever σ may be: either $EQ\sigma < \sigma$ or $E(\int Q\sigma)^2 = \infty$. If $u < d$, (1) occurs whenever σ has finite u-energy; in particular Q lives on the Lebesgue measure of $\mathbb{R}^d$.

However the L^1-theory leads to quite different results, as we shall see later. Q dies on every measure in $\mathbb{R}^d$ when $u \geq 2d$. Q lives on the Lebesgue measure in $\mathbb{R}^d$ when $u < 2d$.

The L^1-theory shows the same kind of phenomenon as the self-similar cascades. When σ is absolutely continuous with respect to Lebesgue measure, we have

$$E(\int Q\sigma)^h < \infty \Leftrightarrow uh < 2d.$$

The first question we shall consider is how Q depends on the $p_n(t,s)$ (always assumed ≥ 0). The fundamental result is that the law of Q depends only on the kernel $q(t,s)$. We call $q(s,t)$ a kernel of σ-positive type (here σ means countable sum). The multiplicative chaos (as Q is called) has the same relation with kernels of σ-positive type as gaussian processes have with respect to kernels of positive type.

A companion will be the comparison principle: if $q(t,s) \lesssim q'(t,s)$, Q is "better" than Q′ in every sensible respect.

Here is a useful version of the comparison principle. Suppose that $\mathcal{S}$ is a contraction of T, and we consider two measures σ and $\sigma' = \mathcal{S}\sigma$. Suppose moreover that $q(t,s)$ is a decreasing function of $|t - s|$. Then Q is better on σ than on σ'.

Assuming that we know what happens for self-similar cascades, this is a way to derive the results we just mentioned.

Fifth lecture

The multiplicative chaos

The program was indicated in lecture 4. We assume

$$p_n(t,s) \geq 0 \qquad (n = 1,2,\ldots;s,t \in T)$$

and write

$$q(t,s) = \sum_1^\infty p_n(t,s).$$

We say that $q(t,s)$ is a kernel of σ-positive type if it can be expressed in this way. Here is an important example of a kernel of σ-positive type. Suppose that the distance $d(t,s)$ in the metric space (T,d) satisfies the condition: for each $\lambda > 0$, $\exp(-\lambda d^2(t,s))$ is a kernel of positive type (when this is true $d^2(t,s)$ has "negative type", by definition). Then

$$\ell(d(t,s)) = \frac{1}{2}\int_1^\infty \exp(-\lambda d^2(t,s))\,\frac{d\lambda}{\lambda}$$

is of σ-positive type. Let us remark that

$$\ell(d(t,s)) = \log^+ \frac{1}{d(t,s)} + O(1)$$

and that $d^2(t,s) = |t-s|^2$ has negative type when $T = \mathbb{R}^d$ and $|\ |$ denotes the euclidean norm. In the sequel we shall consider kernels of the form

(1) $$q_u(t,s) = u \log^+ \frac{1}{|t - s|} + O(1)$$

$(t,s \in \mathbb{R}^d, u > 0)$.

Here is what we shall prove.

Theorem 1. The law of Q depends only on the kernel $q(t,s)$.

Theorem 2. Suppose that we are given two kernels of σ-positive type, $q(t,s)$ and $q'(t,s)$, with

$$q'(t,s) \leq q(t,s),$$

and let Q and Q' be the corresponding operators. Then, whenever Q lives on a measure $\sigma \in M^+(T)$, Q' does the same. Whenever we have $0 < E(\int Q\sigma)^h < \infty$, with $h > 1$, the same holds for Q' instead of Q. If we suppose

$$q'(t,s) = q(t,s) + O(1),$$

then Q and Q' are equivalent in the sense that they live and die on the same measures.

Corollary. Suppose that $q(t,s)$ is a decreasing function of the distance $d(t,s)$, and that $\mathcal{S}$ is a contraction of the space (T,d). Consider $\sigma \in M^+(T)$ and $\sigma' = \mathcal{S}\sigma$ (image of σ by $\mathcal{S}$). If Q dies on σ, it dies on σ'. If $E(\int Q\sigma)^h = \infty$, then $E(\int Q\sigma')^h = \infty$.

Theorem 3. Consider $T = \mathbb{R}^d$ and the operators Q_u given by (1). If $u \geq 2d$ Q_u dies on the Lebesgue measure λ of $\mathbb{R}^d$. If $u < 2d$ Q_u lives on λ and moreover

$$E\left(\int Q_u\lambda\right)^h < \infty \Leftrightarrow uh < 2d$$

when the integral is taken on a domain of finite λ-measure.

The proof of Theorem 1 relies on the following lemma.

Lemma. Suppose

$$p_0(t,s) \leq p_1(t,s)$$

and consider the corresponding weights $P_0(t)$ and $P_1(t)$, together with a measure $\sigma \in M^+(T)$ and a convex function f defined on $\mathbb{R}^+$. Then (omitting parentheses)

$$Ef\int P_0\sigma \leq Ef\int P_1\sigma.$$

Let us prove the lemma, using some familiar ideas about comparisons of gaussian processes, such as Slepian's lemma (but Slepian's lemma does not provide the result, as far as I can see). We consider the gaussian processes $X_0(t)$ and $X_1(t)$ associated with $p_0(t,s)$ and $p_1(t,s)$, and without restriction we suppose that they are independent (taking values in orthogonal gaussian Hilbert spaces). We consider

$$X_\lambda(t) = \sqrt{1-\lambda}\, X_0(t) + \sqrt{\lambda}\, X_1(t)$$

when $0 \leq \lambda \leq 1$ (the previous notation is consistent with this when $\lambda = 0$ or 1), and

$$P_\lambda(t) = \exp(X_\lambda(t) - \frac{1}{2} E\, X_\lambda^2(t))$$
$$= \exp(X_\lambda(t) - \frac{1}{2}(1-\lambda)p_0(t,t) - \frac{1}{2}\lambda\, p_1(t,t)).$$

Writing

$$h(\lambda) = Ef\int P_\lambda \sigma$$

it is enough to prove that $h'(\lambda)) \geq 0$. To make the computation easier let us assume that f is continuously differentiable. We have

$$h'(\lambda) = E(f'(\int P_\lambda\sigma)\int P'_\lambda\sigma)$$

where P'_λ means $\frac{\partial}{\partial\lambda}P_\lambda$, that is

$$P'_\lambda(t) = \frac{1}{2}P_\lambda(t)\left[\frac{-X_0(t)}{\sqrt{1-\lambda}} + \frac{X_1(t)}{\sqrt{\lambda}} + p_0(t,t) - p_1(t,t)\right].$$

From now on we fix t and write

$$U = \frac{-X_0(t)}{\sqrt{1-\lambda}} + \frac{X_1(t)}{\sqrt{\lambda}}\ .$$

We have

$$E(UX_\lambda(s)) = -p_0(t,s) + p_1(t,s) \geq 0.$$

Therefore

$$X_\lambda(s) = \alpha(s)U + W(s)$$

where $W(s)$ is independent of U, and $\alpha(s) \geq 0$. Writing α for $\alpha(t)$ and τ^2 for EU^2, $P'_\lambda(t) = \frac{1}{2} e^{\alpha U - \frac{1}{2}\alpha^2\tau^2} e^{W(t) - \frac{1}{2}EW^2(t)}(U - \alpha\tau^2)$.

We shall prove

$$E(P'_\lambda(t) f'(\int P_\lambda \sigma)) \geq 0.$$

Let us write

$$f'(\int P_\lambda \sigma) = f'(\int e^{\alpha(s)U - \frac{1}{2}\alpha^2(s)\tau^2} e^{W(s) - \frac{1}{2}EW^2(s)} d\sigma(s))$$

and consider first

$$E(P'_\lambda(t) f'(\int P_\lambda \sigma) \,|\, U = u).$$

Since the W's are independent of U we simply integrate the product with respect to the W's and we obtain

$$(u - \alpha\tau^2) e^{\alpha u - \frac{1}{2}\alpha^2\tau^2} g(u)$$

where $g(u)$ is some increasing function of u (because $f'(u)$ is increasing). Writing $\gamma(u)du$ for the distribution of U, what we have to prove is

$$\int (u - \alpha\tau^2) e^{\alpha u - \frac{1}{2}\alpha^2\tau^2} g(u)\gamma(u)du \geq 0$$

when $g(u)$ is an increasing function. Now

$$\int_{-\infty}^{\infty}(u - \alpha\tau^2)e^{\alpha u - \frac{1}{2}\alpha^2\tau^2}\gamma(u)du = 0$$

because it is the derivative of

$$\int_{-\infty}^{\infty} e^{\alpha u - \frac{1}{2}\alpha^2\tau^2}\gamma(u)du \ (= E(e^{\alpha U - \frac{1}{2}\alpha^2\tau^2}) = 1)$$

with respect to α. Since the function

$$\int_{-\infty}^{x}(u - \alpha\tau^2)e^{\alpha u - \frac{1}{2}\alpha^2\tau^2}\gamma(u)du$$

vanishes at $\pm\infty$ and its derivative vanishes once only, passing from negative to positive values, it is negative. Integrating by parts gives the desired inequality. Therefore $h'(\lambda) \geq 0$ and the lemma is proved.

<u>Corollary</u>. Suppose $\varepsilon > 0$, $\alpha > 1$, and f is a positive, increasing and convex function on $\mathbb{R}^+$ such that $f(0) = 0$ and

$$f(\lambda x) \leq \lambda^a f(x)$$

for $x \geq 0$ and $\lambda \geq 1$. Suppose moreover

$$p_1(t,s) \leq p_0(t,s) + \varepsilon.$$

Then

$$Ef\int P_1\sigma \leq (1 + b(\varepsilon,a))Ef\int P_0\sigma$$

with $\lim_{\varepsilon\to 0} b(\varepsilon,a) = 0$.

Let us prove the corollary. Using the lemma we can assume

$$p_1(t,s) = p_0(t,s) + \varepsilon.$$

Therefore we can assume

$$X_1(t) = X_0(t) + Y$$

where Y is gaussian and independent of $(X_0(t))_{t\in T}$, with $EY^2 = \varepsilon$. Then we have

$$Ef\int P_1\sigma = Ef(e^{Y-\frac{1}{2}\varepsilon}\int P_0\sigma),$$
$$E1_{Y\geq\varepsilon/2}f(e^{Y-\frac{1}{2}\varepsilon}\int P_0\sigma) \leq E(1_{Y\geq\varepsilon/2}e^{a(Y-\frac{1}{2}\varepsilon)})Ef\int P_0\sigma,$$
$$E1_{Y<\varepsilon/2}f(e^{Y-\frac{1}{2}\varepsilon}\int P_0\sigma) \leq P(Y<\tfrac{\varepsilon}{2})Ef\int P_0\sigma.$$

Hence the conclusion, with

$$1 + b(\varepsilon,a) = P(Y<\tfrac{\varepsilon}{2}) + E(1_{Y\geq\varepsilon/2}e^{a(Y-\frac{1}{2}\varepsilon)}).$$

<u>Proof of Theorem 1</u>. Let us suppose $p_n(t,s) \geq 0$, $p'_n(t,s) \geq 0$

$$\sum_1^\infty p_n(t,s) = \sum_1^\infty p'_n(t,s).$$

We introduce $q_n(t,s)$, $q'_n(t,s)$, Q and Q' as usual. Suppose for simplicity that T is compact. Choosing an integer ν we certainly have

$$q'_\nu(t,s) \leq q(t,s) + \varepsilon$$

when n is large enough. Therefore

$$Ef \int Q'_\nu\sigma \leq (1 + b(\varepsilon,a))Ef \int Q_n\sigma$$

if f satisfies the assumptions of the corollary.

Suppose now that Q lives on σ. Then, for some function f such that $\frac{f(x)}{x} \to \infty$ as $x \to \infty$, we have

$$Ef \int Q_n\sigma = O(1)$$

and we can assume that f is as in the corollary. It follows that $Ef \int Q'_\nu\sigma = O(1) \quad (\nu \to \infty)$, therefore Q' lives on σ.

Moreover,

$$Ef \int Q'\sigma = Ef \int Q\sigma$$

(because we have $\leq$ and also $\geq$). Since there are plenty of functions f, it means that $\int Q'\sigma$ and $\int Q\sigma$ have the same distribution.

Now, given a finite set of measures σ_j, we have the same distribution for $\int Q \sum u_j\sigma_j$ and $\int Q' \sum u_j\sigma_j$ whatever may be the $u_j \geq 0$. Therefore the joint distributions of the random vectors $(\int Q\,\sigma_j)$ and $(\int Q'\sigma_j)$ are the same. This proves that Q and Q' have the same law and ends the proof of Theorem 1.

The proof of Theorem 2 is still simpler, for $q'(t,s) \leq q(t,s)$ implies $Ef \int Q'\sigma \leq Ef \int Q\sigma$ for all functions f satisfying the assumptions of the corollary of the lemma. Taking $f(x) = x^h$ provides the first result. The second is already proved (choosing a fixed ε).

In order to obtain the corollary of Theorem 2 we write

$$q'(t,s) = q(\varphi t, \varphi s)$$
$$Q_n'(t) = Q_n(\varphi s)$$
$$\int Q_n \sigma' = \int Q_n' \sigma.$$

The assumption on $q(t,s)$ gives $q'(t,s) \geq q(t,s)$, therefore, exchanging q and q', Theorem 2 applies, and gives the conclusion.

Before going to the proof of Theorem 3 we consider again the self similar cascades. Choosing $T = \{0,1,\dots,c-1\}^{\mathbb{N}}$ we consider the c-adic distance

$$d(t,s) = c^{-n}$$

when $t_1 = s_1, \dots, t_n = s_n, t_{n+1} \neq s_{n+1}$, and the weight

$$W = e^{\alpha\xi - ½\alpha^2}$$

where ξ is a standard normal variable. We then define

$$X_n(t) = \alpha \sum_{(n)} \xi_C 1_C(t)$$

where (n) means that we sum over all sets C of the n-th step and all ξ_C are standard normal and independent. Then

$$P_n(t) = \sum_{(n)} W_C 1_C(t)$$

and

$$p_n(t,s) = \begin{cases} 0 & \text{if } d(t,s) > c^n; \\ \alpha^2 & \text{if } d(t,s) \le c^n. \end{cases}$$

Therefore

$$q(t,s) = \alpha^2 \log_c \frac{1}{d(t,s)} .$$

The results of Kahane-Peyrière apply, by noticing that

$$EW^h = e^{\frac{1}{2}\alpha^2(h^2-h)},$$

$$EW \log W = \alpha^2/2.$$

The operator Q dies on Haar measure of T (say, σ) if $\alpha^2 \ge 2 \log c$ and lives on σ if $\alpha^2 < 2 \log c$. Moreover

$$E(Q\sigma(T))^h < \infty \Leftrightarrow \alpha^2 h < 2 \log c.$$

Proof of Theorem 3. The idea is to take $c = 2^d$ and consider a Cantor set in $\mathbb{R}^d$ as a version of T. We construct a symmetric Cantor set of positive Lebesgue measure on the interval $[0,1]$ by taking away intervals of lengths

$\varepsilon_1, \varepsilon_2$(twice), ε_3(4 times),

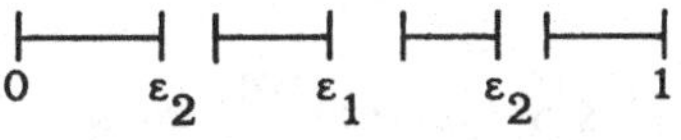

in such a way that

$$\sum_1^\infty \varepsilon_n 2^{n-1} = 1 - \gamma < 1$$

while $\lim \varepsilon_n n^2 = \delta > 0$. Denoting this Cantor set by K we choose

$$T = K^d .$$

This is a version of the abstract space we just considered, and now the cells of the n-th step are cubes with edges $\approx \gamma 2^{-n}$, whose minimal distance is $\approx \delta n^{-2} 2^{-n}$. We equip T with the c-adic distance $d(t,s)$ and with the euclidean distance $|t - s|$. Then $d(t,s) = 2^{-dn}$ implies $\delta' n^{-2} 2^{-n} \lesssim |t - s| \lesssim \gamma' 2^{-n}$. Let us remember that we just considered

$$q(t,s) = \frac{\alpha^2}{\log c} \log \frac{1}{d(t,s)} .$$

and we have to consider

$$q_u(t,s) = u \log^+ \frac{1}{|t - s|} + O(1).$$

If $u \geq 2d$ we choose

$$\alpha^2 = u \log 2.$$

Since $|t - s| \lesssim \gamma' (d(t,s))^{1/d}$ we have

$$q_u(t,s) \geq q(t,s) + O(1).$$

Moreover $\alpha^2 \geq 2 \log c$, and therefore Q dies on the Lebesgue measure on T. According to the comparison principle Q_u dies on the Lebesgue measure on T too. Therefore Q_u dies on the Lebesgue measure of $\mathbb{R}^d$.

If $u < 2d$ we choose

$$u \log 2 < \alpha^2 < 2 \log c.$$

Now we have

$$q_u(t,s) \lesssim q(t,s) + O(1)$$

and Q lives on the Lebesgue measure on T. Therefore the same occurs for Q_u.

We reckon in the same way when $uh \gtrsim 2d$ (no moment of order h) or $uh < 2d$ (finite moment of order h).

This ends the proof of Theorem 3.

Theorem 3 confirms a remarkable intuition of B. Mandelbrot (1972), that is, 6 is the critical value of the parameter u in 3-dimensional space.

It is possible to apply the comparison principle in different ways: for example, if we take for T the ordinary Cantor set, the euclidean distance is equivalent to some power of the c-adic distance, and the translation of Kahane-Peyrière results is obvious.

However another theory has to be developed for more general situations. It involves three ideas: (1) the L^2-theory, already considered; (2) the formula

$$Q_{u+v} = Q_u Q_v$$

where Q_u and Q_v are considered as independent; (3) the use of Peyrière's probability, defined on $\Omega \times T$ as

$$F(\omega,t) \longrightarrow \iint F(\omega,t)dS(t)d\omega$$

where σ is a given probability measure on T and $S = Q\sigma$ as usual; the basic fact is that the $X_n(t,\omega)$ are independent r.v. with this new probability. The idea of Peyrière's probability is exactly what worked beautifully for Riesz products (see lecture 1). All this is developed in my Quebec paper (1985) and my paper in Chinese Annals of Mathematics (1987) (to be read first).

The results are essentially what can be expected: Q_u dies on every measure $\sigma \in M^+(T)$ if $u > 2 \dim T$, where dim denotes the Hausdorff dimension. When $u < 2 \dim T$, Q_u lives on σ when σ is a-regular and $u < 2a$. In this case $Q_u\sigma$ is a.s. b-regular whenever $u < 2(a + b)$.

As a conclusion we may consider that the operators Q associated with a kernel $q(t,s)$ of σ-positive type justify the "log-normal hypothesis" of A. Kolmogorov (1961) and the idea of a "limit-log-normal process" developed by B. Mandelbrot in 1972, in order to give a random model of turbulence. Actually Q (the multiplicative chaos) could very well be called a limit-log-normal operator. However it does not seem appropriate to call $Q\sigma$ a limit-log-normal measure (simply because a sum of log-normal variables is not log-normal any more). A study of the cone Γ generated by e^X($X \in \mathcal{H}$, a gaussian Hilbert space) may lead to another introduction of the multiplicative chaos (maybe, extremal elements in the cone of additive and isometric operators from $L^1_+(\sigma)$ to $\Gamma^1(\mathcal{H})$, the closure of Γ in $L^1(\Omega)$).

This is the end of the lecture, not the end of the topic. The quality of my audience was a stimulation for providing here a few new results (in lectures 3 and 4) and for trying to explain in a proper way the basic ideas surrounding random multiplications and their applications. The studious atmosphere in Altgeld Hall was best possible for this purpose. The constant initiative, encouragement, and help of Earl Berkson were decisive. I wish to thank him for everything - including correcting my English mistakes.

WAVELETS AND OPERATORS

Yves Meyer

CEREMADE, Université de Paris Dauphine

INTRODUCTION

Wavelets were born about a year ago but similar constructions were already used in pure mathematics or in theoretical physics. It is interesting and surprising that some people in geometry of Banach spaces, in constructive quantum field theory and even some people working on certain problems in signal analysis occurring in oil prospecting were looking for about the same thing at about the same time.

They were interested in finding new and more tractable orthogonal expansions of general functions in terms of a system of well behaved explicit functions combining the advantages of the Haar system and of the trigonometrical system.

The Fourier series expansion has quite a long history and precisely for that reason we know what can be easily done and what is quite hard to attain using Fourier series.

What can be done with a Fourier series (without a zero order term) is to integrate as many times as one likes. One still gets a Fourier series whose coefficients are multiplied by $-i/k$ for each integration. No other type of expansion would do that since the only 2π-periodic eigenfunctions of $\frac{d}{dx}$ are precisely e^{ikx}, $k = 0, \pm 1, \pm 2, \ldots$

One can also take the distributional derivative of a Fourier series and one still gets a Fourier series. Finally L^2 norms are easily computed by Fourier series since the functions $(2\pi)^{-1/2}e^{ikx}$, $k \in \mathbb{Z}$, form an orthonormal basis of $L^2(0,2\pi)$.

Putting all those things together we attain all Sobolev norms. But that is about all one can reach with Fourier series expansions. One cannot decide if a function is Hölder continuous by the size of its Fourier coefficients. One also cannot decide if it belongs to L^p for $p \neq 2$.

Let us make this last point more precise. A collection of vectors $e_0, e_1, e_2, \ldots$ in some Banach space B forms an unconditional basis if the following properties are satisfied:

(1) for each x in B there exists a unique sequence $\alpha_0, \alpha_1, \ldots, \alpha_k, \ldots$ of coefficients such that $x = \alpha_0 e_0 + \alpha_1 e_1 + \cdots + \alpha_k e_k + \cdots$ which means that the partial sums of this series converge to x with respect to the B norm.

(2) there exists a constant C such that for each integer m and for each sequence $\alpha_0, \alpha_1, \ldots, \alpha_m$ of coefficients, we have

$$\left\|\sum_0^m \beta_k e_k\right\| \leq C\left\|\sum_0^m \alpha_k e_k\right\|$$

for all sequences $\beta_0, \ldots, \beta_m$ such that $|\beta_k| \leq |\alpha_k|$ for $0 \leq k \leq m$.

If property (1) is satisfied, the sequence $e_0, e_1, \ldots$ is called a Schauder basis for the Banach space B. Property (2) means that as in the case of an orthonormal basis for a Hilbert space, we can decide if a given formal series $\gamma_0 e_0 + \gamma_1 e_1 + \cdots$ will converge to some element x in B by simply checking the sizes $|\gamma_0|, |\gamma_1|, |\gamma_2|, \ldots$ of the coefficients.

The trigonometric system written $1, \cos x, \sin x, \ldots, \cos kx, \sin kx, \ldots$ is a Schauder basis for $L^p(0,2\pi)$ when $1 < p < \infty$ but it is not an unconditional basis unless $p = 2$ (Paley, R.E.A.C. and Zygmund, A. On some series of functions, Proc. of the Cambridge Phil. Soc. 26 (1930), 337-357, 458-474 and 28 (1932), 190-205).

The situation is even worse in the case of Hölder functions since the trigonometric system is not even a Schauder basis for C^r.

On the other extreme, the Haar system provides an orthonormal basis for L^2, and unconditional basis for L^p when $1 < p < \infty$ but is of no help when smoothness plays a role even in a disguised form. For example the Haar system is not an unconditional basis of the Stein and Weiss real variable version of the Hardy space H^1. And the Haar system cannot be used for the Banach space C^r of Hölder functions.

For later purposes let us remind the reader that the Haar system also makes sense on the real line. In that case it is the orthonormal basis consisting of all functions $2^{j/2}h(2^j x - k)$ where $j \in \mathbb{Z}$, $k \in \mathbb{Z}$, and where $h(x) = 1$ on $[0,\frac{1}{2})$, -1 on $[\frac{1}{2},1)$ and 0 elsewhere.

The Haar functions have a good localization with respect to the x variable but their Fourier transforms do not have a good localization. This is both due to the lack of smoothness and to the lack of higher order vanishing moments in the Haar system.

The case of the trigonometric system on $[0,2\pi)$ is the opposite since the functions e^{ikx}, $k = 0,\pm1,\pm2,\ldots$ do not have any localization with respect to the x variable but have a perfect localization in frequencies. The Fourier inversion formula on the line reads $f(x) = (2\pi)^{-1} \int e^{ix\xi}\hat{f}(\xi)d\xi$ and the values of ξ belonging to the support of $\hat{f}$ are called the frequencies of the function f. A similar terminology is used in the periodic case.

The new orthonormal basis of wavelets for $L^2(\mathbb{R})$ combines the good properties of the Haar system with the good properties of the trigonometrical system.

We start with a large integer m. A function ψ is called an analyzing wavelet if we can impose upon ψ the competing (but fortunately compatible!)

properties (1)-(4)

(1) $\psi(x)$ is a function of class C^m

(2) $\psi(x)$ has a rapid decay at infinity together with all its derivatives of order $k \le m$

(3) all moments of ψ of order $k \le m$ vanish

(4) the collection $2^{j/2}\psi(2^j x - k)$, $j \in \mathbb{Z}$, $k \in \mathbb{Z}$, is an orthonormal basis of $L^2(\mathbb{R})$.

If $m = 0$, the Haar system satisfies these conditions with L^∞ instead of C^0 in (1).

Let us denote by $\mathcal{T}$ the collection of all dyadic intervals $I = [k2^{-j},(k+1)2^{-j})$, $k \in \mathbb{Z}$, $j \in \mathbb{Z}$, and write more simply $\psi_I(x) = 2^{j/2}\psi(2^j x - k)$. The functions ψ_I, $I \in \mathcal{T}$, are called <u>wavelets</u> and the expansion of a general function into wavelets is given by

$$(5) \qquad f(x) = \sum_{I\in\mathcal{T}} \alpha(I)\psi_I(x) \quad \text{where} \quad \alpha(I) = \langle f,\psi_I\rangle.$$

The validity of (5) is not limited to the L^2 setting. A smooth function of class C^r where $r < m$ or a nasty distribution S can be written in the form (5) as long as m is larger than the order of the distribution S, since the wavelets are playing the role of testing functions.

The wavelet ψ_I is "almost supported" by the corresponding interval I: for $M > 1$, the L^2 norm of ψ_I outside the interval MI is less than $\epsilon(M)$ where MI is centered at the center of I and has M times the length of I, and $\epsilon(M)$ tends to 0 as M tends to infinity.

The smoothness and the vanishing moment conditions permit one to differentiate or to integrate the wavelets series (5) term by term. The new series will no longer be given by the same function ψ but by its derivative or its primitive and will be unconditionally convergent in the Sobolev spaces H^{s-1} or H^{s+1} when the given series (5) represents a function in H^s.

Indeed the wavelets form a universal unconditional basis for most of the classical spaces (if we except L^1 or L^∞). This is due to the fact that the operators which are diagonal in the wavelets basis (with bounded eigenvalues) are Calderón-Zygmund operators which are precisely bounded on these functional spaces by the general theory.

One can attempt to go further and try to understand the local behavior of a function or of a distribution at a point x_0 in terms of size conditions on the wavelet coefficients $\alpha(I)$ as I shrinks to x_0. The extremely good properties (decay, vanishing moments, and smoothness) of the analyzing wavelet make this kind of local Fourier analysis possible.

Let us give an example. The Zygmund class Λ_* is the collection of all continuous functions $f(x)$ on the line such that

$$| f(x + h) + f(x - h) - 2f(x) | \leq Ch ,$$

for some constant C, each $x \in \mathbb{R}$, and each $h > 0$. The affine functions in Λ_* correspond to $C = 0$, and the quotient space of Λ_* modulo affine functions is a Banach space which (through some abuse of notation) will also be denoted by Λ_*. Then the condition $|\alpha(I)| \leq C|I|^{3/2}$, $I \in \mathcal{J}$, characterizes the Zygmund class. It implies that any series $\sum \alpha(I)\psi_I(x)$ for which $C_1 |I|^{3/2} \leq |\alpha(I)| \leq C_2 |I|^{3/2}$ represents a function in the Zygmund class which is nowhere differentiable. If $f(x)$ is differentiable at x_0, then the limit of $|I|^{-3/2}\alpha(I)$ when $x_o \in 2I$ and $|I|$ tends to 0 is 0. For example, any function f in Λ_* can be written $f = g + h$, where both g and h belong to the Zygmund class, and both g and h are nowhere differentiable.

Many authors have previously used some expansions which are similar in nature to (5). For example A. Chang, R. Fefferman, and A. Uchiyama have decomposed BMO functions in series like (5), but the pieces were "almost orthogonal" instead of being orthogonal. The missing uniqueness of their decompositions was irrelevant for their goals. A similar remark applies to the algorithms developed by M. Frazier and B. Jawerth in their paper "Decomposition of Besov spaces".

By contrast, L. Carleson and P. Wojtaszczyk wanted to construct a specific unconditional basis for the Stein and Weiss real variable version of the Hardy space H^1. L. Carleson used a family of functions of the form $2^{j/2}\psi(2^j x - k)$. But since his collection was not orthogonal, he had to construct the dual basis and to prove some subtle estimates. Wojtaszczyk proved that the Franklin system was an unconditional basis for H^1. The Franklin system is an orthonormal basis for L^2 by virtue of its construction but does not have the simple structure given by (4). Z. Ciesielski in joint work with T. Figiel constructed a basis (φ_k) in $C(X)$, where X is a smooth compact manifold (with boundary or without) such that (φ_k) is simultaneously a basis in all Besov and Sobolev spaces of order not exceeding some integer m. Bočkarev proved that the Franklin system could be used in order to construct a Schauder basis for the disc algebra $A(D)$ or for the polydisc algebra $A(D^2)$.

The smooth wavelets treated in Chapter V throw considerable light on this subject since they form a basis which is unconditional for all Besov, Sobolev, etc. ...spaces whatever be the indices. Moreover all the technicalities in Bočkarev's theorems simply disappear.

Nevertheless we first want to describe what we call the "Franklin wavelets" since their relation to linear and cubic splines is so simple that the "Franklin wavelets" cannot be overlooked in any harmonic analysis approach to splines.

The Franklin wavelets are unique if the sharp localization is expressed in terms of the least square deviation and they provide a very elegant path to Bočkarev and Wojtaszczyk's results.

Once the Franklin wavelets' construction is understood, the same recipe is applied to build the smooth wavelets for which $m = \infty$ in (1)-(3). Sitting in between are the wavelets constructed with splines of order m.

The second part of these notes is devoted to the interplay between the wavelets and the theory of Calderón-Zygmund operators.

We denote by CZO the collection of linear continuous operators $T : L^2(\mathbb{R}^n) \to L^2(\mathbb{R}^n)$ which have the following property. There exist an exponent $\gamma \in (0,1]$ and a constant C such that the off-diagonal part of the distributional kernel of T is a function $K(x,y)$ satisfying the following estimates

(6) $|K(x,y)| \le C|x - y|^{-n}$ if $x \neq y$

(7) $|K(x',y) - K(x,y)| \le C|x' - x|^{\gamma}|x - y|^{-n-\gamma}$ if $x \neq y$ and $|x' - x| \le \frac{1}{2}|x - y|$.

(8) $|K(x,y') - K(x,y)| \le C|y' - y|^{\gamma}|x - y|^{-n-\gamma}$ if $x \neq y$ and $|y' - y| \le \frac{1}{2}|x - y|$.

The real variable methods of Calderón and Zygmund apply to such operators. Natural examples show up when the method of the double layer potential is applied to elliptic partial differential equations in Lipschitz domains. But this collection of operators has been extremely mysterious for a very long time since there were no readily accessible ways for obtaining the basic L^2-estimate when one is given a possibly unbounded operator T whose kernel satisfies (6), (7) and (8).

The first criterion was obtained in 1983 by G. David and J. L. Journé and some lengthy and painful preceding papers were immediately clarified. Chapter VI will cover this material.

But what is the interplay between Calderón-Zygmund operators and wavelets?

Suppose we are given an operator T, acting on $L^2(\mathbb{R})$. Since $L^2(\mathbb{R})$ has the orthonormal basis ψ_I, $I \in \mathcal{T}$, the most natural idea to try is to compute the matrix M of the operator T in our basis ψ_I, $I \in \mathcal{T}$. The entries of M are $m(I,J) = \langle T\psi_I, \psi_J\rangle$ and M is an infinite matrix indexed by $\mathcal{T} \times \mathcal{T}$.

A pedestrian approximation T_A to the operator T is given by the following cut off procedure. We let A be a finite subset of $\mathcal{T}$ and we define the "cut off operator" T_A by $\langle T_A\psi_I, \psi_J\rangle = \langle T\psi_I, \psi_J\rangle$ when $(I,J) \in A \times A$ and 0 if $(I,J) \notin A \times A$. The matrix M_A of T_A is obtained from M by deleting all rows and columns which are not indexed by elements of A.

If T is going to be bounded on $L^2(\mathbb{R})$, obviously the corresponding "cutoff operators" T_A are uniformly bounded and vice versa.

But is it true that when T is a Calderón-Zygmund operator, the cut off operators T_A will satisfy uniform estimates of Calderón-Zygmund type? This uniformity means that both the exponent $\gamma \in (0,1]$ and the constant C can be independent of A in (6), (7), (8). The answer is surprisingly no!

The collection $\mathcal{A}$ of $T \in CZO$ for which the answer is yes forms an operator algebra which has been discovered and studied by P. G. Lemarié in his Ph.D. dissertation.

The Hilbert transform belongs to $\mathcal{A}$ as well as any Calderón-Zygmund operator given by a convolution. In contrast, Calderón's commutators, the Cauchy operator on a Lipschitz curve or the double layer potential on a Lipschitz surface do not belong to $\mathcal{A}$.

The operators $T \in \mathcal{A}$ are characterized by the off-diagonal decay of the entries $m(I,J)$ of the matrix M of T in the wavelet basis. In the one dimensional case, these entries satisfy

(9) $|m(I,J)| \le C|I|^{\frac{1}{2}+\gamma}|J|^{\frac{1}{2}+\gamma}(|I| + |J|)^{-\gamma}(|I| + |J| + \text{dist}(I,J))^{-1-\gamma}$.

Conversely it is an exercise to check the ℓ^2-boundedness of such a matrix and to verify (6), (7) and (8) for $0 < \gamma' < \gamma$ when the entries $m(I,J)$ satisfy (9).

But $\mathcal{A}$ is strictly contained in CZO. For example an operator of pointwise multiplication with a function in L^∞ is certainly a Calderón-Zygmund operator which never belongs to $\mathcal{A}$ (unless the function is a constant).

The operators which are missing are exactly the sums $R_b + R_\beta^*$ where R_b is the paraproduct with a BMO function b and R_β^* is the adjoint of a paraproduct with a BMO function β.

The paraproduct operation is for the classical analyst what a stochastic integral is for the probabilist. Paraproducts might also be compared to Toeplitz or Hankel operators, but without any doubt the best approach to the paraproducts is given by the wavelets.

The paraproduct operator R_b applied to $f \in L^2$ will be written $\pi(b,f)$ and viewed as a bilinear mapping from BMO × L^2 into L^2. If one fixes f in L^2, then the linear mapping which takes b in BMO to $\pi(b,f)$ in L^2 is actually diagonal in the wavelet basis and this remark easily yields the basic estimate. On the other hand, if one fixes b in BMO, the mapping that takes f to $\pi(b,f)$ is a Calderón-Zygmund operator R_b. The David-Journé theorem says that any $T \in$ CZO can be written in a unique way as a sum of three terms T_1, T_2 and T_3 where T_1 belongs to $\mathcal{A}$, $T_2 = R_b$ for some BMO function b and T_3 is the adjoint of R_β for some BMO function β. The next step is to write the <u>atomic decomposition of a Calderón-Zygmund operator</u> T.

The tensor product $u \otimes v$ between two functions u and v in L^2 defines an operator by $(u \otimes v)(f) = \langle f,v\rangle u$.

We now let Δ_I be the triangle function based on the dyadic interval I, normalized by the condition $\int \Delta_I(x)dx = 1$. Then, in the one dimensional case, the atoms which are the building blocks for all Calderón-Zygmund operators, are simply $\psi_I \otimes \psi_J$, $I \in \mathcal{J}$, $J \in \mathcal{J}$, $\psi_I \otimes \Delta_I$, $I \in \mathcal{J}$ and $\Delta_I \otimes \psi_I$, $I \in \mathcal{J}$. They form an unconditional basis for the topological vector space of all Calderón-Zygmund operators.

More precisely any $T \in CZO$ can be written in a unique way

$$(10) \qquad T = \sum_I \sum_J \alpha(I,J)\, \psi_I \otimes \psi_J + \sum_I \beta(I)\Delta_I \otimes \psi_I + \sum_I \gamma(I)\psi_I \otimes \Delta_I$$

where for some constant C and for some exponent $\gamma \in (0,1]$ the coefficients $\alpha(I,J)$ satisfy (9) while the coefficients $\beta(I)$ and $\gamma(I)$ satisfy the Carleson-measure-type condition

$$(11) \qquad \sum_{I \subset I'} (|\beta(I)|^2 + |\gamma(I)|^2) \le C|I'|, \quad I' \in \mathcal{J}.$$

The first term in the right-hand side of (10) belongs to the Lemarié algebra $\mathcal{A}$ and the two others are the paraproduct operators R^*_β and R_b.

The notes end with T Murai's proof for the L^2-boundedness of the Cauchy operator on a Lipschitz curve, and the Appendix contains some references which the reader might find useful.

I. THE FRANKLIN WAVELETS.

Yves Meyer

Introduction. The Franklin wavelets form a new orthonormal basis of the space $L^2(0,1)$. This basis is reminiscent of the Franklin system but has a much simpler and richer structure. In particular the Franklin wavelets are completely explicit special functions. They are, in fact, generated by one function and all estimates on Franklin wavelets are trivial.

The Franklin wavelets from an unconditional basis for the Stein and Weiss real variable version of the Hardy space H^1.

Franklin wavelets have been discovered simultaneously and independently by P. G. Lemarié (Tunis, May 30, 1986) and G. Battle (Cornell, August 24, 1986).

1. Construction of the Franklin wavelets.

Let B be the Banach space of all continuous 1-periodic functions. As usual we shall also view B as the space of continuous function on $[0,1]$ such that $f(1) = f(0)$. Then the Fourier series of a function $f \in B$ is

$$(1.1) \qquad f(x) = \sum_{-\infty}^{\infty} c_k e(kx)$$

where $e(kx) = \exp(2\pi ikx)$. The Fourier coefficients $c_k = \int_0^1 f(x)e(-kx)dx$ and will be sometimes denoted by $\hat{f}(k)$. The Fourier tranform of a function $g \in L^1(\mathbb{R})$ will be defined by

(1.2) $$\hat{g}(\xi) = \int_{-\infty}^{\infty} \exp(-2\pi i\xi x)g(x)dx.$$

We shall often use the observation that the Fourier coefficients of the periodic function $f(x) = \sum_{-\infty}^{\infty} g(x + \ell)$ are precisely $\hat{g}(k)$ when $g \in L^1(\mathbb{R})$. For $j \geq 0$, we define the subspace $B_j \subset B$ of piecewise affine functions with nodes at $k2^{-j}$, $0 \leq k < 2^j$. Then B_0 contains only the constant functions and we have $B_0 \subset B_1 \subset B_2 \subset \cdots \subset B_j \subset \cdots$. The space B_j has dimension 2^j and elements of B_j are spline functions. The construction will be carried over for general splines (of any given order) in section 4. We now imbed B_j in $L^2(0,1)$, the ordinary L^2 space with respect to the Lebesgue measure. Then we define $C_{j+1} \subset B_{j+1}$ to be the orthogonal complement of B_j in B_{j+1}. Therefore the dimension of C_{j+1} is also 2^j. We will also write $C_0 = B_0$.

Then $L^2(0,1)$ splits into a direct orthogonal sum

$$L^2(0,1) = C_0 \oplus C_1 \oplus C_2 \oplus \cdots .$$

We then define Γ_j to be the cyclic group of $k2^{-j}$ (mod.1) where $0 \leq k < 2^j$. We then have

$$\{0\} = \Gamma_0 \subset \Gamma_1 \cap \Gamma_2 \subset \cdots \subset \mathbb{R}/\mathbb{Z}.$$

Let $U_j : B_j \to B_j$ be the translation by 2^{-j}. Then Γ_j acts on B_j by the unitary operators U_j^k, $0 \leq k < 2^j$. There is a cyclic vector for this group action, for example the triangle function which is 1 at 2^{-j} and 0 at the other modes $k2^{-j}$, $k \neq 1$. Therefore the eigenvalues of U_j are simple and are exactly $e(k2^{-j})$, $0 \le k < 2^j$, since the cardinality of Γ_j is the dimension of B_j.

We then define two linear suspaces B'_{j+1} and B''_{j+1} of B_{j+1}. The functions in B'_{j+1} are those functions in B_{j+1} which vanish at $k2^{-j}$, $0 \le k < 2^j$ and similarly functions in B''_{j+1} are those that vanish at $k2^{-j} + 2^{-j-1}$, $0 \le k < 2^j$. Therefore B_{j+1} is the direct (but not orthogonal) sum of B'_{j+1} and B''_{j+1}. The operator U_j acting either on B'_{j+1} or B''_{j+1} has also a cyclic vector (a triangle function with base $[0,2^{-j}]$ or $[2^{-j-1},2^{-j} + 2^{-j-1}]$) and the eigenvalues of U_j acting on B'_{j+1} or B''_{j+1} are simple and are $e(k2^{-j})$, $0 \le k < 2^j$. That proves that the eigenvalues of $U_j : B_{j+1} \to B_{j+1}$ are double and are $e(k2^{-j})$, $0 \le k < 2^j$. Since the eigenvalues of the action of U_j on B_j are also a string of the 2^j-th roots of unity, the same is true for the action of U_j on the orthogonal complement C_{j+1} of B_j in B_{j+1}. Therefore there exists a basis and even an orthonormal basis of C_{j+1} which is an orbit of a vector $g_j \in C_{j+1}$ under the unitary action of Γ_j. There is no uniqueness but all the other vectors g'_j whose orbits are an orthonormal basis of C_{j+1} are obtain from g_j by $g'_j = (\alpha_0 1 + \alpha_1 U_j + \cdots + \alpha_{2^j-1} U_j^{2^j-1})g_j$ where this operator

$\alpha_0 + \alpha_1 U_j + \cdots + \alpha_{2^j-1} U_j^{2^j-1}$ acts unitarily on $L^2(0,1)$. Then our task will be to find the choice of g_j which provides the best estimates like $|g_j(x)| \le C2^{3j/2}(1 + 2^j|x|)^{-2}$.

For doing it, we first want to characterize the Fourier coefficients $c(k)$ of any $f \in B_j$. We have the following simple characterization: the sequence $c(k)k^2$, $-\infty < k < +\infty$ has to be 2^j-periodic in k. The factor k^2 comes from the fact that the second derivative of f is a sum of Dirac masses at $m2^{-j}$, $0 \le m < 2^j$ whose Fourier coefficients have the required periodicity.

It is now easy to write a characterization for the Fourier coefficients of any $f \in C_{j+1}$. The orthogonality relations will be written $\Sigma' \, c(k) \, \gamma(k)k^{-2} = 0$, where $f(x) = \sum_{-\infty}^{\infty} \gamma(k)e^{ikx}$ and where $c(k)$ is any 2^j-period sequence vanishing on $2^j\mathbb{Z}$. The symbol Σ' means that there is no zero order term in the sum. This is not a restriction since f needs to be orthogonal to 1 which gives $\gamma(0) = 0$. It is easily seen that this is equivalent to

$$(1.3) \qquad \sum_{\ell=-\infty}^{\infty} \gamma(k+\ell 2^j)(k+\ell 2^j)^{-2} = 0, \ \gamma(0) = 0, \ k = 1, 2, \ldots, 2^j-1.$$

We also have that $k^2\gamma(k)$ has to vanish on $2^{j+1}\mathbb{Z}$ and is 2^{j+1}-periodic in k. We want to find the simplest function $g_j(x) = \sum_{-\infty}^{\infty} \gamma(j,\ell)e(\ell x)$ belonging to C_{j+1} whose Γ_j-orbit is an orthonormal basis. Since the dimension of C_{j+1} is equal to the cardinality of Γ_j we have only to require that this orbit be an orthonormal family. It is quite simple to write the orthogonality conditons in terms of the coefficients $\gamma(j,\ell)$. It gives

$$(1.4) \qquad \sum_{-\infty<m<+\infty} |\gamma(j,\ell + 2^j m)|^2 = 2^{-j}.$$

For the sake of simplicity we shall look for a specific function $\gamma(x)$, continuous over the real line, such that $x^2\gamma(x)$ be periodic of period 2 and with the property that $\gamma(j,\ell) = 2^{-j/2}\gamma(2^{-j}\ell)$. Then (1.4) is equivalent to

$$(1.5) \qquad \sum_{-\infty}^{\infty} |\gamma(x + m)|^2 = 1.$$

The condition (1.3) will also take a simpler form if we assume that $c(k) = f(2^{-j}k)$ where f is a smooth function, periodic of period 1, such that $f(0) = f'(0) = 0$. Then, passing to the limit as $j \to +\infty$, we obtain $\int_{-\infty}^{\infty} \gamma(x)f(x)x^{-2}dx = 0$. Assuming that $\gamma(x)$ is also smooth with $\gamma(0) = \gamma'(0) = 0$, this condition gives

$$(1.6) \qquad \sum_{-\infty}^{\infty} \gamma(x + m)(x + m)^{-2} = 0.$$

We are now ready for the final explicit computations. Observe that if $\gamma(x)$ satisfies (1.5) and (1.6) the same will be true for the product of $\gamma(x)$ and any unimodular continuous and 1-periodic function.

Our last demand will be

$$\gamma(x) = \exp(-\pi ix)\omega(x) \frac{\sin^2 \pi x/2}{(\pi x/2)^2}$$

where $\omega(x + 2) = \omega(x)$ and $\omega(x)$ is real analytic with zeros at even integers with multiplicity 2. Then our conditions become

$$(1.7) \qquad \omega(x) \sin^2 \pi x/2 \sum_{-\infty}^{\infty} (x + 2\ell)^{-4} = \omega(x + 1) \cos^2 \pi x/2 \sum_{-\infty}^{\infty} (x + 2\ell + 1)^{-4}$$

and

$$(1.8) \qquad |\omega(x)|^2 \sin^4 \pi x/2 \left(\frac{\pi}{2}\right)^{-4} \sum_{-\infty}^{\infty} (x + 2\ell)^{-4}$$
$$+ |\omega(x + 1)|^2 \cos^4 \pi x/2 \left(\frac{\pi}{2}\right)^{-4} \sum_{-\infty}^{\infty} (x + 2\ell + 1)^{-4} = 1.$$

For solving these equations we shall restrict to the case where $\omega(x) \geq 0$ (which is compatible with (1.7)). then we use the following well-known identity

$$\sum_{-\infty}^{\infty} (t + \ell)^{-4} = \pi^4(1 - \frac{2}{3}\sin^2\pi t)(\sin \pi t)^{-4}$$

which implies

$$\sum_{-\infty}^{\infty} (t + 2\ell)^{-4} = (\frac{\pi}{2})^4(1 - \frac{2}{3}\sin^2\pi t/2)(\sin \pi t/2)^{-4}$$

and

$$\sum_{-\infty}^{\infty} (t + 2\ell + 1)^{-4} = (\frac{\pi}{2})^4(1 - \frac{2}{3}\cos^2\pi t/2)(\cos \pi t/2)^{-4}.$$

Easy computations give

$$(1.9) \qquad \omega(x) = \frac{\sin^2\frac{\pi x}{2}}{1 - \frac{2}{3}\sin^2\frac{\pi x}{2}}\left[\frac{\sin^4\frac{\pi x}{2}}{1 - \frac{2}{3}\sin^2\frac{\pi x}{2}} + \frac{\cos^4\frac{\pi x}{2}}{1 - \frac{2}{3}\cos^2\frac{\pi x}{2}}\right]^{-1/2}$$

and we define $g(x) \in L^1(\mathbb{R})$ by its Fourier transform

$$(1.10) \qquad \hat{g}(x) = e^{-i\pi x}\omega(x)\,\frac{\sin^2\frac{\pi x}{2}}{(\frac{\pi x}{2})^2}\,.$$

Here is a picture of $g(x)$.

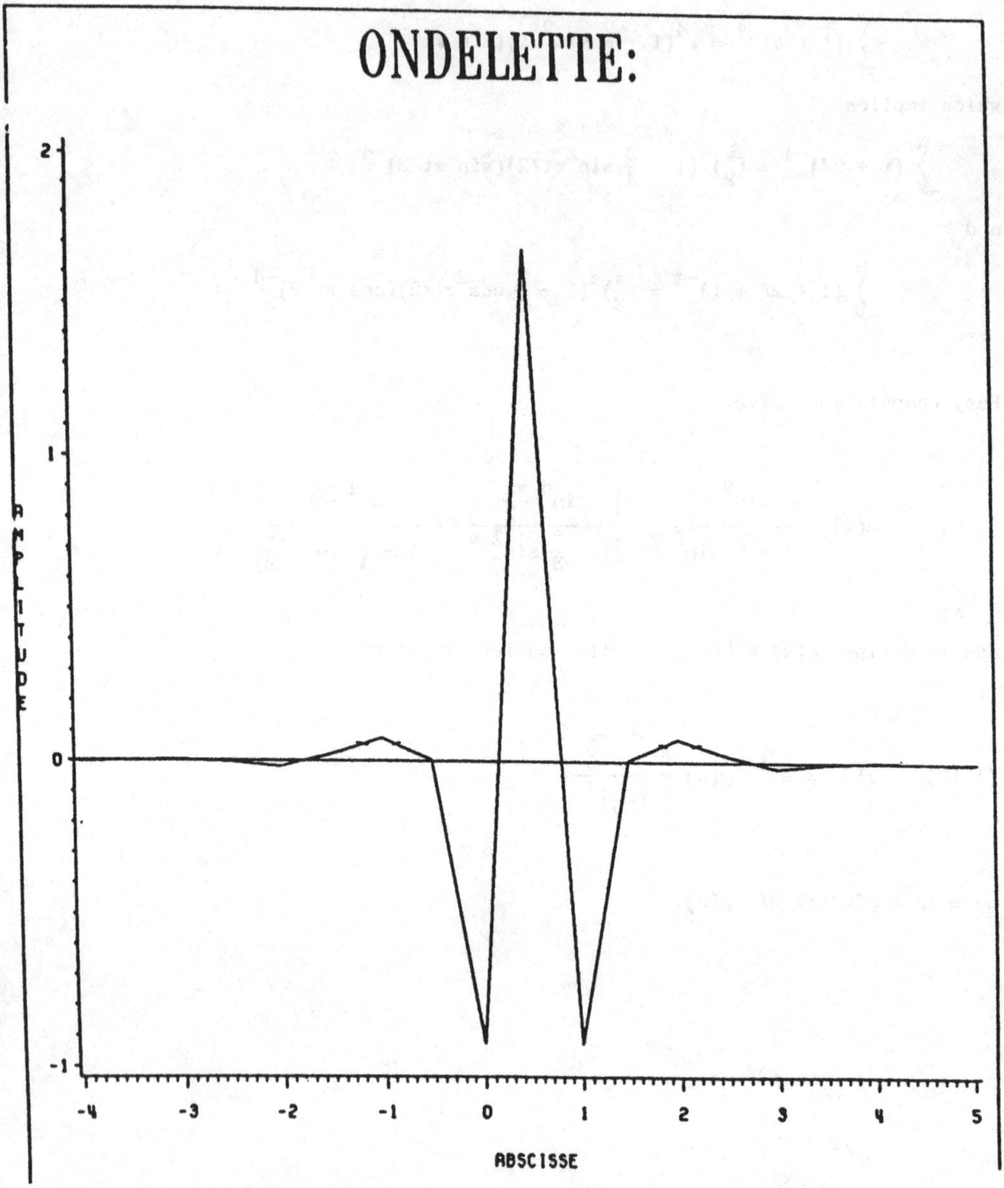
ONDELETTE:
AMPLITUDE
ABSCISSE
2
1
0
-1
-4
-3
-2
-1
0
1
2
3
4
5

Our function $g(x)$ is a Lipschitz function on the real line, piecewise linear with nodes at $m/2$, $m = 0,\pm 1,\pm 2,\ldots$ and has an exponential decay at infinity.

Coming back to the basis $g_{j,k}$ of C_{j+1}, it is defined by $g_{j,k}(x) = g_j(x - 2^{-j}k)$ where

$$g_j(x) = 2^{j/2} \sum_{-\infty}^{\infty} g(2^j x + 2^j m).$$

Estimates on $g_j(x)$ are truly transparent. Since $g_j(x)$ is 1-periodic we can restrict x to $[-\frac{1}{2},\frac{1}{2}]$. If, say, x belongs to $[-\frac{1}{4},\frac{1}{4}]$ $g_j(x) = 2^{j/2}g(2^j x)$ plus an error term whose Lipschitz norm is $O(\exp(-\epsilon 2^j))$ for some $\epsilon > 0$. If x does not belong to $[-\frac{1}{8},\frac{1}{8}]$ the Lipschitz norm of $g_j(x)$ itself decays as fast as $O(\exp(-\epsilon 2^j))$.

2. The function φ and the dyadic partial sums operators.

We now want to compute explicitely the orthogonal projection $\Pi_j : L^2(0,1) \to B_j$. A first observation is that $g_{j,k}$, $0 \le k < 2^j$ together with 1 form when $0 \le j < m$ an orthonormal basis of B_m. Therefore

$$\Pi_m(f) = c_0 + \sum_{\{0\le k<2^j,\,0\le j\le m-1\}} \langle f,g_{j,k}\rangle g_{j,k}. \tag{2.1}$$

The operator Π_m appears as a partial sum operator of order 2^m with respect to the Franklin wavelets.

We are now going to produce a function φ_m in B_m such that the orbit of φ_m under the action of the cyclic group Γ_m be an orthonormal basis of B_m. Since the dimension of B_m equals the order of Γ_m, it suffices that the functions $\varphi_m(x - k2^{-m})$ be an orthonormal family when $0 \le k < 2^j$. As before the Fourier coefficients of φ_m will be sought with the hint that $\hat{\varphi}_m(k) = 2^{-m/2}\hat{\varphi}(2^{-m}k)$ where $\varphi \in L^1(\mathbb{R})$. Then the condition on φ is $\sum_{-\infty}^{\infty}|\hat{\varphi}(x+m)|^2 = 1$ together with the requirement that $x^2\hat{\varphi}(x)$ be periodic of period 1. The simplest way to achieve those properties is to assume that $\hat{\varphi}(x) \geq 0$ which immediately gives

$$\hat{\varphi}(x) = \left(\frac{\sin \pi x}{\pi x}\right)^2\left(1 - \frac{2}{3}\sin^2 \pi x\right)^{-1/2}.$$

Therefore $\varphi(x)$ is a continuous function on $\mathbb{R}$, $\varphi(x) = O(e^{-\epsilon|x|})$ at infinity where $0 < \epsilon < \epsilon_0$ with ch $\epsilon_0 = \sqrt{\frac{3}{2}}$ and φ is piecewise linear with nodes located on the integers $0, \pm 1, \pm 2, \ldots$ As above $\varphi_j(x) = 2^{j/2}\sum_{-\infty}^{\infty}\varphi(2^j x + 2^j\ell)$ and $\varphi_{j,k}(x) = \varphi_j(x - k2^{-j})$.

Then $\Pi_m(f) = \sum_{0 \leq k < 2^m}\langle f, \varphi_{m,k}\rangle\varphi_{m,k}$.

We prefer to write this sum as a triple sum. Indeed, since f is periodic of period 1,

$$\begin{aligned}\langle f, \varphi_{m,k}\rangle &= 2^{m/2}\sum_{\ell}\int_0^1 f(y)\varphi(2^m y + 2^m\ell - k)\,dy \\ &= 2^{m/2}\int_{-\infty}^{\infty} f(y)\varphi(2^m y - k)\,dy = \alpha(m,k).\end{aligned}$$

Therefore $\Pi_m(f)$ is given by

$$\Pi_m(f)(x) = \sum_{0 \leq k < 2^m,\ \ell}\sum \beta(m,k)\varphi(2^m x + 2^m\ell - k) \tag{2.2}$$

where $\beta(m,k) = 2^m\int_{-\infty}^{\infty} f(y)\varphi(2^m y - k)\,dy$.

These coefficients should be thought as mean values of f over the interval centered at $2^{-m}k$ of size 2^{-m}. Let us prove that these operators Π_m are umiformly bounded on L^1 and on L^∞.

If f belongs to L^∞, then $|\beta(m,\ell)| \le \|f\|_\infty \|\varphi\|_1$

and $$\|\Pi_m(f)\|_\infty \le \|f\|_\infty \|\varphi\|_1 \ \Big\| \sum_{-\infty}^{\infty} |\varphi(x+\ell)| \Big\|_\infty .$$

If f belongs to $L^1(0,1)$ we use the fact that Π_m is self-adjoint and we are reduced to the preceding case.

The last estimate we would like to give is a uniform Lipschitz norm estimate for Π_m.

If f is Lipschitz we write $f(y) = f(2^{-m}k) + f(y) - f(2^{-m}k)$ which gives $\beta(m,k) = f(2^{-m}k) + \epsilon(m,k)$ where $|\epsilon(m,k)| \le C2^{-m}$, C being related to the Lipschitz constant of f.

Therefore $\Pi_m f(x) = \sum_{-\infty}^{\infty} f(k2^{-m})\varphi(2^m x - k)$ + error (f is extended by periodicity to the real line). The Lipschitz norm of the error term can be controlled by the L^∞ norm of $\sum_{-\infty}^{\infty} |\varphi'(x-k)|$. For estimating the lipschitz norm of the main term, my student El Hodaibi, applied the following trick. There exists a function $\beta(x)$ with exponential decay at ∞ such that $\varphi'(x) = \beta(x-1) - \beta(x)$. This function $\beta(x)$ is a step function with jump discontinuities at $0, \pm 1, \pm 2, \ldots$ Its Fourier transform is given by

$$\hat{\beta}(x) = \frac{e^{i\pi x}}{2} \frac{\sin \pi x}{\pi x} \left(1 - \frac{2}{3}\sin^2 \pi x\right)^{-1/2}.$$

Then $$\frac{d}{dx}\Big(\sum_{-\infty}^{\infty} f(k2^{-m})\varphi(2^m x - k)\Big) = 2^m\Big(\sum_{-\infty}^{\infty} f(k2^{-m})(\beta(2^m x - k - 1) - \beta(2^m x - k)\Big)$$
$$= 2^m \sum_{-\infty}^{\infty} (f((k-1)2^{-m}) - f(k2^{-m}))\beta(2^m x - k).$$

Finally $2^m|f((k-1)2^{-m} - f(k2^{-m})| \le C$ and $\sum_{-\infty}^{\infty} |\beta(x-k)|$ belongs to $L^\infty(\mathbb{R})$.

3. What are the Franklin wavelets doing for you?

A quite simple exercise is to check that the Franklin wavelets form an unconditional basis for the Stein and Weiss real variable version of the Hardy space H^1.

This fact comes from some general theorems about singular integral operators.

Let $T : L^2(\mathbb{R}/\mathbb{Z}) \to L^2(\mathbb{R}/\mathbb{Z})$ be a bounded operator. Then T is called a Calderón-Zygmund operator if there exists a function $K(x,y)$, $x \in \mathbb{R}/\mathbb{Z}$, $y \in \mathbb{R}/\mathbb{Z}$ such that $|K(x,y)| \le C|x-y|^{-1}$ (where $|x|$ is the distance from x to the nearest integer),

$$\left|\frac{\partial}{\partial x} K(x,y)\right| + \left|\frac{\partial}{\partial y} K(x,y)\right| \le \frac{C}{|x-y|^2}$$

and $(Tf)(x) = \int K(x,y)f(y)dy$ for all continuous functions f and x not belonging to the support of f. (Then the integral is not singular).

If we now assume the T enjoys the following property

$$(3.1) \quad f \in L^2(0,1) \text{ and } \int_0^1 f(x)dx = 0, \text{ imply } \int_0^1 (Tf)(x)dx = 0,$$

then T extends to an operator (still denoted by T) which is bounded on the Stein and Weiss version of the Hardy space H^1. The proof of this simple observation is an exercise if H^1 is defined by its atomic decomposition. Then T maps "atoms" into "molecules".

Let us return to the Franklin wavelets.

We want first to check the fundamental inequality: there exists C such that for all finite sums $\sum\sum \alpha(j,k)g_{j,k}(x) = f(x)$ we have

$$(3.2) \qquad \sup_{|\lambda(j,k)|\le 1} \|\sum\sum \alpha(j,k)\lambda(j,k)g_{j,k}(x)\|_{H^1} \le C\|f\|_{H^1}.$$

The proof of (3.2) is very simple. We introduce the kernel $K(x,y) = \sum\sum \lambda(j,k)g_{j,k}(x)g_{j,k}(y)$, the sum being finite. Then the size and smoothness properties of the Franklin wavelets easily provide the Calderon-Zygmund type estimates.

Property (3.1) is trivial for the following two reasons. The sum which defines $K(x,y)$ should begin with a constant term λ (since the Franklin wavelets begin with the constant function 1). This term disappears once we have integrated against $f(y)$ (since the integral of $f(y)$ vanishes). All other terms have a vanishing integral.

The L^2-continuity of T is also trivial since T is diagonal in some orthonormal basis. Therefore (3.2) is proved.

The span of the Franklin wavelets is dense in $L^2(0,1)$. Therefore it is dense in H^1. But this, together with (3.2) imply that the Franklin wavelets form an unconditional basis for H^1.

Obviously the Franklin wavelets form an unconditional basis for all L^p spaces when $1 < p < +\infty$. They form a Schauder basis for the limiting space L^1 and for B once they are ordered the canonical way as a sequence $1, g_{0,0}, g_{1,0}, g_{1,1}, g_{2,0}, g_{2,1}, g_{2,2}, g_{2,3} \dots$ These two last properties come from the fact that the operators Π_m are uniformly bounded on L^1 and on B. These operators give only the partial sums of order 2^m but the remaining cases follow from the size estimates (sharp localization) of the wavelets: $|g_{j,0}(x)| + |g_{j,1}(x)| + \cdots + |g_{j,2^j-1}(x)| \le C2^{j/2}$.

It is also interesting to observe that the Franklin wavelets are a weak-Schauder basis for the space of Lipschitz functions.

4. Generalizations to higher order splines.

For $m \ge 1$ we define $B_j^{(m)}$ as the linear subspace of B consisting of functions of class C^{m-1} on the real line, 1-periodic, which coincide with a polynomial of degree less than or equal to m on each interval $[k2^{-j},(k+1)2^{-j}]$, $0 \le k < 2^j$.

The Fourier coefficients $c(k)$ of $f \in B_j^{(m)}$ are easily characterized by the condition that $k^{m+1}c(k)$ has to be 2^j-periodic in k. This forces $c(2^jk)$ to vanish for $k = \pm1, \pm2, \pm3, \ldots$ but does not provide any information on $c(0)$.

Therefore the dimension of $B_j^{(m)}$ is 2^j. As above we imbed all the spaces $B_j^{(m)}$ in $L^2(0,1)$ and we define $C_{j+1}^{(m)}$ to be the orthogonal complement of $B_j^{(m)}$ in $B_{j+1}^{(m)}$. We therefore have a unitary group action of the cyclic group Γ_j of order 2^j on $C_{j+1}^{(m)}$. There is an orbit of this group-action which is an orthonormal basis of $C_{j+1}^{(m)}$ and the union of all those basis will be the spline-wavelets (of order m). We take advantage of the non uniqueness of our orthonormal basis to select carefully a basis with good localization. It means that we want

$$|g_j(x)| \le C2^{j/2}(1 + 2^j|x|)^{-2}, \ldots, |g_j^{(m)}(x)| \le C2^{j/2}2^{jm}(1 + 2^j|x|)^{-2}$$

(where the basis of $C_j^{(m)}$ will be $g_j(x - 2^{-j}k)$). This can be achieved by exactly the same calculation as above. We replace everywhere ξ^2 coming from the second derivative by ξ^{m+1}.

II. BOČKAREV'S THEOREMS

The Franklin-wavelets have a remarkable symmetric structure. This structure can be used to avoid most of the technicalities in Bočkarev's proof of the existence of a Schauder basis for the disc algebra.

1. Construction of holomorphic wavelets.

We denote by $S : L^2(0,1) \to L^2(0,1)$ the symmetry defined by $Sf(x) = f(1-x)$. Observing that $g(1-x) = g(x)$ for real x, we immediately deduce that $S(g_{j,k}) = g_{j,k^*}$ where $k^* = 2^j - k - 1$. If $j = 0$, $k = k^* = 0$ while if $j \geq 1$, k and k^* are distinct integers.

We denote by $E \subset L^2(0,1)$ the closed subspace defined by $f(x) = f(1-x)$ and $\int_0^1 f(x)dx = 0$.

Then it is an easy exercise to check that $\sqrt{2}\, P : E \to H_0^2$ is an isomorphic isometry, where P is the projection onto H^2. We have denoted by $H_0^2 \subset H^2$ the closed subspace defined by $\int_0^1 f(x)dx = 0$.

Let us use the Franklin wavelets to construct an orthonormal basis of E. We claim that $g_{0,0}$ and $\frac{1}{\sqrt{2}}(g_{j,k} + g_{j,k^*})$ is such a basis (for $j \geq 1$, $0 \leq k \leq 2^{j-1} - 1$).

Indeed if $f \in E$, f can be expanded into a series of Franklin wavelets

$$f = \alpha(0,0)g_{0,0} + \sum_{\{0 \leq k < 2^j,\, j \geq 1\}} \sum \alpha(j,k) g_{j,k}(x).$$

Applying the symmetry operator S, we get

$$S(f) = \alpha(0,0)g_{0,0} + \sum_{\{0\le k<2^j,\, j\ge 1\}} \sum \alpha(j,k)g_{j,k^*}(x).$$

Identifying the two expansions gives $\alpha(j,k^*) = \alpha(j,k)$ which is what we have claimed.

Now we apply our isometry $\sqrt{2}\,P : E \to H_0^2$. We therefore obtain the following basis of H_0^2

$$(1.1)\qquad h_{0,0} = P(g_{0,0}),\ h_{j,k} = P(g_{j,k}) + P(g_{j,k^*}) \\ 0 \le k \le k(j),\quad k(j) = 2^{j-1} - 1.$$

For getting a better understanding of what this basis looks like, we compute $P(g_{j,k})$. Since P is translation invariant, we have only to compute $P(g_j)$. An obvious computation gives

$$(1.2)\qquad P(g_j) = H_j \quad \text{where} \quad H_j(x) = 2^{j/2} \sum_{-\infty}^{\infty} H(2^j x + 2^j \ell)$$

and $\qquad \hat{H}(x) = \hat{g}(x)$ if $x \ge 0$, $\hat{H}(x) = 0$ if $x \le 0$.

We have the following informations about the size and the smoothness properties of $H(x)$:

$$(1.3)\qquad |H(x)| \le C(1 + x^2)^{-1}$$

$$(1.4)\qquad |H(x') - H(x)| \le C|x' - x|^{\alpha}(1 + x^2)^{-1}$$

for any $\alpha \in (0,1)$, $C = C(\alpha)$, $|x' - x| \le 1$.

This function $H(x)$ is no longer a lipschitz function: its first derivative has logarithmic singularities at $\ell/2$, $\ell = 0, \pm 1, \pm 2, \ldots$ Using splines of higher order, the corresponding function $H(x)$ will be as smooth as we like.

In G. Weiss' terminology, $H(x)$ is called a smooth molecule.

Now $P(g_{j,k}) = H_{j,k} = H_j(x - 2^{-j}k)$. Therefore $H_{j,k}$ is concentrated around $2^{-j}k$. For the same reason H_{j,k^*} is concentrated around $2^{-j}k^*$ and these two points are symmetrical with respect to $\frac{1}{2} - 2^{-j-1}$.

The conclusion of our discussion is that $1, \sqrt{2}h_{0,0}, h_{1,0}, h_{2,0}, h_{2,1}, h_{3,0}, h_{3,1}, h_{3,2}, h_{3,3}, \ldots$ is an orthonormal basis of H^2. We observe that this basis as a very simple translation invariance structure. If we let the cyclic group Γ_j act on $L^2(0,1)$ by translations, we have, as usual

$$U_j(h_{j,k}) = h_{j,k+1}$$

where here $k + 1$ has to be taken modulo 2^{j-1}.

2. Proof of Bočkarev's and Wojtaszczyk's theorems.

Theorem 1. The basis $1, \sqrt{2}\, h_{0,0}, h_{1,0}, h_{2,0}, h_{2,1}, h_{3,0}, h_{3,1}, \ldots$ of H^2 is a Schauder basis for the disc-algebra $A(D)$.

Theorem 2. This basis is an unconditional basis for the holomorphic H^1 space.

For proving Theorem 1, we shall first compute some partial sum operators with respect to our basis. They are

$$\sigma_m(f) = \langle f,1\rangle 1 + 2\langle f,h_{0,0}\rangle h_{0,0} + \sum_{\{1\le j<m,\, 0\le k\le k(j)\}} \langle f,h_{j,k}\rangle h_{j,k}$$

These partial sum operators will be called dyadic partial sum operators.

We assume that $f \in H^2$. Therefore $f = P(f)$. Since P is self-adjoint, we always have $\langle f,P(g_{j,k})\rangle = \langle f,g_{j,k}\rangle$. Therefore

$$\sigma_m(f) = \langle f,1\rangle 1 + 2\langle f,h_{0,0}\rangle h_{0,0} + P\left\{ \sum_{\{1\le j<m,\, 0\le k\le k(j)\}} \langle f,g_{j,k} + g_{j,k^*}\rangle (g_{j,k} + g_{j,k^*})\right\}.$$

Observe next that $g_{j,k^*} = S(g_{j,k})$ and that S is also self-adjoint.

Therefore

(2.1) $$\sigma_m(f) = P(\Pi_m(f) + \Pi_m(S(f))) - \langle f,1\rangle 1.$$

We shall now accept the following crucial lemma (whose proof will be given in the next section).

<u>Lemma 1</u>. <u>There exists a constant</u> C <u>such that the operator norm of the commutator</u> $[\Pi_m,P]$ <u>acting on the space</u> B <u>of 1-periodic continuous functions be bounded by</u> C. <u>Moreover for each</u> $f \in B$, $[\Pi_m,P]f$ <u>converges uniformly to</u> 0 <u>as</u> m <u>tends to infinity</u>.

We now apply our lemma.

Since f belongs to the disc algebra, $f = Pf$ and we have $P\Pi_m(f) = \Pi_m P(f)$ + error term = $\Pi_m(f)$ + error term = f + error terms. The two terms tend uniformly to 0.

For the other term, we first write $f = c + F$ where $c = \int_0^1 f(x)dx$. Then $S(c) = c$ and $\Pi_m(c) = c$ which finally cancells out with $\langle f,1\rangle 1$.

We now return to the main term $P\Pi_m S(F) = \Pi_m PS(F)$ + error term = error term. Indeed $PS(F) = 0$ since $S(F)$ is anti-holomorphic.

3. Proof of the main lemma.

We begin with a few observations on operators and kernels.

Lemma 2. Let $a(x)$ and $b(x)$ be two 1-periodic continuous functions. For integers N we consider the kernels

$$K_N(x,y) = \sum_{0\le k<N} a(x - kN^{-1})b(y - kN^{-1})$$

and the operator T_N defined by these kernels. Then we have, for $m \in \mathbb{Z}$, $T_N(e(mx)) = \tau_m(x)e(mx)$ where the symbol τ_m of the operator T_N is given by

$$\tau_m(x) = N\,\hat{b}(-m) \sum_{-\infty}^{\infty} \hat{a}(m + \ell N)e(\ell Nx).$$

The proof is quite simple and left to the reader.

Lemma 3. We now assume that there exists a C^1-function $f(x)$ which is $O(x^{-2})$ at infinity together with its first derivative and such that $\hat{a}(m) = N^{-1/2}f(mN^{-1})$. We furthermore assume that $f(m) = 0$ for $m = \pm1, \pm2, \ldots$ We also assume that $\hat{b}(m) = N^{-1/2}g(mN^{-1})$ where g has a jump discontinuity at the origin, is C^1 away from the origin and with the same behavior at infinity as f. Then

$$T_N = M_N + R_N$$

where M_N is the convolution operator defined by $M_N(e(mx)) = f(mN^{-1})\bar{g}(mN^{-1})$ and where the norm of $R_N : L^\infty(0,1) \to L^\infty(0,1)$ is uniformly bounded.

The operator of pointwise multiplication with the function $e(\ell x)$ will be denoted χ_ℓ and we then have

$$R_N = \sum_{-\infty}^{\infty}{}' \chi_{\ell N} R_N^{(\ell)}.$$

For ending the proof we want to show that $R_N^{(\ell)}$ is bounded on L^∞ with a norm which is $O(\ell^{-2})$.

For doing it we denote $A(\mathbb{R})$ the Wiener algebra of Fourier transforms of $L^1(\mathbb{R})$ functions equipped with the L^1-norm. We know that if $m(x)$ belongs to $A(\mathbb{R})$, then for any $\epsilon > 0$, $m(\epsilon k)$ is the sequence of Fourier coefficients of a function in $L^1(0,1)$ whose L^1-norm does not exceed $\|m(x)\|_{A(\mathbb{R})}$. Furthermore we know that $\|m(x)\|_{A(\mathbb{R})} \le C\|m\|_2^{1/2}\|m'\|_2^{1/2}$ where m' is the distributional derivative of m. Returning to our specific situation, $\epsilon = N^{-1}$, $m(x) = m_\ell(x) = f(x + \ell)\bar{g}(x)$ and finally $\|m_\ell\|_{A(\mathbb{R})} \le C\ell^{-2}$ which gives the required estimate.

Let us return to the commutator $[\Pi_m, P]$. We have

$$\Pi_m(f) = \sum_{0 \le k < 2^m} \langle f, \varphi_{m,k} \rangle \varphi_{m,k}.$$

Hence

$$[\Pi_m, P](f) = \sum_{0 \le k < 2^m} \langle f, \tilde{\varphi}_{m,k} \rangle \varphi_{m,k} - \sum_{0 \le k < 2^m} \langle f, \varphi_{m,k} \rangle \tilde{\varphi}_{m,k}.$$

We have $\tilde{\varphi}_{m,k} = \tilde{\varphi}_m(x - 2^{-m}k)$ and the Fourier coefficients of $\tilde{\varphi}_m$ are 0 if $\ell < 0$, $2^{-m/2}\hat{\varphi}(2^{-m}k)$ if $\ell \ge 0$, where $f(x) = \hat{\varphi}(x) = (\frac{\sin \pi x}{\pi x})^2(1 - \frac{2}{3}\sin^2 \pi x)^{-½}$. If, in Lemma 2, $a(x)$ is $\varphi_m(x)$ and $b(x)$ is $\tilde{\varphi}_m$, then the main term will be the same as when $a = \tilde{\varphi}_m$ and $b = \varphi_m$. The main terms cancell and the error terms are uniformly bounded.

Returning to the last assertion of Lemma 1, we restrict our attention to the case where f is a C^1 function. Then Pf is certainly a continuous function and $\Pi_m(Pf)$ converges uniformly to Pf. On the other hand, $\Pi_m(f)$ converges uniformly to f in Lipschitz norm and $P\Pi_m(f)$ converges uniformly to Pf.

4. Proof of Theorem 2. As in Chapter I we study the kernel

$$\sum \sum \lambda(j,k)(h_{j,k}(x) + h_{j,k^*}(x))(g_{j,k}(y) + g_{j,k^*}(y))$$

which is obviously a sum of four terms. The kernels

$$K_1(x,y) = \sum \sum \lambda(j,k) h_{j,k}(x) g_{j,k}(y) \quad \text{or}$$

$$K_2(x,y) = \sum \sum \lambda(j,k) h_{j,k^*}(x) g_{j,k^*}(y)$$

satisfy the assumptions of Section 3, Chapter I. For the H^1-boundedness only regularity with respect to the y variable is needed.

The operators defined by the kernels $K_3(x,y)$ and $K_4(x,y)$ are not of Calderón-Zygmund type. Nevertheless they will belong to that class, once we perform the change of variable $y \to 1 - y$. This remark ends the proof of Theorem 2.

III. CUBIC SPLINES AND FRANKLIN WAVELETS

We first show how to use the tools we have developed in Chapter 1 to give an elementary approach to cubic splines. Then we characterize the Zygmund class Λ_* by the rate of convergence measured in L^∞-norm of its cubic-splines approximations. The result is interesting since the Zygmund class cannot be characterized by the rate of convergence of its polygonal approximations.

1. Cubic splines on the circle group.

We are given a sequence $y_0, \ldots, y_{2^j-1}$ of real or complex numbers together with an integer $j \geq 0$. We are looking for the nicest function f such that $f(k2^{-j}) = y_k$, $0 \leq k < 2^j$ and by that we mean that we want to minimize the functional

(1.1) $$\int_0^1 |f''(x)|^2 dx$$

where f'' is the distributional second derivative of the 1-periodic function f.

Without loss of generality we can assume that $y_k = F(k2^{-j})$ where F belongs to the 1-periodic Sobolev space H^2.

Theorem 1. The optimal function f exists, is unique and has the following equivalent characterization

(1.2) f is a 1-periodic function of class C^2 such that $f(x) = a_k + b_k x + c_k x^2 + d_k x^3$ on each interval $[k2^{-j}, (k + 1)2^{-j}]$, $0 \leq k < 2^j$, and such that $f(k2^{-j}) = y_k$. Moreover

(1.3) $$f(x) = \sum_{0}^{2^j-1} f(k2^{-j})\theta_j(x - k2^{-j})$$

<u>where, for a universal constant</u> C,

$$|(\tfrac{d}{dx})^q\theta_j(x)| \le C2^{qj}(1 + 2^j\|x\|)^{-2}, \quad q = 0,1,2, \text{ and } 3,$$

where, for $0 \le x < 1$, $\|x\| = \inf\,\{x, 1-x\}$. ,Let us denote by $\Delta_{j,k}$ the triangle function with height 1 and base $[(k - 1)2^{-j}, (k + 1)2^{-j}]$.

Then $f(k2^{-j}) = F(k2^{-j})$ implies $\langle f'', \Delta_{j,k}\rangle = \langle F'', \Delta_{j,k}\rangle$ which means that $f'' - F''$ is orthogonal to B_j. Since $\|f''\|_2$ should be minimal f'' has to be the orthogonal projection of F'' on B_j. Therefore we have (1.2).

Conversely $\langle f'', \Delta_{j,k}\rangle = \langle F'', \Delta_{j,k}\rangle$ yields $f(k2^{-j}) = F(k2^{-j})$ after fixing the mean value of f on $[0,1]$.

For proving the converse implication, the uniqueness of the solution f to (1.2) is needed. This uniqueness will follow immediately from the fact that the dimension of the linear space $S_3(j)$ of all cubic splines with nodes at $k2^{-j}$ is precisely 2^j.

But the Fourier coefficients $\gamma(\ell)$ of $f \in S_3(j)$ are characterized by $(\ell + 2^j)^4\gamma(\ell + 2^j) = \ell^4\gamma(\ell)$, $\ell \in \mathbb{Z}$. Hence the dimension of $S_3(j)$ is 2^j. Therefore for $0 \le k < 2^j$ there exists a unique function $\theta_{j,k} \in S_3(j)$ such that $\theta_{j,k}(k'2^{-j}) = 0$ if $k' \ne k$, 1 if $k' = k$. Using the Γ_j group action defined in Chapter 1 together with the uniqueness of $\theta_{j,k}$, we obtain $\theta_{j,k}(x) = \theta_j(x - k2^{-j})$. The function $\theta_j(x)$ belongs to $S_3(j)$ and satisfies

(1.4) $$\theta_j(0) = 1, \ \theta_j(2^{-j}k) = 0 \ \text{ for } \ 1 \le k < 2^j.$$

Therefore its Fourier coefficients $\gamma_j(\ell)$ satisfy

(1.5) $$\sum_{-\infty}^{\infty} \gamma_j(\ell_0 + 2^j \ell) = 1, \ 0 \le \ell_0 < 2^j$$

and

(1.6) $$(\ell + 2^j)^4 \gamma_j(\ell + 2^j) = \ell^4 \gamma_j(\ell).$$

The coefficients $\gamma_j(\ell)$ can be computed explicitly and we obtain

(1.7) $$\theta_j(x) = \sum_{-\infty}^{\infty} \beta(2^j x + 2^j k)$$

where $\hat{\beta}(\xi) = (\frac{\sin \pi\xi}{\pi\xi})^4 (1 - \frac{2}{3} \sin^2 \pi\xi)^{-1}$.

Then the estimates on θ_j are straightforward.

2. Characterization of the Zygmund class.

For a 1-periodic continuous function F, $S_j(F)$ will denote the cubic spline which belongs to $S_3(j)$ and coincides with F at the sampling points $k2^{-j}$, $0 \le k < 2^j$.

We want to show that $S_j(F)$ is a good approximation of F in the sense that the smoothness properties of F are correctly reflected by size properties of $S_{j+1}(F) - S_j(F) = \Delta_j(F)$.

Theorem 2. The two following properties are equivalent

(2.1) $F(x)$ belongs to the Zygmund class Λ_*

(2.2) there exists a constant C such that $\|\Delta_j(F)\|_\infty \le C2^{-j}$ for all j's.

The Zygmund class is defined by

$$(2.3)\qquad |F(x+h)+F(x-h)-2F(x)| \le Ch \qquad (0 \le x \le 1,\ 0 \le h \le 1).$$

Before proving this theorem, let us make a few remarks and comments. The Hölder class C^r is defined by $|F(x+h)-F(x)| \le Ch^r$ when $0 < r < 1$ (x and h are real) and by requiring that F be of class C^1 with $F' \in C^{r-1}$ when $1 < r < 2$. Finally if $2 < r < 3$, we similarly define $F \in C^r$ by $F' \in C^{r-1}$. But for $r = 1$ we will have to replace the usual C^1 by the Zygmund class Λ_* and similarly when $r = 2$, we shall replace the usual C^2 by the 1-periodic primitives of the functions in the Zygmund class.

Then, for $0 < r < 3$, F belongs to C^r (with the given substitutes when $r = 1,2$) if and only if $\|\Delta_j(F)\|_\infty \le C2^{-rj}$.

The proof which will be given yields the general case. Moreover if $0 < r < 1$, $S_j(F)$ could be replaced by the polygonal approximation (splines of order 3 could be replaced by splines of order 1). But it is no longer the case when the Zygmund class is concerned.

We first assume that F belongs to Λ_*.

Then $S_j(F) = \sum F(k2^{-j})\theta_j(x - k2^{-j})$.

Therefore $\frac{d^2}{dx^2} S_j(F) = \sum F(k2^{-j})\frac{d^2}{dx^2}\theta_j(x - k2^{-j})$.

We now use the fundamental observation that $\beta''(x) = \gamma(x+1) + \gamma(x-1) - 2\gamma(x)$ where $\gamma(x)$ is a piecewise linear function with nodes at $k \in \mathbb{Z}$, continuous on the real line, with exponential decay at infinity and whose Fourier transform is given by $\hat{\gamma}(\xi) = \left(\frac{\sin \pi\xi}{\pi\xi}\right)^2\left(1 - \frac{2}{3}\sin^2 \pi\xi\right)^{-1}$.

Therefore $\frac{d^2}{dx^2}(\theta_j(x)) = 4^j\{\gamma_j(x + 2^{-j}) + \gamma_j(x - 2^{-j}) - 2\gamma_j(x)\}$ where γ_j is defined in terms of γ exactly the same way as θ_j is defined in terms of β.

Finally $\frac{d^2}{dx^2} S_j(F) = 4^j \sum \{F((k+1)2^{-j}) + F((k-1)2^{-j}) - 2F(k2^{-j})\}$ $\gamma_j(x - k2^{-j})$ and we then can apply the fundamental inequality defining the Zygmund class. It gives $\|\frac{d^2}{dx^2}(S_j(F))\|_\infty \le C2^j$.

Therefore $\|\frac{d^2}{dx^2}(\Delta_j(F))\|_\infty \le 3C2^j$ but we also know that $\Delta_j(F)(k2^{-j}) = 0$ for $0 \le k < 2^j$. Simple calculus yields $\|\Delta_j(F))\|_\infty \le C'2^{-j}$. There is an other and more natural way to compute $\|\frac{d^2}{dx^2}(\Delta_j(F))\|_\infty$. It is to observe that this second derivative $\frac{d^2}{dx^2}(\Delta_j(F))$ is indeed the orthonormal projection of F'' onto the space C_{j+1}. Therefore $\frac{d^2}{dx^2}(\Delta_j(F)) = \sum_k \langle F'', \psi_{j,k}\rangle$ $\psi_{j,k}$.

We now want to show that $|\langle F'', \psi_{j,k}\rangle| \le C2^{j/2}$ whenever F belongs to the Zygmund class. That estimate will give $|\frac{d^2}{dx^2}(\Delta_j(F))| \le C2^{j/2} \sum|\psi_{j,k}(t)|$ $\le C'2^j$ which is enough to ensure that $\|\Delta_j(F)\|_\infty \le C''2^{-j}$. But $\langle F'', \psi_{j,k}\rangle =$ $\langle F, d\mu_{j,k}\rangle$ where $d\mu_{j,k}$ is the second derivative in the distributional sense of $\psi_{j,k}$. Since we know $\psi_{j,k}$ explicitly, we have

(2.4) $d\mu_{j,k}$ is a sum of Dirac masses at the points $\ell 2^{-j-1}$, $0 \le \ell \le 2^{j+1}$

(2.5) $d\mu_{j,k}$ is symmetric with respect to the point $(2k+1)2^{-j-1}$.

(2.6) $\int d\mu_{j,k} = 0$ and $\int |t - (2k+1)2^{-j-1}| d|d\mu_{j,k}| \le C2^{j/2}$.

We then have $|\langle F, d\mu_{j,k}\rangle| \le C'2^{j/2}\|F\|_*$.

Now we prove that $\|\Delta_j(F)\|_\infty \le C2^{-j}$ for all j's implies $F \in \Lambda_*$.

For doing that we use the following Bernstein's type inequalities lemma. There is a universal constant C such that, when $q = 0,1,2,$ or $3,$

$$\|(\frac{d}{dx})^q f(x)\|_\infty \le C2^{jq} \sup_{0\le k<2^j} |f(k2^{-j})|$$

for all $j \ge 0$ and all cubic splines $f \in S_3(j)$.

These estimates are straightforward since $f(x)$ is given by the explicit representation of Theorem 1.

We are reduced now to study the smoothness property of the sum of $f(x)$ of a series $g_0(x) + g_1(x) + g_2(x) + \cdots$ where $\|g_j\|_\infty \le C2^{-j}$ and $\|g_j''\|_\infty \le C2^j$.

For proving that $f(x)$ belongs to the Zygmund class, we have to show that $|f(x + h) + f(x - h) - 2f(x)| \le C|h|$. We define m by $2^{-m} \le |h| < 2^{-m+1}$ and split f into $g + h$ where $g = g_0 + \cdots + g_m$ and $h = g_{m+1} + \cdots$ We obviously have $\|h\|_\infty \le C'2^{-m}$ and the three terms where h shows up have trivial estimates. Applying Taylor's formula yields $|g_j(x + h) + g_j(x - h) - 2g_j(x)| \le C2^jh^2$ and $h^2 \sum_0^m 2^j \le C'h$.

IV. FRANKLIN WAVELETS ON THE LINE

Our goal will be to explain the connection between the construction of the Franklin wavelets on the circle group with the similar construction on the line which is somehow more simple and transparent since dilations act in a simpler way.

Afterwards we will show how the Franklin wavelets on the line are connected with the usual interpolation with cubic splines. The main difference with the situation on the circle group comes from the fact that a very specific growth condition is needed to ensure the uniqueness of the cubic-spline approximation to a function f given by the sampling $f(k2^{-j})$, $k = 0,\pm 1,\ldots$ This growth condition fits exactly with the decay of the specific functions we have constructed so far.

1. Construction of the Franklin wavelets on the line.

Our reference Hilbert space will be the usual $L^2(\mathbb{R})$. We first will try to understand the subspace $B_0 \subset L^2(\mathbb{R})$ whose elements are continuous piecewise linear functions with nodes at $0,\pm 1,\pm 2,\pm 3,\ldots$ We construct an orthonormal basis for B_0 whose elements have the form $\varphi(x - k)$, $k = 0,\pm 1,\pm 2,\ldots$ Then the function $\varphi(x)$ will be unique (up to a multiplication by a complex number of modulus 1) if we impose the following extra condition

(1.1) among all other possible choices of a function $g \in B_0$ for which $g(x - k)$, $k = 0,\pm 1,\pm 2,\ldots$ is orthonormal basis of B_0, our special choice φ minimizes the norm $\left(\int_{-\infty}^{\infty} x^2 |g(x)|^2 dx\right)^{1/2}$.

We first want to compute the Fourier transform of a function $f \in B_0$. Such a function is completely defined by the sequence $f(k)$, $k = 0, \pm 1, \pm 2, \ldots$ which belongs to $\ell^2(\mathbb{Z})$. Moreover if $\Delta(x) = \sup(0, 1 - |x|)$ is the standard triangle function whose base is $[-1,1]$, we have $f(x) = \sum_{-\infty}^{\infty} f(k)\Delta(x-k)$.
Therefore

$$\hat{f}(\xi) = \int_{-\infty}^{\infty} \exp(-2\pi i\xi x) f(x)\,dx = \frac{\sin^2 \pi\xi}{(\pi\xi)^2} \sum_{-\infty}^{\infty} f(k)e(-k\xi) = \frac{\sin^2 \pi\xi}{(\pi\xi)^2}\, m(\xi)$$

where $m(\xi) \in L^2(0,1)$ and is extended by periodicity.
We can now list our requirements about the function φ. We should first have

$$\hat{\varphi}(\xi) = \frac{\sin^2 \pi\xi}{(\pi\xi)^2}\, m(\xi) \quad \text{with} \quad m(\xi) \in L^2(0,1) \tag{1.2}$$

and $m(\xi)$ being 1-periodic and then have

$$\sum_{-\infty}^{\infty} |\hat{\varphi}(\xi + \ell)|^2 = 1. \tag{1.3}$$

The second condition comes from the orthogonality relations between the $\varphi(x-k)$'s.
We will write down a specific solution which generates all the other solutions. This specific solution is

$$\hat{\varphi}(\xi) = \frac{\sin^2 \pi\xi}{(\pi\xi)^2} \left(1 - \frac{2}{3}\sin^2 \pi\xi\right)^{-\frac{1}{2}}. \tag{1.4}$$

It is clear from (1.3) that all other solutions are obtained from this special solution by multiplication by unimodular 1-periodic functions.

Let us prove that the collection of functions $\varphi(x - k)$, $-\infty < k < +\infty$, is complete in B_0. By the Fourier transform it amounts to proving that the collection of $\hat{\varphi}(\xi)e(k\xi)$ $k \in \mathbb{Z}$, is complete in $\mathcal{F}(B_0)$ where $\mathcal{F}$ denote the Fourier transform. As it was obvious from the very beginning, B_0 is isomorphic to $\ell^2(\mathbb{Z})$, the isomorphism being given by $f \to (f(k))_{-\infty<k<+\infty}$ or equivalently by $f \to m(\xi) \in L^2(0,1)$. It means that we can "erase $\frac{\sin^2 \pi\xi}{(\pi\xi)^2}$ " and we are reduced to proving that the functions $(1 - \frac{2}{3}\sin^2\pi\xi)^{-½}e(k\xi)$ are complete in $L^2(0,1)$ which is clear.

Finally we want to find the optimal solution. We observe that $\int x^2|g(x)|^2dx$ being finite, $\hat{g}$ has to belong to the Sobolev space H^1. But $\hat{g}(\xi) = \hat{\varphi}(\xi)\chi(\xi)$ where $\chi(\xi)$ is 1-periodic and unimodular. It immediately implies that $\chi(\xi)$ should be a 1-periodic function which locally belongs to H^1. We therefore can differentiate $\hat{\varphi}(\xi)\chi(\xi)$ and minimize the L^2 norm of the derivative. We have $\chi'(\xi) = i\lambda(\xi)\chi(\xi)$ where $\lambda(\xi)$ is real valued, 1-periodic and belongs to $L^2(0,1)$.

Therefore

$$\|\frac{d}{d\xi}(\hat{\varphi}(\xi)\chi(\xi))\|_2 = \|\hat{\varphi}'(\xi) + i\lambda(\xi)\hat{\varphi}(\xi)\|_2 = (\int_{-\infty}^{\infty}|\hat{\varphi}'(\xi)|^2d\xi + \int_{-\infty}^{\infty}|\lambda^2(\xi)||\hat{\varphi}|^2d\xi)^{½}.$$

The optimal choice is the one for which $\lambda(\xi) = 0$. Therefore $\chi(\xi)$ is a constant complex number of modulus 1.

Next we define $B_j \subset L^2(\mathbb{R})$ as the Hilbert space of continuous functions on the line which are piecewise linear with nodes at $k2^{-j}$, $k \in \mathbb{Z}$.

We then would like to find an orthonormal basis $\psi_{j,k}$ of $L^2(\mathbb{R})$ with the following properties

(1.5) $\psi_{j,k}$ belongs to B_{j+1} and is orthogonal to B_j

(1.6) $\psi_{j,k} = \psi_j(x - k2^{-j})$

(1.7) $\psi_{j,k}$ has a sharp localization on the dyadic interval $I(j,k) = [k2^{-j},(k+1)2^{-j}]$.

By (1.7) we mean that $\int_{-\infty}^{\infty}(x - x(j,k))^2|\psi_{j,k}|^2dx$ is minimal with respect to all the other possible solutions of (1.5), (1.6), $x(j,k)$ being the center of the interval $I(j,k)$.

Moreover it will be proved that such a basis is unique.

For that purpose, we need the following lemma

Lemma 1. <u>There is a unique function</u> $\psi \in B_1$ <u>with the following properties</u>

(a) ψ <u>is orthogonal to</u> B_0

(b) ψ <u>is orthogonal to</u> $\psi(x - k)$ <u>for</u> $k = \pm1, \pm2, \ldots$

(c) $\|\psi\|_2 = 1$ <u>and</u> $\int(x - \frac{1}{2})^2|\psi(x)|^2dx$ <u>is minimal</u>.

Let us first prove the lemma. For the sake of simplifying the calculation, let us write $\omega(x)$ instead of $\dfrac{\sin^2\pi\xi}{(\pi\xi)^2}$. First $\psi \in B_1$ gives $\hat{\psi}(\xi) = \omega(\xi/2)m(\xi/2)$ where $m(\xi)$ is periodic of period 1. Then the orthogonality between ψ and B_0 yields $\int_{-\infty}^{\infty} e(k\xi)\omega(\xi)\omega(\xi/2)m(\xi/2)d\xi = 0$ for $k \in \mathbb{Z}$. It means

$\sum_{-\infty}^{\infty} \omega(\xi + k)\omega(\xi/2) + k/2)m(\xi/2 + k/2) = 0$. Since $m(\xi)$ is periodic of period 1, this sum can be rewritten

$$(1.8) \qquad A(\xi)m(\xi) + A(\xi + \tfrac{1}{2})m(\xi + \tfrac{1}{2}) = 0 \quad \text{where} \quad A(\xi) = \sum_{-\infty}^{\infty} \omega(2\xi + 2k)\omega(\xi + k).$$

Finally the condition (b) yields $\sum_{-\infty}^{\infty} |\omega(\xi/2 + k/2)|^2 |m(\xi/2 + k/2)|^2 = 1$ which can be rewritten

$$(1.9) \qquad B(\xi)|m(\xi)|^2 + B(\xi + \tfrac{1}{2})|m(\xi + \tfrac{1}{2})|^2 = 1.$$

We immediately recognize the equations (1.7) and (1.8) in Chapter 1. Therefore a particular solution is given by

$$(1.10) \qquad \hat{\psi}(\xi) = e^{-i\pi\xi}C(\xi)$$

where $C(\xi) \geq 0$ is 2-periodic and real analytic.

The same argument we have used to prove the uniqueness of φ gives the uniqueness of ψ when we impose condition (c). Uniqueness means up to a multiplication by a complex number of modulus 1.

Returning to the construction of the full basis $\psi_{j,k}$ we use the simple fact that $f(x) \in B_0$ is equivalent to $f(2^j x) \in B_j$. This observation and the uniqueness obtained in Lemma 1 give $\psi_{j,k}(x) = c_j 2^{j/2}\psi(2^j x - k)$ where c_j is a constant and $|c_j| = 1$.

What remains to be done is to prove that this orthonormal family is a basis. Let us denote by C_{j+1} the orthogonal complement of B_j in B_{j+1}. Then we want to prove that, for any fixed j, the collection $\psi_{j,k}$ is an orthonormal basis for C_{j+1}. By rescaling everything we are reduced to the case where $j = 0$. Then it suffices to prove that if f belongs to B_1 and is orthogonal both to B_0 and to all $\psi(x - k)$, necessarily f is 0. Writing $\hat{f}(\xi) = \omega(\xi/2)\mu(\xi/2)$ where $\mu \in L^2(0,1)$ and is 1-perodic, we have as above

$$A(\xi)\mu(\xi) + A(\xi + \tfrac{1}{2})\mu(\xi + \tfrac{1}{2}) = 0.$$

The orthogonality between f and all $\psi(x -k)$ yields

$$D(\xi)\mu(\xi) - D(\xi + \tfrac{1}{2})\mu(\xi + \tfrac{1}{2}) = 0$$

where both $A(\xi)$ and $D(\xi)$ are real analytic and non-negative. Therefore the determinant $A(\xi)D(\xi + \frac{1}{2}) + D(\xi)A(\xi + \frac{1}{2})$ cannot vanish identically and $\mu(\xi) = 0$ (a.e.).

2. <u>The partial sums operators and the relation with the Franklin wavelets on the circle group</u>.

Let us denote by Π_j the orthogonal projection from $L^2(\mathbb{R})$ onto B_j. Then $D_j = \Pi_{j+1} - \Pi_j$ is the orthogonal projection from $L^2(\mathbb{R})$ onto C_{j+1} and we obviously have

$$(2.1) \qquad \sum_{-m}^{m} D_j = \Pi_{m+1} - \Pi_{-m}.$$

But $D_j(f) = \sum_{-\infty}^{\infty} \langle f,\psi_{j,k}\rangle\psi_{j,k}$ and

$$\Pi_j(f) = \sum_{-\infty}^{\infty} \langle f,\varphi_{j,k}\rangle \varphi_{j,k}$$

where both ψ and φ have an exponential decay at infinity. It means that a very simple limiting argument allows us to write, for any $f \in L^\infty(\mathbb{R})$,

$$(2.2) \qquad \sum_{-m}^{m} D_j(f) = \Pi_{m+1}(f) - \Pi_{-m}(f)$$

We would like to pass to the limit as m tends to infinity. (we first replace f by $f_N(x) = f(x)$ when $|x| \le N$, $f_N(x) = 0$ elsewhere and we next observe that the error terms are uniformly bounded in L^∞-norm and converge uniformly to 0 on compact intervals). It is easy to observe that $\|\Pi_m(\varphi)\|_\infty \le C\|f\|_\infty$ where $C = \|\varphi\|_1 \| \sum_{-\infty}^{\infty} |\varphi(x+k)| \|_\infty$. Moreover $\Pi_m(f) \to f$ a.e. as m tends to $+\infty$. When m tends to $-\infty$ we still have uniform boundedness for the L^∞ norms of $\Pi_m(f)$. Moreover the L^∞-norm of the derivative of $\Pi_m(f)$ tends to 0. But in general $\Pi_m(f)$ does not converge (the only possibility for the limit would be to be a constant).

Nevertheless if f belongs to $L^\infty(\mathbb{R})$ and is 1-periodic, this problem does not exist since we already have $\Pi_0(f) = c = \int_0^1 f(x)dx$.

For proving this remark we simply observe that $\sum_{-\infty}^{\infty} \varphi(x+k) = 1$ (as Poisson's formula shows) and therefore $\int_{-\infty}^{\infty} f(t)\varphi(t)dt = \int_0^1 f(t)dt$.

We have, in the 1-periodic case, $\Pi_{m+1}(f) = \int_0^1 f(x)dx + \sum_0^m \sum_{-\infty}^{\infty} \langle f,\psi_{j,k}\rangle \psi_{j,k}$.

We will now reduce the sum in k to be a finite one. We have

$$2^{-j/2}\langle f,\psi_{j,k}\rangle = \int_{-\infty}^{\infty} f(t)\psi(2^j t - k)dt = \sum_{-\infty}^{\infty}\int_0^1 f(t)\psi(2^j t + 2^j \ell - k)dt$$
$$= \int_0^1 f(t)\tilde{\psi}_{j,k}(t)dt$$

where now $\tilde{\psi}_{j,k}$ is 2^j-periodic in k and 1-periodic in t. Since our coefficients $\langle f,\psi_{j,k}\rangle$ happen to be 2^j-periodic in k we can regroup all terms for which $k \bmod 2^j$ is a given k_0 in $[0,2^j)$ and we finally get the Franklin wavelets of Chapter 1.

3. <u>Franklin wavelets and cubic splines</u>.

Given a sequence y_k, $-\infty < k < +\infty$, we want to find the nicest function $y(x)$ such that $y(x) = y_k$, $-\infty < k < +\infty$. Here nicest means that we want the best control on the second derivative. To be more precise we would like to minimize

(3.1) $$\int_{-\infty}^{\infty} |y''(x)|^2 dx$$

over all $y(x)$ whose second derivative in the distributional sense belongs to $L^2(\mathbb{R})$ and such that

(3.2) $$y(k) = y_k.$$

Notice that (3.2) makes sense since $y'(x)$ belongs to the Hölder space $C^{1/2}$ and $y(x)$ to the Hölder space $C^{3/2}$.

Our purpose is to explain the simple relation that exists between this classical problem (whose solution is given by cubic splines) and our orthonormal basis.

We first denote by $f(x)$ the second derivative of y in the distributional sense and rewrite (3.2) in the seemingly equivalent form

$$(3.3) \qquad \int_{-\infty}^{\infty} f(x)\Delta(x-k)dx = y_{k+1} + y_{k-1} - 2y_k = z_k.$$

We immediately see from (3.3) that a necessary condition is $z_k \in \ell^2(\mathbb{Z})$. We are now going to show that for any $z_k \in \ell^2(\mathbb{Z})$, the infimum of $\|f\|_2$ with the constraints (3.3) is attained and f belongs to B_0. Going back to a second primitive y of f, we will have $y(k) = \tilde{y}_k$ where $\tilde{y}_{k+1} + \tilde{y}_{k-1} - 2\tilde{y}_k = z_k = y_{k+1} + y_{k-1} - 2y_k$. Therefore $\tilde{y}_k = y_k + ak + b$. But then the second primitive y can be adjusted so that (3.2) holds.

With those precautions (3.1) and (3.2) is equivalent to minimizing $\|f\|_2$ under the constraints (3.3).

Observe now that, $\varphi(x)$ being defined by (1.4), we have

$$(3.4) \quad \varphi(x) = \sum_{-\infty}^{\infty} \omega_j \Delta(x-j)$$

where the coefficients ω_j are the Fourier coefficients of the 1-periodic function $(1 - \frac{2}{3}\sin^2\pi\xi)^{-1/2}$. Similarly we also have

$$(3.5) \qquad \Delta(x) = \sum_{-\infty}^{\infty} \omega_j' \varphi(x-j)$$

where, this time, ω_j' are the Fourier coefficients of

$$(1 - \tfrac{2}{3}\sin^2\pi\xi)^{1/2}.$$

Therefore (3.3) is equivalent to

(3.6) $$\int_{-\infty}^{\infty} f(x)\varphi(x - k)dx = u_k$$

where $$u_k = \sum_{-\infty}^{\infty} \omega_j z_{k-j}$$

and similarly

$$z_k = \sum_{-\infty}^{\infty} \omega'_j u_{k-j}.$$

Now the solution to (3.6) is obvious since $\varphi(x - k)$, $k \in \mathbb{Z}$, is an orthonormal basis of B_0. Therefore the minimum is attained when

(3.7) $$f(x) = \sum_{-\infty}^{\infty} u_k \varphi(x - k).$$

Returning to the z_k, (3.7) can be rewritten

(3.8) $$f(x) = \sum_{-\infty}^{\infty} z_k \Delta^*(x - k)$$

where

(3.9) $$\Delta^*(x) = \sum_{-\infty}^{\infty} \omega_k \varphi(x - k).$$

Therefore the Fourier transform of Δ^* is given by

$$\mathcal{F}(\Delta^{*}) = \frac{\sin^2 \pi\xi}{(\pi\xi)^2} \left(1 - \frac{2}{3} \sin^2 \pi\xi\right)^{-1}. \tag{3.10}$$

In other words $\Delta(x - k)$, $-\infty < k < \infty$, is an unconditional basis of the Hilbert space B_0 and the dual basis is precisely given by $\Delta^{*}(x - k)$, $-\infty < k < \infty$.

We now define the "fundamental cubic spline" $\beta(x)$ by its Fourier transform

$$\hat{\beta}(\xi) = \left(\frac{\sin \pi\xi}{\pi\xi}\right)^4 \left(1 - \frac{2}{3} \sin^2 \pi\xi\right)^{-1}. \tag{3.11}$$

It is clear that $\beta(x)$ belongs to $C^2(\mathbb{R})$, coincides with a cubic polynomial on each interval $[k, k + 1]$, $k \in \mathbb{Z}$ and that

$$\beta(x) = O(\exp(-c|x|)) \text{ at infinity where} \tag{3.12}$$

$$c = \log(2 + \sqrt{3}). \tag{3.13}$$

The function $\beta(x)$ and $\Delta^{*}(x)$ are related by

$$\beta''(x) = \Delta^{*}(x - 1) + \Delta^{*}(x + 1) - 2\Delta^{*}(x) \tag{3.14}$$

and therefore one has the fundamental identity

$$y(x) = \sum_{-\infty}^{\infty} y_k \, \beta(x - k) \tag{3.15}$$

which gives the solution to the optimal problem (3.1), (3.2).

There are some cases where (3.15) can be understood as giving the solution of another problem. Given a sequence y_k, $-\infty < k < \infty$ with some growth condition, is it possible to find a function $y(x)$ with the following three properties

(3.16) $y(x)$ belongs to $C^2(\mathbb{R})$ and coincides with a cubic polynomial $P_k(x)$ on each interval $[k, k+1]$, $k \in \mathbb{Z}$

(3.17) for some $c < \log(2+\sqrt{3})$, we have $y(x) = O(e^{c|x|})$ at infinity

(3.18) $y(k) = y_k$ for $k \in \mathbb{Z}$?

Let us first prove the existence of a solution under the condition $|y_k| = O(e^{c|k|})$ for some $c < \log(2+\sqrt{3})$.

For doing it we just plug those coefficients into (3.15). The series converge since the decay of β matches the growth of the coefficients. Moreover $|y(x)| = O(e^{c|x|})$ where c is the same constant that gives the exponential growth of the coefficients.

Finally we have $\beta(0) = 1$, $\beta(\pm 1) = 0$, $\beta(\pm 2) = 0, \ldots$ This is obvious from the fact that (3.15) gives the solution to the interpolation by cubic splines when say $y_k = 1$ for $k = 0$ and $y_k = 0$ for $k \neq 0$.

Let us prove the uniqueness. We therefore assume that (3.16) and (3.17) hold together with $y(k) = 0$ for $k \in \mathbb{Z}$. We want to prove that y vanishes identically.

For doing it we consider the two by two matrix M that takes $(y'(k), y''(k))$ to $(y'(k+1), y''(k+1))$. This matrix does not depend on k and is

$-\begin{bmatrix} 2 & \frac{1}{2} \\ 6 & 2 \end{bmatrix}$. The eigenvalues of M are $2 \pm \sqrt{3}$ and therefore $|y'(k)| + |y''(k)| + |y'(-k)| + |y''(-k)|$ is equivalent to $C(2+\sqrt{3})^{|k|}$ when $|k|$ tends to $+\infty$ where $c > 0$ unless y vanishes identically.

We can summarize what we have obtained and state a theorem.

Theorem 1. Let y_k, $k = 0, \pm1, \pm2, \ldots$ be a sequnce of real or complex numbers such that for some $r < 2 + \sqrt{3}$ and some positive constant C, we have

$$|y_k| \le Cr^{|k|}.$$

Then there is a unic cubic-spline function f(x) of class C^2 on the line, with nodes at $0, \pm1, \pm2, \ldots$, such that

$$f(k) = y_k, \ k = 0, \pm1, \pm2, \ldots$$

and $$|f(x) \le C'r^{|x|}$$ for some constant C'.

Moreover $f(x) = \sum_{-\infty}^{\infty} y_k \beta(x-k)$ where $\beta(x)$ is defined by (3.11).
If F(x) is a continuous function on the line whose second derivative in the distributional sense F''(x) belongs to $L^2(\mathbb{R})$, then the cubic-spline solution to $f(k) = F(k)$, $k = 0, \pm1, \pm2, \ldots$ has the property that its second derivative is the orthogonal projection of F'' on B_0.
This last observation can be easily proved. The equalities $F(k) = f(k)$ can be rewritten $\int F''(x)\Delta(x-k)dx = \int f''(x)\Delta(x-k)dx$. Since B_0 is the closed

linear span of the functions $\Delta(x - k)$, $k \in \mathbb{Z}$, $F'' - f''$ has to be orthogonal to B_0. But f'' belongs to B_0 since f is a cubic spline. Therefore f'' is the orthogonal projection of F'' onto B_0.

5. Why quadratic splines should be disregarded?

Let us return to the case of the circle group. If the linear space $S_2(j)$ consists of all C^1 functions on the circle group which are piecewise quadratic polynomials on each interval $[k2^{-j},(k + 1)2^{-j}]$, then the remarquable interpolation properties we have obtained for cubic-splines simply disappear. For example there does not exist an $f \in S_2(j)$ such that $f(0) = 1$ and $f(k2^{-j}) = 0$ for $1 \leq k \leq 2^j - 1$. The mapping that takes $f \in S_2(j)$ to the sampling $f(k2^{-j})$, $0 \leq k < 2^j$ is neither injective nor surjective. Therefore quadratic splines cannot be used for recovering an approximation to a function which is only given by a sampling. But quadratic splines can be used without any restriction for the construction of an orthonormal basis of wavelets.

6. The Franklin wavelets in two variables.

For a better understanding of the structure of the Franklin wavelets in two variables we should first describe the Haar system in two variables.

Let $h(x)$ be the step-function which is supported by $[0,1)$ and which takes the value 1 on the first half of this interval and -1 on the second half. Then the Haar system in one variable is the orthonormal basis of $L^2(\mathbb{R})$ consisting of all $2^{j/2}h(2^jx - k)$ $j = 0,\pm1,\pm2,\ldots,k = 0,\pm1,\pm2,\ldots$

In two variables one might take the tensor product of this basis with itself. But the functions we are constructing are now supported by arbitrarily eccentric rectangles (instead of squares).

If we want that the underlying geometrical setting be given by balls or squares instead of eccentric rectangles, we shall do something distinct. Let us denote by $\chi(x)$ the characteristic function of $[0,1)$. Then it is a simple exercise to check that the three strings

$$2^j h(2^j x_1 - k_1) h(2^j x_2 - k_2),\ j \in \mathbb{Z},\ k_1 \in \mathbb{Z},\ k_2 \in \mathbb{Z}$$
$$2^j h(2^j x_1 - k_1) \chi(2^j x_2 - k_2) \ \ldots\ldots$$
$$2^j \chi(2^j x_1 - k_1) h(2^j x_2 - k_2) \ \ldots\ldots$$

all together form an orthonormal basis of $L^2(\mathbb{R}^2)$. We shall mimic this construction.

Theorem 2. The functions $2^j\psi(2^j x_1 - k_1)\psi(2^j x_2 - k_2)$, $2^j\psi(2^j x_1 - k_1)\varphi(2^j x_2 - k_2)$, $2^j\varphi(2^j x_1 - k_1)\psi(2^j x_2 - k_2)$ all together form an orthonormal basis of $L^2(\mathbb{R}^2)$.

The orthogonality is straightforward since it reduces in all cases to the case of one real variable.
The completeness comes from the following trivial identity

$$(6.1) \qquad \Pi_{j+1} \otimes \Pi_{j+1} - \Pi_j \otimes \Pi_j = D_j \otimes \Pi_j + \Pi_j \otimes D_j + D_j \otimes D_j .$$

Each one of the three terms in the right-hand side of (5.1) corresponds to the orthogonal projection on one of our strings.
But $\Pi_j \otimes \Pi_j \uparrow 1$ as j tends to $+\infty$, and to 0 as j tends to $-\infty$, and this remark ends the proof of Theorem 2.

In more than two variables, the construction is identical.

V. THE SMOOTH WAVELETS

Taking advantage of our preceding experiences, we will construct another family of wavelets. This other family will be better behaved as much as smoothness is concerned, and as much as the localization of Fourier coefficients is concerned.

This new basis could be used to give a proof of Bockarev's theorem in which the technicalities in Lemma 1, Chapter II, would disappear.

1. Construction of the smooth wavelets on the real line

The construction is completely similar to the one used for the Franklin wavelets. The only slight modification is the definition of the functions ψ and φ. The new ψ is defined, as above by its Fourier transform

$$\hat{\psi}(x) = e^{-i\pi x}\theta(x) \tag{1.1}$$

where $\theta(x)$ is a $C_0^\infty(\mathbb{R})$ function which is supported by $[-1 - 2\epsilon, -\frac{1}{2} + \epsilon] \cup [\frac{1}{2} - \epsilon, 1 + 2\epsilon]$ and even. Here $\epsilon > 0$ is fixed and we impose that $\epsilon \leq \frac{1}{2}$. The function $\theta(x)$ should in fact be written $\theta_\epsilon(x)$. Some other properties of $\theta(x)$ are needed

$$\theta(x) \geq 0 \text{ everywhere and } \theta(x) = 1 \text{ on } [\tfrac{1}{2} + \epsilon, 1 - 2\epsilon] \tag{1.2}$$

$$\theta^2(1 - x) + \theta^2(x) = 1 \text{ on } [\tfrac{1}{2} - \epsilon, \tfrac{1}{2} + \epsilon] \tag{1.3}$$

$$\theta(2x) = \theta(1 - x) \text{ on } [\tfrac{1}{2} - \epsilon, \tfrac{1}{2} + \epsilon] \tag{1.4}$$

It is an easy exercise to construct such functions $\theta(x)$.
We also need a function φ as before. The Fourier transform $\hat{\varphi}$ of φ will be even, non negative, in $C_0^\infty(\mathbb{R})$ and related to θ by

$$(1.5) \qquad \theta^2(x) = (\hat{\varphi}(x/2))^2 - (\hat{\varphi}(x))^2$$

which implies that

$$(1.6) \qquad \hat{\varphi}(x) = 1 \quad \text{on} \quad [-\tfrac{1}{2} + \epsilon, \tfrac{1}{2} - \epsilon] \quad \text{and}$$

$$(1.7) \qquad \hat{\varphi}(x) = 0 \quad \text{outside} \quad [-\tfrac{1}{2} - \epsilon, \tfrac{1}{2} + \epsilon].$$

We denote by $\varphi_{j,k}$ and $\psi_{j,k}$ the functions $2^{j/2}\varphi(2^j x - k)$ and $2^{j/2}\psi(2^j x - k)$. We then have

Theorem 1. With the preceding notations, the collection $\psi_{j,k}$, $j \in \mathbb{Z}$, $k \in \mathbb{Z}$, is an orthonormal basis of $L^2(\mathbb{R})$. We furthermore have the following identity

$$(1.8) \qquad S_m(f) = \sum_{-\infty<k<+\infty} \langle f, \varphi_{m,k}\rangle \varphi_{m,k} = \sum_{\{j<m, -\infty<k<+\infty\}} \langle f, \psi_{j,k}\rangle \psi_{j,k}.$$

An easy computation shows that the functions $\psi_{j,k}$ are orthonormal.
The completeness follows from the identity (1.8) and from the easy observation that $S_m \uparrow 1$ and $S_m \downarrow 0$ as m tends to $+\infty$ or $-\infty$.
We shall denote by $D_{j,k}$ the orthonormal projection on the vector $\psi_{j,k}$ and Δ_j will be the convolution operator with $2^j\psi(2^j x)$. Finally M_j will denote the operator of pointwise multiplication with the function $\exp(2\pi i 2^j x)$.
We will denote by $C(\mathbb{R})$ the Frechet space of continuous functions on

the line, by $M(\mathbb{R})$ the Frechet space of Borel measures and for a given $\mu \in M(\mathbb{R})$, by $\{\mu\} : C(\mathbb{R}) \to M(\mathbb{R})$ the operator mapping $f \in C(\mathbb{R})$ onto the measure $d\nu = f d\mu$. Denoting by δ_a the Dirac mass at the point a, we then have

$$(1.9) \qquad D_{j,k} = 2^{-j}\Delta_j\{\delta_{2^{-j}k}\}\Delta_j^*.$$

Let us check this trivial remark. If we apply to f the right-hand side of (1.9), it gives first $\int 2^j \bar{\psi}(2^jy - 2^jx)f(y)dy$ which then becomes $(\int 2^j \bar{\psi}(2^jy - k)f(y)dy)\delta_{2^{-j}k} = 2^{j/2}\langle f,\psi_{j,k}\rangle\delta_{2^{-j}k}$ and finally we obtain $\psi_{j,k}\langle f,\psi_{j,k}\rangle$.

We then calculate $\sum_{-\infty}^{\infty} D_{j,k} = D_j$ by the Poisson formula which gives $2^{-j}\sum_k \delta_{2^{-j}k} = \sum_k \exp(2\pi i 2^j kx)$. We therefore have

$$(1.10) \qquad D_j = \sum_k \Delta_j(M_j)^k\Delta_j^* = \Delta_j\Delta_j^* + \Delta_jM_j\Delta_j^* + \Delta_jM_{j+1}\Delta_j^* + \Delta_jM_j^*\Delta_j^* + \Delta_jM_{j+1}^*\Delta_j^*.$$

The last identity comes from the fact that $\Delta_j(M_j)^k\Delta_j^* = 0$ for $|k| \geq 3$.

This is due to the following observation. If T is any convolution operator, with symbol $\tau(\xi)$ and if M is the pointwise the multiplication by a character $e(ax)$ then

$$(1.11) \qquad MTM^{-1} = T_a$$

is still a convolution operator whose symbol τ_a is $\tau(\xi - a)$. For proving (1.11), it suffices to apply the left- and the right-hand side to any character $e(\xi x)$ and to know that the relation between a pseudo-differential operator and its symbol is given by $T(e(\xi x)) = \tau(\xi)e(\xi x)$.

This observation and straightforward computations give (1.10).

Finally we call P_j the convolution operator with $2^j\varphi(2^jx)$. Then the following identities are easily checked by (1.11)

$$\Delta_j\Delta_j^* = P_{j+1}P_{j+1}^* - P_jP_j^* \tag{1.12}$$

$$\Delta_jM_j\Delta_j^* = -P_jM_jP_j^* \tag{1.13}$$

$$\Delta_jM_{j+1}\Delta_j^* = P_{j+1}M_{j+1}P_{j+1}^*. \tag{1.14}$$

Finally we have obtained the remarkable identity

$$\begin{cases} D_j = S_{j+1} - S_j \quad \text{where} \\ S_j = P_jP_j^* + P_jM_jP_j^* + P_j^*M_j^*P_j \end{cases} \tag{1.15}$$

We have indeed $P_j^* = P_j$ but we wanted to stress the self-adjointness of S_j in the right-hand side. The observation which plays the role of Lemma 1, Chapter II is the identity

$$[S_j,H] = 2(P_jM_jP_j^* - P_j^*M_j^*P_j). \tag{1.16}$$

The right-hand side of (1.16) is obviously uniformly bounded on L^∞ and that takes care of the technicalities we have been facing in Lemma 1, Chapter II. For constructing the Schauder basis of the disc algebra, we first periodize the smooth wavelets following Chapter III's recipe. Then we pair these new wavelets by couples in order to construct the even wavelets which form a basis for the subspace of $L^2(0,1)$ composed of even functions. Finally we project those even wavelets on H^2.

The smooth wavelets on the circle group will be denoted $\psi^{\#}_{j,k}$, $j \geq 0$, $0 \leq k < 2^j$ (and we have to add the constant function 1 to obtain a basis).

2. What's good or bad with smooth wavelet?

There is a striking similarity between the Franklin wavelets (on the circle group) and the smooth wavelets.

To the advantage of the Franklin wavelets is the fact that the spaces B_m generated $g_{j,k}$, $0 \leq k < 2^j$, $0 \leq j < m$, together with the function 1 have a very simple geometrical meaning: B_0 is the space of all continuous piecewise linear functions on the circle with nodes at $k2^{-m}$, $0 \leq k < 2^m$. You do not get anything similar with the smooth wavelets. The corresponding subspaces would consist of trigonometric polynomials of order less than or equal to $2^m(\frac{1}{2} + \epsilon)$. The dimension of this subspace of trigonometric polynomials is $2^m + 2\epsilon 2^m + 1$. It means that there is a gap between the number of vectors $\psi^{\#}_{j,k}$ you have been using and the dimension of the natural space they sit in. On the other hand, ϵ could not be replaced by 0 without immediately loosing the localization properties of the wavelets.

Another wrong point with the smooth wavelets is that the numerical localization of the function ψ is not quite good compared to the one of the function g.

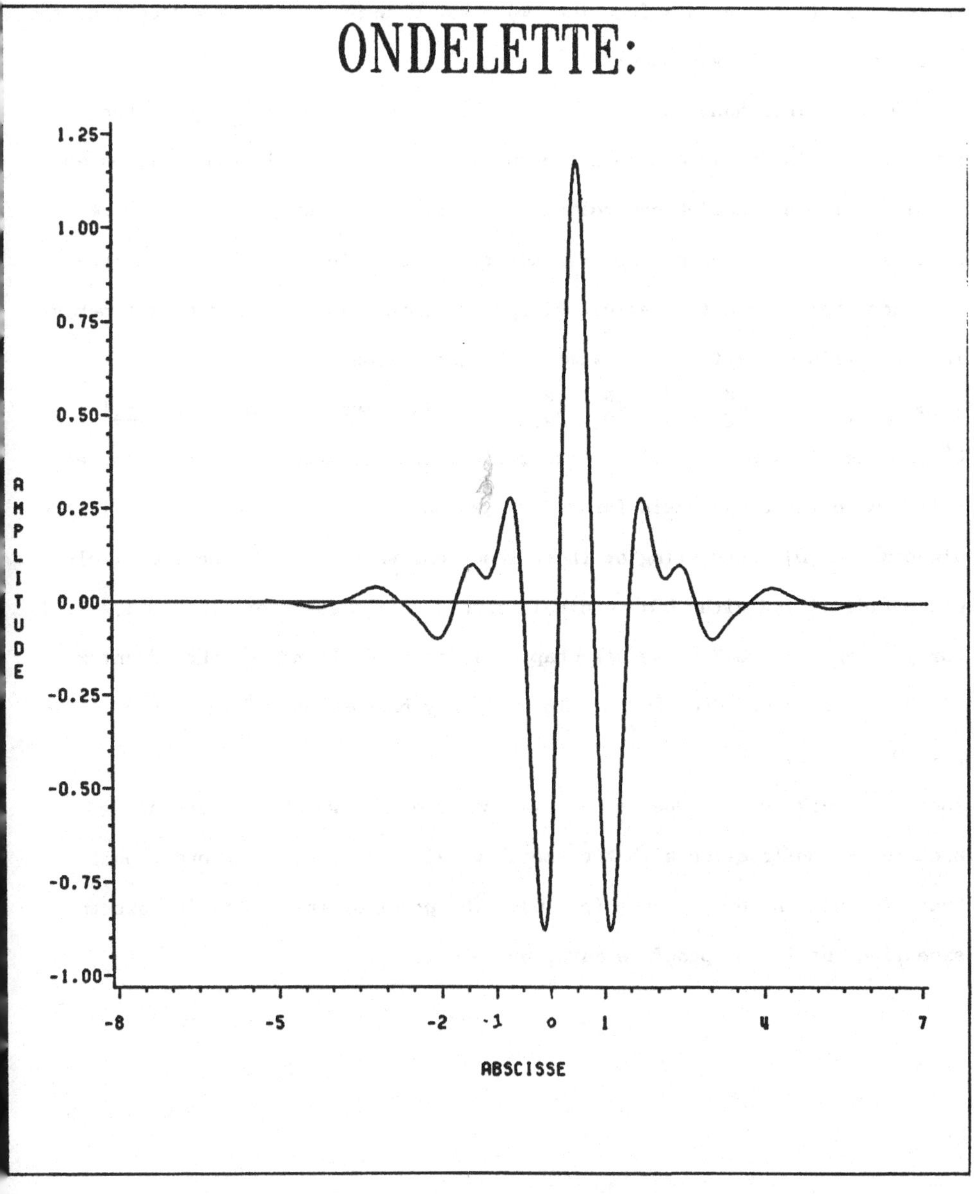
ONDELETTE:
AMPLITUDE
1.25
1.00
0.75
0.50
0.25
0.00
-0.25
-0.50
-0.75
-1.00
-8
-5
-2
-1
o
1
4
7
ABSCISSE

It looks as if the computer were making a difference between a function with exponential decay (like the function g in Chapter I) and a function with rapid decay (like our new function ψ). Rapid decay is not very good from the numerical analyst's viewpoint.

On the other hand, smooth wavelets have two related qualities. Each wavelet of this family can be differentiated or integrated as many times as you like without loosing control on the estimates (except that there is a scaling factor 2^j which appears each time you perform a differentiation). It means that from a theoretical viewpoint, smooth wavelets are better behaved then Franklin-wavelets. For example the collection $\{1,\psi^{\#}_{0,0},\psi^{\#}_{1,0},\psi^{\#}_{1,1},\psi^{\#}_{2,0},\psi^{\#}_{2,1},\psi^{\#}_{2,2},\psi^{\#}_{2,3},\ldots\}$ is a Schauder basis for <u>all</u> the C^r spaces $(r = 0,1,2,\ldots)$. If we perform the Bockarev's construction, we obtain an orthonormal basis for $H^2(D)$ which is a Schauder basis for all the algebras $A^m(D)$ consisting of those functions which are C^m on the circle group (identified with ∂D) and holomorphic on the interior $(m = 0,1,2,\ldots)$. For proving this remark, we use Chapter II's approach and we take advantage of (1.16). The error term $[S_j,H]$ is uniformly bounded on C^m (m fixed, $j = 0,1,2,\ldots$).

Another remark is that the orthonormal basis of H^2 which is given by Bockarev's construction with the smooth wavelets is also an unconditional basis for all the H^p spaces $(p > 0)$. The proof of the remark follows the same lines as in the proof we gave for $p = 1$.

VI. CALDERON ZYGMUND OPERATORS

1. Introduction.

The definition of Calderón-Zygmund operators has changed during the past decades. In the fifties, Calderón-Zygmund operators were still convolution operators with quite specific distributions. Nowadays Calderón-Zygmund operators include completely different operators as the Cauchy kernel on a Lipschitz curve, the double layer potential on the boundary of a Lipschitz domain, etc.

We would like in this introduction to describe this evolution in order to motivate the contemporary problems.

The Calderón-Zygmund program began in the fifties with the systematic study of those singular integral operators of convolution type which naturally appear in linear partial differential equations, mostly in the elliptic case.

Examples are the Riesz transforms $R_j = -i \frac{\partial}{\partial x_j} (-\Delta)^{-½}$ where $\Delta = \frac{\partial^2}{\partial x_1^2} + \cdots + \frac{\partial^2}{\partial x_n^2}$, $1 \leq j \leq n$. The Riesz transforms are the n-th dimensional generalization of the Hilbert transform.

Let Ω be homogeneous of degree 0 in $\mathbb{R}^n$ (Ω is not defined in 0) and let us suppose that Ω, restricted to the unit sphere $S^{n-1} \subset \mathbb{R}^n$ is C^∞ and that

$$(1.1) \qquad \int_{S^{n-1}} \Omega(x) d\sigma(x) = 0$$

where $d\sigma$ is the usual surface measure on S^{n-1}.

Then the "first generation" of singular integral operators studied by Calderón and Zygmund was given by

(1.2) $$Tf(x) = \text{p.v} \int_{\mathbb{R}^n} f(x - y) \frac{\Omega(y)}{|y|^n} dy.$$

This principal value is clearly defined when f is a smooth function with compact support and one can prove that the Fourier transform of Tf is given by the product of the Fourier transform of f by a multiplier $m(\xi)$ which is the Fourier transform of the distribution $\text{p.v.} \frac{\Omega(x)}{|x|^n}$. This multiplier $m(\xi)$ satisfies the following estimates

(1.3) $$|m(\xi)| \le C_0$$

(1.4) $$|\partial^\alpha m(\xi)| \le C_\alpha |\xi|^{-|\alpha|}$$

when $\alpha = (\alpha_1, \dots, \alpha_n)$ and $|\alpha| = \alpha_1 + \dots + \alpha_n$

(1.5) $$\int_{S^{n-1}} m(\xi) d\sigma(\xi) = 0.$$

From (1.3) one can see that T is bounded on $L^2(\mathbb{R}^n)$. But Calderón and Zygmund wanted to know if (1.2) was making sense for any $f \in L^2(\mathbb{R}^n)$. This problem leads one to study the maximal operator T^* given by

(1.6) $$T^*f(x) = \sup_{\epsilon > 0} \left| \int_{|y| \ge \epsilon} f(x - y) \frac{\Omega(y)}{|y|^n} dy \right|.$$

Two natural problems have been raised and solved by Calderón and Zygmund

(1.7) what are the other continuity properties of those singular integral operators?

(1.8) do these singular integral operators form an algebra?

Singular integral operators had already a long history when Calderón and Zygmund made their fundamental contribution. These operators appear in 1936 in G. Giraud and Mihlin. It is interesting to observe that multipliers for trigonometric series (i.e. operators which are diagonalized in the basis e^{ikx}, $k = 0, \pm 1, \pm 2, \ldots$) appear around the same years. Marcinkiewicz proved in 1938 his celebrated multiplier theorem: if $M(e^{ikx}) = m_k e^{ikx}$, $k = 0, \pm 1, \pm 2, \ldots$ if $|m_k| \le C_0$ and $|k||m_{k+1} - m_k| \le C_1$, then M is bounded on $L^p(0,2\pi)$ for $1 < p < \infty$. Marcinkiewicz's proof used a theorem by Littlewood and Paley (1931) giving the equivalence between the L^p norm of the boundary values of a holomorphic function $F(z) = c_0 + c_1 z + c_2 z^2 + \cdots$ and the L^p norm of the Littlewood-Paley functional

$$g(\theta) = |c_0| + \left(\sum_0^\infty |\Delta_j(F)|^2\right)^{1/2} \quad \text{where}$$

$$\Delta_j(F) = \sum_{2^j \le k < 2^{j+1}} c_k e^{ik\theta}.$$

The reduction to the holomorphic case was obtained by the celebrated F. and M. Riesz's theorem. These two steps require $1 < p < \infty$.

The complex method does not exist in several real variables. It was then a major breakthrough when Calderón and Zygmund discovered that the "real variable methods" could give L^p estimates as well. One main new discovery that has opened the gate to the real variable methods was the Calderón-Zygmund decomposition of an L^1 function into a sum of an L^2 function and of a series of oscillating terms carried by a set of small measure. This set is a (countable) union of disjoint cubes which support each of the oscillating terms.

The L^2 estimate for an operator T together with a mild hypothesis like $\int_{|x-y|\geq 2|y'-y|} |K(x,y') - K(x,y)|dx \leq C$ on the kernel of T provide the fundamental weak-L^1 estimate for Tf when f belongs to L^1. The Marcinkiewicz's interpolation theorem applies and gives L^p-estimates for $1 < p \leq 2$. The L^q estimates are obtained by duality.

Calderón and Zygmund noticed that the Marcinkiewicz's multipliers m_k were the discrete analogues of the ones given by (1.3), (1.4) and that the real variable methods could therefore give Marcinkiewicz's multiplier theorem.

The singular integral operators given by (1.2) are of convolution type. Therefore the symbolic calculus is transparent. In order to compose two operators, it suffices to multiply their symbols. Therefore, once the identity operator is included into the collection (1.2), we obtain a commutative operator algebra.

This operator algebra can only be used to give estimates for partial differential equations of elliptic type with constant coefficients. For dealing with elliptic P.D.E.'s with smooth coefficients, Calderón and Zygmund created what I would like to call "the second generation of Calderón-Zygmund operators. This "second generation" has been given another name by Kohn and Nirenberg (pseudo-differential operators).

These singular integral operators are no longer of convolution type but are still given by

$$Tf(x) = \text{p.v.}\int K(x,y)f(y)dy \tag{1.9}$$

where $K(x,y) = L(x,x-y)$ and $L(x,z)$ satisfies the following three properties

(1.10) $L(x,\lambda z) = \lambda^{-n}L(x,z)$ for $\lambda > 0$, $x \in \mathbb{R}^n$, $0 \neq z \in \mathbb{R}^n$

(1.11) $|\partial_z^\alpha \partial_x^\beta L(x,z)| \le C$ for $|z| = 1$ and $0 \le |\alpha| \le N$, $0 \le |\beta| \le m$, m and N being as large as it is required for the other estimates

(1.12) $\int_{S^{n-1}} L(x,z)d\sigma(z) = 0$ for every $x \in \mathbb{R}^n$.

For proving L^2 and L^p estimates for the new class, Calderón and Zygmund used the method of separation of variables. By means of expansion in spherical harmonics, the kernel $L(x,z)$ can be written as the sum of a series $m_0(z)H_0(z) + \cdots$ and the convergence is good enough to provide any estimate we want. Therefore it suffices to analyze each term of the series, which reduces the problem to convolution operators followed by multiplications by smooth functions.

These classical pseudo-differential operators form a commutative Banach algebra modulo the operators S which are smoothing of order 1, that is to say which are bounded in L^2 as well as $\partial/\partial x_j\, S$ and $S\, \partial/\partial x_j$. This fact was proved by Calderón and Zygmund, assuming m and N are quite large.

A few years later A. Calderón succeeded in proving

(1.13) $\|[T_1,T_2]\partial/\partial x_j f\|_2 \le C\|f\|_2$, $1 \le j \le n$

and

(1.14) $\|\partial/\partial x_j[T_1,T_2]f\|_2 \le C\|f\|_2$

with minimal smoothness assumptions with respect to the x variable on the kernels $L_1(x,z)$ and $L_2(x,z)$ of T_1 and T_2. In the applications he had in mind, he wanted to get rid of unnecessary smoothness assumptions on the coefficients of the partial differential equations he studied.

An equivalent formulation is to describe the functions $A(x)$ such that the commutator between the operator A of pointwise multiplication by $A(x)$ and any classical pseudo-differential operator of order 1 is bounded on L^2.

In one variable the problem reduces to studying the commutator $[A,\Lambda]$ between the pointwise multiplication by the function A and the operator $\Lambda = DH$ where $D = -i\frac{d}{dx}$ and H is the Hilbert transform.

In 1965 Calderón proved that this commutator is bounded on $L^2(\mathbb{R})$ if and only if $A(x)$ is Lipschitz that is to say if there exists a constant C such that $|A(y) - A(x)| \le C|y - x|$. The necessity of this condition is easy. The other implication was deep and relied on the theorem that the holomorphic H^1 space is characterized by the condition that the Lusin area functional should belong to L^1.

Then using the method of rotations of Calderón and Zygmund, Calderón proved that the collection of singular integral operators of type (1.9) with $m = 1$ and N large in (1.11) was indeed a commutative Banach algebra modulo operators S which are smoothing of order 1.

Even if the L^2 estimates for these commutators $[A,\Lambda]$ are quite involved, nevertheless their kernel p.v. $\frac{A(x) - A(y)}{(x - y)^2}$ satisfies the smoothness and size estimates that permit one to apply the "real variable methods". Therefore the L^p estimates follow easily when $1 < p < \infty$.

A. Calderón made the fundamental observation that similar operators show up in seemingly unrelated problems. The boundary values of the Cauchy integral of an L^2 function on a Lipschitz curve Γ are given by

$$(1.15) \qquad Tf(x) = \lim_{\epsilon \downarrow 0} \frac{1}{\pi i} \int_{-\infty}^{\infty} (z(y) - z(x) - i\epsilon)^{-1} f(y)dz(y)$$

where $z(x) = x + iA(x)$ with $A'(x) \in L^\infty(\mathbb{R})$.
If the L^∞ norm of A' is small, one is tempted to write the operator T as a Neumann series of perturbations of the Hilbert transform. The term which shows up after the Hilbert transform is precisely our friend $[A,\Lambda]$.
A third operator of similar nature is given by the double layer potential of an L^2-function on the boundary of a Lipschitz domain. In local coordinates, such an operator is given by

$$(1.16) \qquad Tf(x) = \frac{1}{\omega_n} \text{ p.v.} \int_{\mathbb{R}^n} [|x - y|^2 + (A(x) - A(y))^2]^{-\frac{n+1}{2}} (A(x) - A(y) - (x - y)\cdot\nabla A(y))f(y)dy.$$

Here ω_n is the area of the unit sphere S^{n-1} and solving the Dirichlet problem in Lipschitz domains by the method of the double layer potential requires to prove the boundedness of T and furthermore the invertibility of $1 \pm \frac{1}{2} T$.
Calderón developed algorithms relating (1.15) to (1.16) and the celebrated method of rotations of Calderón and Zygmund is one of these. He then proved in 1977 the boundedness of the Cauchy operator (1.15) for small Lipschitz norms. Then the boundedness of the double layer potential followed with the same limitation and E. Fabes, M. Jodeit and N. Rivière could solve the Dirichlet and the Neumann problem in C^1-domains using the double layer potential approach.

Our modern understanding of this theory originates with a problem raised by A. Calderón in 1976. If $K(x,y)$ is a function of two real variables x and y, defined away from the diagonal $y = x$ and satisfying $K(y,x) = -K(x,y)$ together with the estimates

$$(1.17)\qquad |K(x,y)| \le C|x,y|^{-1},\ |\partial/\partial x\, K(x,y)| \le C|x-y|^{-2},$$

then one can certainly define an operator T taking test-functions to distributions by

$$(1.18)\qquad \langle Tf,g\rangle = \lim_{\epsilon\downarrow 0} \iint_{\{|x-y|\ge\epsilon\}} K(x,y)f(y)g(x)dydx.$$

Here f and g are two test functions, smooth with compact supports and the formal singular integral can be rewritten $\frac{1}{2}\iint K(x,y)(f(y)g(x) - f(x)g(y))dydx$. Calderón wanted to show that such an operator T is always bounded on $L^2(\mathbb{R})$ and this result would immediately imply the boundedness of the Cauchy kernel. A counter-example was found during one of my visits to Chicago and that convinced Calderón to give up the real variable approach to the Cauchy kernel and to use the full power of complex methods (conformal mapping, etc.) instead.

The real variable approach to the generalized singular integral operators has been J. L. Journé's program and the first outstanding result along this direction is the celebrated $T(1)$ theorem by Journé and David.

The David-Journé $T(1)$ theorem is a striking result since it gives immediately, by a simple integration by parts, the deep results obtained by A. Calderón during the two last decades.

Moreover this $T(1)$ theorem together with rather simple real variable methods yields the boundedness of the Cauchy kernel for general Lipschitz curves.

This fact was first discovered by G. David but the proof has been considerably simplified by T. Murai.

2. Definition of Calderón-Zygmund operators

We do not want to define Calderón-Zygmund operators by the principal values of some singular integrals. The reason is the following. If we take, for example, a function $K(x,y)$ of two real variables, defined if $y \neq x$ and satisfying the conditions $K(x,y) = -K(y,x)$ and $|K(x,y)| \leq C|x-y|^{-1}$, $|\partial/\partial x\, K(x,y)| \leq C|x-y|^{-2}$, then the principal value approach does not work. The limit as ϵ tends to 0 of $\int_{|x-y|\geq\epsilon} K(x,y)f(y)dy$ does not exist even if $f(y)$ is a smooth function. A very simple example is $K(x,y) = \sum_0^\infty \exp(i2^k(x+y))2^k\theta(2^k(x-y))$ where $\theta(t)$ is C^∞, supported by $[-\frac{4}{3},-\frac{2}{3}] \cup [\frac{2}{3},\frac{4}{3}]$, odd, and where $\int_{-\infty}^{\infty} \sin t\, \theta(t)dt = 1$. If $f(y)$ is also C_0^∞ and is, say, identically 1 on $[-10,10]$, then the existence of the limit of $\int_{|x-y|\geq\epsilon} K(x,y)f(y)dy$ would imply the existence of the limit of the series $\sum_0^\infty e^{i2^k x}$ which is known to diverge everywhere.

As is suggested by this example, we have to give up convergence almost everywhere and replace it by the convergence in the distributional sense. Indeed, if $g(x)$ is a second testing function and if $J_\epsilon(f,g) = \iint_{|x-y|\geq\epsilon} K(x,y)f(y)dy\, g(x)dx$, then $\lim_{\epsilon\downarrow 0} J_\epsilon(f,g)$ exists when $K(x,y) = -K(y,x)$ and $|K(x,y)| \leq C|x-y|^{-1}$. Let us denote by $J(f,g)$ this limit. It is given by $J(f,g) = \frac{1}{2}\iint K(x,y)\{f(y)g(x) - f(x)g(y)\}dxdy$ and defines an unbounded operator T by $\langle Tf,g\rangle = J(f,g)$ when f and g belong to C_0^∞.

Unbounded operators are common in mathematics. We start with a Hilbert space H and a dense linear subspace $V \subset H$ which is generally a locally convex topological vector space but which can also be a linear space without topology. Then assuming that the inclusion $V \subset H$ is continuous in the former case, this inclusion defines by duality the inclusion $H \subset V'$ where V' is the topological or algebraical dual of V.

An unbounded operator is either defined as a continuous linear mapping $T : V \to V'$ or as a bilinear continuous form J on $V \times V$. The two approaches being the same if J and T are related by $\langle Tu, v\rangle = J(u,v)$, $u \in V$, $v \in V$ and where $\langle \cdot, \cdot \rangle$ expresses the duality between V' and V.

To simplify the notations we shall restrict ourselves to the one-dimensional case. Then the Hilbert space H we have in mind is $L^2(\mathbb{R})$ and V will be the linear subspace (without topology) of compactly supported piecewise linear continuous functions on the line with nodes at $k2^{-j}$ ($k = 0, \pm 1, \pm 2, \ldots,$ and j is fixed but depends on f). In other words V is the linear space of continuous function with compact support belonging to the union of all the spaces B_j spaces of Chapter III.

We are first going to define possibly unbounded operators whose kernels satisfy the Calderón-Zygmund estimates. Then we will be able to raise the problem of how to find a criterion giving the L^2-estimate.

In the definition of a Calderón-Zygmund bilinear form, Ω will denote the open subset of $\mathbb{R}^2$, defined by $y \neq x$ (Ω is the complement of the diagonal in $\mathbb{R}^2$).

Definition 1. A Calderón-Zygmund form is a bilinear form J on $V \times V$ which has the following properties: there exists a (unique) continuous function $K(x,y)$ on Ω such that

$$J(f,g) = \iint K(x,y)f(y)g(x)dydx \tag{2.1}$$

for all $f \in V$ and all $g \in V$ whose supports are disjoint from the support of f. Moreover there exist a constant C and an exponent $\gamma \in (0,1]$ such that

(2.2) $$|K(x,y)| \le C|x - y|^{-1}$$

(2.3) $$|K(x',y) - K(x,y)| \le C|x' - x|^{\gamma}|x - y|^{-1-\gamma}$$

<u>when</u> $|x' - x| \le \frac{1}{2}|x - y|$

(2.4) $$|K(x,y') - K(x,y)| \le C|y' - y|^{\gamma}|x - y|^{-1-\gamma}$$

<u>when</u> $|y' - y| \le \frac{1}{2}|x - y|$.

Observe that the kernel $K(x,y)$ is unique, once the form J is given. We insist that we <u>cannot</u> in general entirely recover the form from its kernel. Let us give a very simple example. If $m(x)$ is any locally integrable function on the line, then the integral $\int f(x)g(x)m(x)dx$ defines a Calderón-Zygmund form for which the corresponding kernel K vanishes identically. Moreover the problem of the existence of a form J, given a kernel K satisfying (2.2), (2.3) and (2.4), is also present since the integral (2.1) does not converge, unless f and g have disjoint supports. As we have seen, this problem has a very simple and canonical solution when $K(y,x) = -K(x,y)$. Then the form will always be defined by
$J(f,g) = \frac{1}{2}\iint K(x,y)\{f(y)g(x) - f(x)g(y)\}dxdy$.

Definition 2. <u>A Calderón-Zygmund operator is a bounded linear operator</u> $T : L^2(\mathbb{R}) \to L^2(\mathbb{R})$ <u>such that</u> $\langle Tf,g\rangle$ <u>is a Calderón-Zygmund bilinear form</u>.

Let us list some properties of Calderón-Zygmund operators. The real variable methods apply to all Calderón-Zygmund operators. Using (2.4) and the L^2-continuity of T, one gets the weak-L^1 estimate. By Marcinkiewicz's interpolation theorem, one obtains the L^p estimates for $1 < p \leq 2$. Then we observe that if T is a Calderón-Zygmund operators, the same is true for its adjoint T^*. Therefore T is bounded on L^p for $2 \leq p < \infty$. Let us study the limiting case.

Lemma 1. <u>A Calderón-Zygmund operators</u> T <u>is bounded from the atomic space</u> H^1 <u>to</u> L^1.

The proof uses simply the atomic decomposition. If $a(x)$ is an atom with supporting interval I, it means that $\|a\|_\infty \leq |I|^{-1}$ and $\int a(x)dx = 0$. Let us denote by x_0 the center of I, ℓ the length of I and $2I$ the interval centered at x_0 with length 2ℓ.

Then if x does not belong to $2I$, we have by (2.1)

$$Ta(x) = \int K(x,y)a(y)dy = \int [K(x,y) - K(x,x_0)]a(y)dy$$

and therefore $|Ta(x)| \leq C\ell^\gamma |x - x_0|^{-1-\gamma}$. This gives $\int_{|x-x_0|\geq 2\ell} |Ta(x)|dx \leq C_\gamma$. The other part of $\int_{-\infty}^{\infty} |Ta(x)|dx$ is $\int_{2I} |Ta|dx$ and is treated by the L^2- continuity of T. Cauchy-Schwarz inequality gives $\int_{2I} |Ta|dx \leq (2\ell)^{½}\|Ta\|_2 \leq (2\ell)^{½}\|T\|\|a\|_2 \leq 2^{½}\|T\|$.

By duality a Calderón-Zygmund operator T defines a mapping from L^∞ to BMO which is continuous when both spaces are equipped with the norm topology but which is also continuous for the weak-star topologies $\sigma(L^\infty,L^1)$ and $\sigma(BMO,H^1)$.

The last definition we need is the weak-boundedness property. As usual, we denote by $\mathcal{F}$ the collection of all dyadic intervals $I = [k2^{-j},(k + 1)2^{-j}]$ where $j = 0,\pm1,\pm2,\ldots,k = 0,\pm1,\pm2,\ldots$ and for each $I \in \mathcal{F}$, $\Delta_I(x)$ will be the triangle function defined by $\Delta_I(x) = 2^j(1 - |2^j x - k|)^+$. The normalization is such that $\int_{-\infty}^{\infty} \Delta_I(x)dx = 1$. Then a bilinear form J on V × V has the weak-boundedness property if there exists a constant C such that, whenever $I \in \mathcal{T}$, $J \in \mathcal{T}$ and $|I| = |J|$, one has

$$(2.5) \qquad |J(\Delta_I, \Delta_J)| \le C|I|^{-1}.$$

Observing that $\|\Delta_I\|_2 = (\frac{2}{3}|I|)^{½}$, we see that (2.5) is a weak formulation of the Cauchy-Schwarz inequality. It is also easily seen that (2.5) is true in the case when J is defined by an antisymmetric kernel $K(x,y)$ satisfying (2.2).

We denote by ψ_I, $I \in \mathcal{T}$, the Franklin wavelets orthonormal basis. We then have

Lemma 2. Let J be a Calderón-Zygmund form on V × V. Then the two following properties are equivalent

(2.6) J has the weak boundedness property

(2.7) there exists a constant C such that $|J(\psi_I, \psi_J)| \le C$ for all $I \in \mathcal{T}$ and $J \in \mathcal{T}$.

In (2.7) we do not suppose $|I| = |J|$.

For proving that (2.6) implies (2.7) we define some new wavelets imitating our orthonormal basis ψ_I, $I \in \mathcal{T}$. We first denote by $W(x)$ the function $T(x) - \frac{1}{2}(T(x - \frac{1}{2}) + T(x + \frac{1}{2}))$ where $T(x)$ is the triangle function, with height one, based on the interval $[0,1]$. Then the Fourier transform of ψ and of W are related by

(2.8) $\hat{\psi}(\xi) = \hat{W}(\xi)\alpha(\xi)$

where $\alpha(\xi)$ is a 2 periodic real-analytic function on the line which is strictly positive. Therefore

(2.9) $\psi(x) = \sum_{-\infty}^{\infty} \alpha_k W(x - k/2)$ where the coefficients α_k have an exponential decay.

Conversely one can also write $W(x) = \sum_{-\infty}^{\infty} \alpha'_k \psi(x - k/2)$ where α'_k have also an exponential decay. It means that all estimates which are translation invariant can be transferred from ψ to W and vice versa. For example it suffices to prove $|J(W_I, W_J)| \le C$ where $W_I(x) = 2^{j/2}W(2^j x - k)$ in order to get (2.6). We can assume $|I| \le |J|$ and decompose $W_J(x) = U_J(x) + V_J(x)$ where V_J. vanishes on the doubled interval $2I$ of I. This decomposition is obtained by the polynomial approximation of W_J with nodes at $k2^{-j}$ (where $2^{-j} = |I|$) and therefore U_J is a sum of a few triangle functions based on intervals I' with length 2^{-j}. The weak-boundedness property applies to $J(W_I, U_J)$ while the estimate on $J(W_I, V_J)$ is obtained by Lemma 1's proof.

Conversely we assume that $|J(\psi_I, \psi_J)| \le C$ and we want to compute $J(\Delta_{I_0}, \Delta_{J_0})$. It suffices to write $\Delta_{I_0} = \sum \alpha(I)\psi_I$ and prove that $\sum|\alpha(I)| \le C|I_0|^{-½}$. In fact $\alpha(I) = \langle\Delta_{I_0}, \psi_I\rangle = 0$ when $|I| \le |I_0|$ by construction of the Franklin wavelets. If $|I| > |I_0|$, we write $|\alpha(I)| \le \int|\Delta_{I_0}(x)|2^{j/2}|\psi(2^j x - k)|dx = \gamma(j,k)$. We first majorize $\sum_k |\psi(u - k)|$ by a constant and we have then $\sum_k \gamma(j,k) \le C2^{j/2}$, which gives the required estimate since $j \le j_0$ and $|I_0| = 2^{-j_0}$.

3. The definition of $T(1)$.

Let us denote by $B^0_{\infty,\infty}$ the space of Bloch functions on the line. The space $B^0_{\infty,\infty}$ is the dual space of the atomic space $B^0_{1,1}$ generated by special atoms. Using the basis of Franklin wavelets, these two spaces are easily characterized as well as all Besov spaces $B^s_{p,q}$. The functions of the space $B^0_{1,1}$ are by definition given by $f(x) = \sum \lambda(I)\psi_I(x)$ where the sum runs over all dyadic intervals $I \in \mathcal{J}$ and where $\sum|\lambda(I)||I|^{1/2} < \infty$. The decomposition is unique and the sum is the norm of f in $B^0_{1,1}$.

Obviously $B^0_{1,1}$ is contained in the usual Hardy space H^1 but these two spaces are distinct. The space $B^0_{1,1}$ has been studied by G. de Souza and O'Neil.

The dual space is a space of distributions modulo constant functions (since all test functions $f \in \mathcal{S}(\mathbb{R})$ with a vanishing integral belong to $B^0_{1,1}$). It is denoted $B^0_{\infty,\infty}$ and a distribution S belongs to $B^0_{\infty,\infty}$ if and only if $|\langle S,\psi_I\rangle| \le C|I|^{1/2}$. It is easily seen that these distributions are exactly the distributional derivatives of the functions belonging to the Zygmund class.

Lemma 3. Let J be a Calderón-Zygmund bilinear form with the weak boundedness property and let $T : V \to V'$ be the operator defined by the form J. If, for $m \ge 0$, $\gamma_m(x) = (1 - 2^{-m}|x|)^+$, then $T(\gamma_m)$ converges in $B^0_{\infty,\infty}$ for the $\sigma(B^0_{\infty,\infty},B^0_{1,1})$ topology to an element $S \in B^0_{\infty,\infty}$ which will be denoted $T(1)$.

It suffices to prove that $\langle T(\gamma_m),\psi_I\rangle = J(\gamma_m,\psi_I)$ converges as m tends to infinity and that we have $\lim|J(\gamma_m,\psi_I)| \le C|I|^{1/2}$.

The proof follows the same lines as the one of Lemma 2. We write $\gamma_m = \gamma'_m + \gamma''_m$ where $\gamma''_m = 0$ on the doubled interval $2I$ of I and where γ'_m is a sum of a few triangle functions based on intervals I' with $|I'| = |I|$

and $I' \subset (10)\ I$. We also replace ψ_I by W_I and then repeat the arguments used in the preceding lemmas.

We similarly define $T^*(1)$ by $\langle T^*(1),\psi_I\rangle = \lim_{m\to+\infty} J(\psi_I,\gamma_m)$.

Another approach to the definition of $T(1)$ is the following. We consider the distribution $T^*(\psi_I)$. It is $O(|x|^{-1-\gamma})$ at infinity. But any distribution S which coincides outside a large compact set with an L^1 function has an integral over the line: $\int_{-\infty}^{\infty} S(x)dx$ makes sense. Therefore $\int_{-\infty}^{\infty} T^*(\psi_I)dx$ exists and is $\langle T(1),\ \psi_I\rangle$.

4. <u>The</u> T(1)-<u>theorem</u>.

In 1983 G. David and J. L. Journé proved the fundamental $T(1)$ theorem which reshaped entirely the theory of Calderón-Zygmund operators. The corresponding conjecture has been formulated by J. L. Journé during his visit to Washington University, St. Louis.

Theorem 1. <u>Let</u> J <u>be a Calderón-Zygmund bilinear form on</u> $V \times V$ <u>and let</u> $T : V \longrightarrow V'$ <u>be the operator which is weakly defined by that form. Then the two following properties are equivalent</u>

(4.1) T <u>is bounded on</u> L^2

(4.2) $T(1)$ <u>belongs to</u> $BMO(\mathbb{R})$, $T^*(1)$ <u>belongs to</u> $BMO(\mathbb{R})$ <u>and</u> J <u>has the weak boundedness property</u>.

We first want to give a very simple form to the condition (4.2). Let ψ_I, $I \in \mathcal{F}$, be the orthonormal basis of Franklin wavelets. Then the wavelet coefficients $\gamma(I)$ of $T(1)$ are given by $\gamma(I) = \langle T(1),\ \psi_I\rangle = \langle 1,\ T^*(\psi_I)\rangle = \int_{-\infty}^{\infty} T^*(\psi_I)dx$. Here T^* is the transposed operator and T, T^* and the Calderón-Zygmund form are related by $\langle Tu,v\rangle = \langle u,T^*v\rangle = J(u,v)$. The pairing

$\langle\cdot,\cdot\rangle$ is the same as the one expressing the duality between distributions and test-functions.

Similarly the wavelet coefficients $\beta(I)$ of $T^*(1)$ are $\beta(I) = \int_{-\infty}^{\infty} T(\psi_I)dx$.

Then the condition (4.2) can be rewritten

(4.3) J has the weak-boundedness property and there exists a constant C such that, for all dyadic intervals $I' \in \mathcal{J}$, $\sum_{I\subset I'} |\gamma(I)|^2 \le C|I'|$ and $\sum_{I\subset I'} |\beta(I)|^2 \le C|I'|$.

These Carleson measure type conditions give the characterization of the wavelet coefficients of the functions in BMO.

The proof of the implication (4.1) ⇒ (4.2) has already been given (lemma 1).

The proof of the deep implication (4.2) ⇒ (4.1) will give a surprisingly simple description of all Calderón-Zygmund operators. But we first would like to introudce a few more notations.

If H is a Hilbert space and e and f are two elements of H then the tensor product $e \otimes f^*$ defines an operator on H by $(e \otimes f^*)(x) = \langle x|f\rangle e$. When H is $L^2(\mathbb{R})$, and when $e = a(x)$, $f = b(x)$ are two functions in $L^2(\mathbb{R})$, then $e \otimes f^*$ is given by the kernel $K(x,y) = a(x)\bar{b}(y)$. Returning to the abstract situation, let us assume that we are given an orthonormal basis e_k, $k \in \mathbb{N}$, of H. It is then natural to try to characterize a bounded operator $T : H \longrightarrow H$ by its entries $\langle Te_j, e_k\rangle = \alpha(j,k)$ and to approach T in the strong operator topology by $T_N = \sum_0^N \sum_0^N \alpha(j,k)e_j \otimes e_k$. We obviously have $\|T_N\| \leq \|T\|$ and $T_N(x) \longrightarrow T(x)$ for $x \in H$.

Since the Franklin wavelets form such a nice basis of $L^2(\mathbb{R})$ it is natural to try to write any Calderón-Zygmund operator T as a limit of

$\sum_{I\in\mathcal{T}_N} \sum_{J\in\mathcal{T}_N} \alpha(I,J)\psi_I \otimes \psi_J = T_N$ where $\mathcal{T}_N$ is the collection of all dyadic intervals $[k2^{-j}, (k+1)2^{-j}]$ such that $|k| \leq N$, $|j| \leq N$.

It is true as before that the operator norms of $T_N : L^2 \longrightarrow L^2$ are bounded by $\|T\|$. But what is no longer true is the fact that T_N would satisfy uniform Calderón-Zygmund estimates (2.2), (2.3) and (2.4) for some $\gamma' \in (0,1)$ and some constant C'.

If the operators T_N have uniform Calderón-Zygmund estimates and if their operator norms are uniformly bounded, then they converge to a Calderón-Zygmund operator T such that $T(1) = 0$ (modulo constant functions) and $T^*(1) = 0$ modulo constant functions. Moreover this happens if and only if for some $\alpha \in (0,1)$ and some constant C

$$(4.4)\qquad |\alpha(I,J)| \leq C\,(\inf(|I|,|J|))^{\alpha}|I|^{1/2}|J|^{1/2}(|I| + |J| + \text{dist}\,(I,J))^{-1-\alpha}.$$

Let us denote by $\mathcal{A}_\alpha$ the collection of all those T for which $\alpha(I,J) = \langle T\psi_I,\psi_J\rangle$ satisfy (4.4). Then any $T \in \mathcal{A}_\alpha$ is a Calderón-Zygmund operator where (2.3) and (2.4) are obtained for $0 < \gamma < \alpha$ and conversely any Calderón-Zygmund operator T such that $T(1) = 0, T^*(1) = 0$ belongs to $\mathcal{A}_\alpha$ for $0 < \alpha \leq \gamma$.

The complete description of all Calderón-Zygmund operator is given by the following theorem. If $I = [k2^{-j}, (k + 1)2^{-j}]$ is a dyadic interval, we denote as before by $\Delta_I(x)$ the function $2^j(1 - |2^jx - k|)^+$.

Theorem 2. <u>The Calderón-Zygmund operators are exactly given by</u>

$$T = \sum_I \sum_J \alpha(I,J)\psi_I \otimes \psi_J + \sum_I \beta(I)\psi_I \otimes \Delta_I + \sum_I \gamma(I)\Delta_I \otimes \psi_I \tag{4.5}$$

<u>where, for some constant</u> C <u>and for some</u> $\alpha \in (0,1)$

$$|\alpha(I,J)| \le C(|I| \wedge |J|)^\alpha \frac{|I|^{1/2}|J|^{1/2}}{(|I|+|J|+\mathrm{dist}(I,J))^{1+\alpha}} \tag{4.6}$$

(4.7) <u>for every dyadic interval</u> $I' \in \mathcal{T}$,

$$\sum_{I \subset I'} |\beta(I)|^2 \le C^2|I'|$$

<u>and for every dyadic interval</u> $I' \in \mathcal{T}$

$$\sum_{I \subset I'} |\gamma(I)|^2 \le C^2|I'|. \tag{4.8}$$

In other words the Calderón-Zygmund operators are decomposed in an unconditional basis which complements the basis $\psi_I \otimes \psi_J$, $I \in \mathcal{T}$, $J \in \mathcal{T}$ of $\mathcal{A}_\alpha$ with two more sequences which are $\psi_I \otimes \Delta_I$ and $\Delta_I \otimes \psi_I$, $I \in \mathcal{T}$.

The decomposition (4.5) is unique as long as the coefficients satisfy (4.6), (4.7) and (4.8).

The organization of the proof of the non-trivial implication (4.2) ⇒ (4.1) runs as follows. We denote by u and v the two BMO functions $T(1)$ and $T^*(1)$ and construct two Calderón-Zygmund operators R_u and S_v such that $R_u(1) = u$, $R_u^*(1) = 0$ and $\|R_u\|_{2,2} \le C\|u\|_{BMO}$. Similarly $S_v(1) = 0$, $S_v^*(1) = u$ and $\|S_v\|_{2,2} \le C\|v\|_{BMO}$. Once this is done, we can write $T = R_u + S_v + G$ where G is given by a Calderón-Zygmund form with the weak boundedness property and $G(1) = 0$, $G^*(1) = 0$ (modulo constant functions).

Then we compute the matrix entries $\langle G\psi_I, \psi_J\rangle$ and prove estimates like (4.6).

From these estimates it is obvious to check the L^2-continuity of G. Schur's lemma applies directly if $\frac{1}{2} < \alpha < 1$. If $0 < \alpha < 1/2$, the L^2-continuity is still quite easy but can also be obtained from the decomposition of G in the basis of Franklin wavelets in two variables. Indeed this decomposition gives G as a series $\sum_{-\infty}^{\infty} G_j$ and Cotlar's lemma applies and yields the L^2 estimate.

The operators R_u and S_v are somehow the same. In fact one is the adjoint of the other. They will be constructed in the next section. Then the operator G will be analyzed and the proof of theorem 1 will be concluded. Theorem 2 will follow from the information gathered in the proof.

5. Paraproducts.

We all know that L^∞ acts on L^2 by pointwise multiplication and that the converse is also true. If an operator acts on L^2 by pointwise multiplication, it has to be given by an L^∞ function.

Therefore it is quite remarkable that BMO also acts on L^2 by an operation which is somehow related to the pointwise multiplication. This operation is the paraproduct. There are in fact several realizations of the paraproduct and it is not entirely clear what is the best one. We shall give a realization based on the Franklin wavelets. Here also, for the sake of simplifying the notations we will consider the one real variable situation.

Then the paraproduct between $u \in BMO(\mathbb{R})$ and $f \in L^2(\mathbb{R})$ is, by definition, the object g given by the series

$$g = \sum_{I \in \mathcal{T}} \langle u, \psi_I \rangle \langle f, \Delta_I \rangle \psi_I . \tag{5.1}$$

We will immediately prove the L^2 estimates. Since $\psi_I, I \in \mathcal{T}$, is an orthonormal basis, we have

$$\|g\|_2 = \sum_{I \in \mathcal{T}} |\langle u, \psi_I \rangle|^2 |\langle f, \Delta_I \rangle|^2 . \tag{5.2}$$

We conclude the proof with an inequality due to Carleson which will be proved later.

Lemma 4. <u>If $\mathcal{T}$ is the collection of all dyadic intervals and if</u> $f(I), p(I)$ <u>are two non-negative functions defined on</u> $\mathcal{T}$, <u>then we have</u>

$$\sum_I p(I) f(I) \le C \int_{-\infty}^{\infty} \omega(x) dx \tag{5.3}$$

where $\omega(x) = \sup_{x\in I} f(I)$

and $C = \sup_{J\in\mathcal{J}} \frac{1}{|J|} \sum_{I\subset J} p(I)$.

We now return to the situation where $p(I) = |\langle u,\psi_I\rangle|^2$ and $f(I) = |\langle f,\Delta_I\rangle|^2$. Then $\omega(x) \le (f^*(x))^2$ where f^* is the Hardy and Littlewood maximal function of f. Therefore by Hardy and Littlewood's theorem, we have $\int_{-\infty}^{\infty} \omega(x)dx \le C\|f\|_2^2$.

The control on C is given by the following lemma.

Lemma 5. <u>There is a constant</u> C <u>such that</u> $\sum_{I\subset J} |\langle u,\psi_I\rangle|^2 \le C|J|\,\|u\|^2_{BMO}$.

The proof runs as follows. We denote by $2J$ the doubled interval and write the BMO function $u(x)$ as a sum of three terms $c(J) + u_1 + u_2$ where $c(J)$ is a constant on which we have no information, u_1 is supported by $2J$ and u_2 vanishes on $2J$. Moreover $\int |u_1|^2 dx \le 2|I|\,\|u\|^2_{BMO}$ and similarly $\int |u_2|^2 \omega_J(x)dx \le C|J|\,\|u\|^2_{BMO}$ with $\omega_J(x) = \dfrac{|J|^2}{(\mathrm{dist}(x,J))^2}$. Then we have

$$\langle c(J).\psi_I\rangle = 0,\ \sum_I |\langle u_1,\psi_I\rangle|^2 = \|u_1\|_2^2 \le C|J|\,\|u\|^2_{BMO} \quad \text{and}$$

$$\sum_{I\subset J} |\langle u_2,\psi_I\rangle|^2 \le \left(\int |u_2|^2\omega_J(x)dx\right)\left(\sum_{I\subset J}\int_{(2J)^c} |\psi_I(x)|^2 \frac{(\mathrm{dist}(x,J))^2}{|J|^2} dx\right)$$

and

$$\sum_{I\subset J}\int_{(2J)^c} |\psi_I(x)|^2 \left[\frac{\mathrm{dist}(x,J)}{|J|}\right]^2 dx \le C.$$

For obtaining this last inequality it is enough to know that $|\psi(x)| \le C(1+x^2)^{-1}$.

We now return to the proof of lemma 1. We introduce the indicator function θ on $\mathcal{T} \times (0,+\infty)$ defined by $\theta(I,t) = 1$ if $0 < t < f(I)$ and $\theta = 0$ elsewhere. Then $\sum_I p(I)f(I) = \int_0^\infty \sum_I \theta(I,t)p(I)dt$. We then fix t and denote by Ω_t the union of all intervals $I \in \mathcal{T}$ for which $f(I) > t$. In other words Ω_t is defined by $\omega(x) > t$. If $\theta(I,t) = 1$ we certainly have $I \subset \Omega_t$ and therefore

$$\sum_I \theta(I,t)p(I) \le \sum_{I\subset\Omega_t} p(I) = \sum_k \sum_{I\subset I_k} p(I)$$

(where I_k are the maximal dyadic intervals contained in Ω_t. Finally $\sum_{I\subset I_k} p(I) \le C|I_k|$, and $\sum_k |I_k| = |\Omega_t|$. But $\int_0^\infty |\Omega_t|dt = \int_{-\infty}^\infty \omega(x)dx$ which ends the proof.

We now study the operator R_u defined by (5.1). If $|\langle u,\psi_I\rangle| \le C|I|^{1/2}$, that is if u belongs to the Bloch space $B^0_{\infty,\infty}$, then $\sum_{I\in\mathcal{T}} \langle u,\psi_I\rangle\langle f,\Delta_I\rangle\langle g,\psi_I\rangle = J_u(f,g)$ is a Calderón-Zygmund form. It is easy to see (and will not be needed in our discussion) that this Calderón-Zygmund form has the weak boundedness property when $u \in B^0_{\infty,\infty}$. Therefore R_u is a Calderón-Zygmund operator when u belongs to BMO.

Jean Lin Journé found another proof of the continuity of the paraproduct. He made the following observation.

Lemma 6. Let $T : V \longrightarrow V'$ be the linear operator given by a Calderón–Zygmund bilinear form. If we assume that there exists a constant C such that, for all $f \in V$, $\|T(f)\|_{BMO} \le C\|f\|_\infty$, then T is a Calderón–Zygmund operator.

This lemma will be proved later on. Returning to (5.1) and assuming $f \in L^\infty$, we know that $\lambda(I) = \langle f, \Delta_I \rangle$ satisfies $|\lambda(I)| \le \|f\|_\infty$. But we also know that the Franklin wavelets ψ_I form an unconditional basis for BMO, equipped with the $\sigma(\mathrm{BMO}, H^1)$ topology. Therefore $\sum_{I \in \mathcal{T}} \lambda(I)\langle u, \psi_I\rangle\psi_I$ belongs to BMO as long as u does. And this is the end of J. L. Journé's remarkable proof.

Let us return to the lemma. We first observe that if I is an interval and if f is supported by I with $\|f\|_\infty \le |I|^{-1}$ and $\int f(x)dx = 0$, then we have $\|Tf\|_1 \le C'$ when the hypotheses of the lemma are satisfied for T.

For proving this remark we let I' be the interval whose length and distance to I equal $|I|$. We know that Tf belongs to BMO and want to check the BMO-norm on $J = 7I'$. We know that the "floating constant" $c(J)$ can also be computed as the mean value of Tf on I'. But if x belongs to I', $Tf(x)$ can be computed using the kernel representation of T, and one obtains $|Tf(x)| \le C|I|^{-1}$. Therefore the BMO condition on J yields $\int_J |Tf - c(J)|dx \le C|J|\|f\|_\infty \le 7C$. Since $|c(J)| \le C|I|^{-1}$ we obtain $\int_J |Tf(x)|dx \le C'$. Outside J the smoothness of the kernel with respect to the y variable gives $\int_{J^c} |Tf(x)|dx < C''$ as in lemma 1.

Now let us assume that f is any H^1 function. Then f has an atomic decomposition $\sum_0^\infty \lambda_k a_k(x)$ where $\sum_0^\infty |\lambda_k|$ is finite and bounded by $2\|f\|_{H^1}$ and where the $a_k(x)$ belong to V, are supported by intervals I_k, and satisfy $\|a_k\|_\infty \le |I|^{-1}$ and $\int a_k(x)dx = 0$.

From the preceding remarks $\|Ta_k\|_1$, $k = 0, 1, 2, \dots$, are uniformly bounded, and therefore T maps H^1 into L^1. But any continuous operator from H^1 into L^1 which is also continuous from L^∞ into BMO is continuous from L^2 into itself. Therefore T was a Calderón-Zygmund operator.

6. <u>The</u> L^2<u>-boundedness of</u> G.

The operator G is given by a Calderón-Zygmund form with the weak-boundedness property and satisfies $G(1) = G^*(1) = 0$. We shall prove that G is bounded on $L^2(\mathbb{R})$. For doing so we simply compute the entries $\langle G\psi_I, \psi_J\rangle$ of the matrix of G in the Franklin-wavelets orthonormal basis. We first suppose $|I| \le |J|$. Then these entries will be small because the integral of $G\psi_I$ vanishes, and the scalar product with ψ_J is like an average over a large interval. Observe that the integral of $G\psi_I$ vanishes since $G^*(1) = 0$. If $|I| > |J|$, we rewrite $\langle G\psi_I, \psi_J\rangle$ as $\langle \psi_I, G^*(\psi_J)\rangle$ and are reduced to the first situation.

The "technical lemma" we need is the following simple observation.

Lemma 7. <u>Let</u> $S \in V'$ <u>be such that</u> $|\langle S, \Delta_k\rangle| \le C(1 + |k|)^{-1-\gamma}$, <u>where</u> $\Delta_k(x) = (1 - |x - k|)^+$ <u>and</u> $\int_{-\infty}^{\infty} S(x)dx = 0$. <u>Then, if</u> $J = [k2^{-j}, (k+1)2^{-j}]$ <u>and</u> $j \le 0$, <u>we have</u> $|\langle S, \psi_J\rangle| \le c2^{j(2^{-1}+\gamma)}(1 + |k|)^{-1-\gamma}$.

After rescaling we apply this lemma to $G\psi_I = S$. Bounds on $\langle S, \Lambda_{I'}\rangle$ with $|I'| = |I|$ are obtained by the weak-boundedness property together with the size and smoothness estimates on the kernel. We then have, if $|I| \le |J|$,

$$(6.1) \qquad |\langle G\psi_I, \psi_J\rangle| \le C\left(\frac{|I|}{|J|}\right)^{2^{-1}+\gamma}\left(\frac{|J|}{|J|+\mathrm{dist}(I,J)}\right)^{1+\gamma}.$$

Now these estimates on the entries provide L^2-boundedness by the well-known Schur lemma.

Lemma 8. <u>If</u> $M = ((m_{p,q}))$ <u>is an infinite matrix indexed by</u> $\mathbb{N} \times \mathbb{N}$ <u>such that, for some constant</u> C <u>we have</u> $\sum_q |m_{p,q}| \le C$ <u>for all</u> p, <u>and</u> $\sum_p |m_{p,q}| \le C$ <u>for all</u> q, <u>then</u> M <u>is bounded on</u> $\ell^2(\mathbb{N})$.

In our case the indices I belong to the denumberable set of dyadic intervals, and checking the two conditions in Schur's lemma for $I = [k'2^{-j'}, (k'+1)2^{-j'}]$ and $J = [k2^{-j}, (k+1)2^{-j}]$ amounts to checking that

$$\sum_{m\ge 0}\sum_{k'} 2^{-m(2^{-1}+\gamma)}(1 + |k - k'2^{-m}|)^{-1-\gamma} \le C,$$

and that

$$\sum_{m\ge 0}\sum_{k} 2^{-m(2^{-1}+\gamma)}(1 + |k - k'2^{-m}|)^{-1-\gamma} \le C.$$

The first requirement holds (uniformly in k) when $\frac{1}{2} < \gamma < 1$, and the second for all $\gamma > 0$.

We have thus proved David-Journé's theorem, at least when $\frac{1}{2} < \gamma < 1$. With a little bit more work it can be proved that for all $\gamma > 0$ any matrix $((\alpha(I,J)))$, $I \in \mathcal{T}$, $J \in \mathcal{T}$, whose entries satisfy the size condition (6.1) is bounded on $\ell^2(\mathcal{T})$.

We prefer to give another proof of the boundedness of G based on Cotlar's lemma.

We denote by $B_j \subset L^2(\mathbb{R})$ the space of all continuous functions which are piecewise linear with nodes at $k2^{-j}$, $k = 0, \pm 1, \pm 2, \ldots$. The orthogonal projection from $L^2(\mathbb{R})$ onto B_j will be denoted Π_j.

We can write G as the telescopic series

$$G = \sum_{-\infty}^{\infty} (\Pi_{j+1} G \Pi_{j+1} - \Pi_j G \Pi_j) = \sum_{-\infty}^{\infty} G_j. \tag{6.2}$$

This series converges in the following sense. If u and v are two finite combinations of the Franklin wavelets, we have $\langle G_j u, u\rangle = 0$ when $|j|$ is large enough and therefore $\langle Gu, v\rangle$ is the finite sum of the right-hand side of (6.2). But those u's and v's are dense in L^2 and suffice for proving the L^2-boundedness of G.

The proof of the L^2-boundedness of G will now be identical to the one given by Cotlar in the case of the Hilbert transform. It consists in proving the following estimates on the kernel $G_j(x,y)$ of the operator G_j

(6.3) $|G_j(x,y)| \le C2^j(1 + 2^j|x - y|)^{-1-\gamma}$

(6.4) $|\partial/\partial x\, G_j(x,y)| + |\partial/\partial y\, G_j(x,y)| \le C4^j(1 + 2^j|\, x - y|)^{-1-\gamma}$

(6.5) $\int G_j(x,y)dy = 0$ for every x

(6.6) $\int G_j(x,y)dx = 0$ for every y.

Then we want to show that for some constant C and some $\epsilon > 0$, we have

(6.7) $$\|G_j^* G_k\|_{2,2} + \|G_j G_k^*\|_{2,2} \le C2^{-\epsilon|j-k|},\ j \in \mathbb{Z},\ k \in \mathbb{Z}.$$

At this stage, Cotlar's lemma provides the L^2-estimate.

We return to the proof of (6.3)-(6.6).

We have $\Pi_{j+1} - \Pi_j = D_j$ and

$$D_j(f) = \sum_k \langle f, \psi_{j,k}\rangle \psi_{j,k}$$

$$\Pi_j(f) = \sum_k \langle f, \varphi_{j,k}\rangle \varphi_{j,k}.$$

We therefore have $G_j = D_j G\Pi_j + \Pi_j G D_j + D_j G D_j$, and the three terms will receive similar treatment. The kernel $A_j(x,y)$ of $D_j G\Pi_j$ is given by

$$A_j(x,y) = \sum_k \sum_\ell \alpha(j,k,\ell)\psi_{j,k}(x)\varphi_{j,\ell}(y)$$

where $\alpha(j,k,,\ell) = \langle G\varphi_{j,\ell}, \psi_{j,k}\rangle$.

The estimates $|\alpha(j,k,\ell)| \le C(1 + 2^{-j}|k - \ell|)^{-1-\gamma}$ come either from the weak-boundedness property or from the smoothness properties of the kernel away from the diagonal (see lemma 2). These estimates immediately give (6.3) and (6.4). The hypothesis $G(1) = 0$ gives $\sum_{\ell} \alpha(j,k,\ell) = \sum_{\ell} \langle G\varphi_{j,\ell}, \psi_{j,k}\rangle = 0$ since $\sum_{\ell} \varphi_{j,\ell} = 2^{j/2}$. Then $\int A_j(x,y)dy = 0$. But $\int A_j(x,y)dx = 0$ by the cancellation properties of the Franklin-wavelets.

Let us sketch the proof of (6.7). We compute directly the kernel $\int \bar{G}_j(u,x)G_k(u,y)du$ of $G_j^*G_k$. Inside this integral, one term carries the cancellation and the other is flat, and the largest index is the one of the oscillating term. Using (6.3)-(6.6) we immediately obtain, if $j \geq k$,

$$\left|\int \bar{G}_j(u,x)G_k(u,y)du\right| \leq C2^{-\epsilon|j-k|}2^k(1 + 2^k|x - y|)^{-1-\gamma}$$

for $0 < \epsilon < \gamma$. That estimate together with all the similar ones ends our proof.

7. Generalizations and applications.

In some applications it is convenient to impose some more regularity on the testing functions that are used to give the weak definition of the possibly unbounded operator T. Let us fix an integer $m \geq 1$ and for all $j \in \mathbb{Z}$, define V_m^j as the linear space of compactly supported functions of class C^m on the line which coincide with polynomials of degree $m + 1$ on each interval $[k2^{-j},(k + 1)2^{-j}]$. There is a fundamental function Δ_m in V_m that plays the role of the ordinary triangle function when $m = 0$. This Δ_m can be defined as the convolution product $\chi * \cdots * \chi$ ($(m + 1)$ times) where χ is the characteristic function of $[0,1]$.

Then V_m^0 is spanned by the functions $\Delta_m(x - k)$, $k \in \mathbb{Z}$, and similarly V_m^j is spanned by $2^j\Delta(2^jx - k)$, $k \in \mathbb{Z}$.

Then theorem 1 remains valid if the space V of testing functions we have been using is replaced by the space V_m (which is the union of all the V_m^j for $j \in \mathbb{Z}$).

We now want to describe the extension to several dimensions. If Q is the dyadic cube of $\mathbb{R}^n$ defined by $0 \leq 2^j x_1 - k_1 < 1, \ldots, 0 \leq 2^j x_n - k_n < 1$, then we denote by $\varphi_Q(x)$ the function $2^{nj}\Delta_m(2^j x_1 - k_1) \cdots \Delta_m(2^j x_n - k_n)$ and by V_m the span of all those functions. Then a linear operator $T : V_m \longrightarrow V_m'$ has the weak-boundedness property if $|\langle T\varphi_Q, \varphi_{Q'}\rangle| \leq C|Q|^{-1}$ when $|Q'| = |Q|$.

We also assume that the off-diagonal part of the distributional kernel of T is an honest function $K(x,y)$ satisfying

$$|K(x,y)| \leq C|x - y|^{-n}, \; |K(x',y) - K(x,y)| \leq$$
$$C|x' - x|^{\gamma}|x - y|^{-n-\gamma} \quad \text{when} \quad |x' - x| \leq \tfrac{1}{2}|x - y|,$$

and $|K(x,y') - K(x,y)| \leq C|y' - y|^{\gamma}|x - y|^{-n-\gamma}$ when $|y' - y| \leq \frac{1}{2}|x - y|$.

Then the L^2-boundedness of T is equivalent to the following condition:

$T(1) \in BMO$, $T^*(1) \in BMO$ and T has the weak-boundedness property.

Let us give a striking application of the n-dimensional version of the T(1) theorem. We consider a classical pseudo-differential operator L of order 1. Its symbol $\lambda(x,\xi)$ satisfies the Hörmander type estimates

$$|\partial_\xi^\alpha \partial_x^\beta \lambda(x,\xi)| \leq C(\alpha,\beta)(1 + |\xi|)^{1-|\alpha|}.$$

Let us denote by A the operator of pointwise multiplication by the functions $A(x)$.

Theorem 3. With the preceding notations, the commutator $[L,A]$ is bounded on $L^2(\mathbb{R}^n)$ if $A(x)$ is a Lipschitz function.

This Lipschitz function $A(x)$ does not have to be bounded and the special case where $L = \partial/\partial x_j$ shows that the Lipschitz condition is necessary.

There are two remarkable examples of theorem 3. If $n = 1$ and if $\lambda(x,\xi) = |\xi|$ (the homogeneous case makes it possible to drop the smoothness condition on the symbol at 0), then the commutator $[L,A]$ is the celebrated Calderón first commutator whose kernel is p.v. $\frac{A(x)-A(y)}{(x-y)}$. Similarly if $L = \sqrt{-\Delta}$, then $[L,A]$ is given by the kernel p.v. $\frac{A(x)-A(y)}{|x-y|^{n+1}}$, and the second case can be obtained from the first one by the method of rotations of Calderón and Zygmund.

However, the general case where L is not given by a homogeneous symbol does not result from those specific examples.

We first observe that the off-diagonal part of the distributional kernel of L is a smooth function $L(x,y)$ satisfying the estimates $|L(x,y)| \le C|x-y|^{-n-1}$, $|\partial/\partial x L| + |\partial/\partial y L| \le C|x-y|^{-n-2}$. Therefore the off-diagonal part of the distributional kernel of T is a function $K(x,y)$ which satisfies estimates of Calderón-Zygmund type.

For proving the weak boundedness property, it can be assumed that $\lambda(x,0) = 0$. Otherwise we call M the operator of pointwise multiplication with $\lambda(x,0)$ and observe that $[M,A] = 0$.

Then we have the following estimate on T

$$\|Tf\|_2 \le C(\|D_1 f\|_2 + \dots + \|D_n f\|_2)$$

where $D_j = -i\partial/\partial x_j$..

We want to majorize $|\langle T\varphi_Q, \varphi_{Q'}\rangle|$ when $|Q| = |Q'|$. The only case where the estimate is not given by the size of the kernel is when the distance between the two cubes is less than Cd where d is the size of Q. Then we denote by x_Q the center of Q ,and we do not change $[L,A]$ if A is replaced by $A_Q(x) = A(x) - A(x_Q)$.

But then both $\|L(A_Q\varphi_Q)\|_2$ and $\|A_QL(\varphi_Q)\|_2$ are easily majorized since L is bounded from the homogeneous Sobolev space, defined by $\nabla f \in L^2$, into L^2.

Finally we have to compute $T(1)$ and $T^*(1)$. The two cases are similar since $T^* = -[L^*,A^*]$ and L^* is also a pseudo-differential operator of order 1.

Then as before we can assume $\lambda(x,0) = 0$. This gives $T(1) = 0$ and $AT(1) = 0$. On the other hand $T = T_1D_1 + \dots + T_nD_n$ where $T_1,\dots,T_n$ are pseudo-differential operators of order 0. Therefore $T(A)$ belongs to BMO. The vindication of this seemingly formal computation is easy. Indeed the smoothness of the symbol with respect to the ξ variable implies that the kernel $L(x,y)$ of L satisfies $|L(x,y)| \le C_N|x - y|^{-N}$ for all $N \ge n + 1$, and similarly $|\partial/\partial xL| + |\partial/\partial yL| \le C_N|x - y|^{-N}$ for all $N \ge n + 2$. This implies that all the integrals used in the computation of $T(1)$ converge at infinity.

Our last example will give a new way of constructing the paraproduct operator $R_u : L^2(\mathbb{R}) \longrightarrow L^2(\mathbb{R})$ for $u \in$ BMO. This construction is probably not very useful but its interest comes from the obvious connection with martingale transforms.

As in Chapter IV, let us denote by $B_j \subset L^2(\mathbb{R})$ the space of continuous piecewise affine functions with nodes at $k2^{-j}$, $k = 0, \pm 1, + 2, \dots$. The operator $E_j : L^2(\mathbb{R}) \longrightarrow B_j$ will then be the orthogonal projection onto B_j and finally $D_j = E_{j+1} - E_j$.

We then have

Theorem 4. <u>For any</u> $u \in BMO$ <u>and any</u> $f \in L^2$, <u>the sum</u> $\sum_{-\infty}^{\infty} D_j(u)E_j(f)$ <u>converges to a function</u> $g \in L^2$, <u>and the operator</u> R_u defined by $R_u(f) = g$ <u>has the following properties</u>:

(7.1) R_u <u>is a Calderón-Zygmund operator</u>;

(7.2) $R_u(1) = u$ <u>and</u> $R_u^*(1) = 0$.

<u>Conversely let us only assume that</u> $u(x)$ <u>is locally square integrable on the line and that</u> $\int_{-\infty}^{\infty} |u(x)|^2 e^{-\epsilon|x|} dx$ <u>is finite for</u> $\epsilon > 0$. <u>Then if</u> R_u <u>is bounded on</u> $L^2(\mathbb{R})$ <u>we have</u> $u(x) = a + bx + c|x| + v(x)$ <u>where</u> a, b, c <u>are three constants, and where</u> $v(x)$ <u>belongs to</u> BMO <u>and we have</u> $R_u = R_v$.

Therefore we could have used R_u for defining the paraproduct and proving theorem 1. But we do not know any short proof for the L^2 boundedness of R_u. Instead we use Theorem 1 for proving this L^2 boundedness.

The converse implication is striking since we do not know in advance that the L^2 bounded operator R_u is a Calderón-Zygmund operator, or, in other words, we do not assume that $u(x)$ belongs to the space $B^0_{\infty,\infty}$ of Bloch functions.

That u is not uniquely defined by R_u comes from the fact that $D_j(f) = 0$ for all $j \in \mathbb{Z}$ is equivalent to $f(x) = a + bx + c|x|$ if we assume a growth condition at infinity like the one satisfied by u.

But let us first assume that $u(x)$ belongs to BMO and let us compute the off-diagonal part $R_u(x,y)$ of the distributional kernel of R_u.

We can simply ignore the convergence problem if we approach the BMO function u by its standard continuous piecewise linear approximations u_N given by the following lemma

Lemma 9. The operators E_j are uniformly bounded on BMO, and, for any $u \in BMO$, $u_N = E_{N+1}(u) - E_{-N}(u)$ converges to u for the $\sigma(BMO, H^1)$ topology as N tends to infinity.

Let us prove this simple remark. The BMO-norm is dilation invariant and therefore a rescaling reduces E_j to E_0. For proving that E_0 maps BMO into itself continuously, we observe that E_0 is a Calderón-Zygmund operator, and its kernel $E_0(x,y)$ satisfies $\int E_0(x,y)dy = 1$.

The second part of lemma 9 comes from the fact that the Franklin wavelets ψ_I are total in H^1. Since the u_N are bounded in BMO-norm, it suffices to check that $\langle u_N, \psi_I\rangle \longrightarrow \langle u, \psi_I\rangle$ as N tends to infinity. But $\langle u_N, \psi_I\rangle = \langle u, E_{n+1}(\psi_I)\rangle - \langle u, E_{-N}(\psi_I)\rangle = \langle u, \psi_I\rangle$ when N is large enough. Once u is replaced by u_N in theorem 4, all sums are finite. We drop the index N and rewrite u for the function u_N.

We now want to compute the off-diagonal part of the distributional kernel of R_u.

We apply the following observation.

Lemma 10. If u belongs to BMO, or if, more generally, u is the derivative of a function in the Zygmund class, we have $\|D_j u\|_\infty \le C\|u\|_*$ and $\|\frac{d}{dx} D_j u\|_\infty \le C2^j\|u\|_*$ where $\|u\|_*$ is the norm of u in the space $B^0_{\infty,\infty}$.

The proof is straightforward. The wavelets with the L^1 normalization are $2^j\ \psi(2^jx - k)$, and are "special atoms" in $B^0_{1,1}$. Therefore $|\langle u, 2^{j/2}\psi_{j,k}\rangle| \le C$. Another method of proof would be to write u as the derivative of a function F in the Zygmund class and to integrate by parts (see theorem 2, Chapter III). Then $D_j(u) = \sum_k \alpha(j,k)\psi_{j,k}$, which yields all the required estimates.

The proof of the weak-boundedness property and of the reverse implication in theorem 4 will both be simplified if one uses the following remark.

Lemma 11. Let $u(x)$ be a locally integrable function on the line. If a constant C exists such that for each $j \in \mathbb{Z}$ one can find a continuous piecewise linear function $u_j(x)$ with nodes at $k2^{-j}$, $k \in \mathbb{Z}$, such that

$$(7.3) \qquad \int_{k2^{-j}}^{(k+1)2^{-j}} |u(x) - u_j(x)|dx \le C2^{-j} \quad \text{for all } k,j \in \mathbb{Z},$$

then $u(x) = a + bx + c|x| + v(x)$, where $v(x)$ belongs to BMO, and a,b,c, are three constants.

We denote by $v_j(x)$ the difference $u_{j+1}(x) - u_j(x)$ and we first would like to prove the estimate

$$(7.4) \qquad \|v_j\|_\infty \le 16C.$$

For obtaining (7.4) we denote by I the interval $[k2^{-j}, (k+1)2^{-j}]$ and by I', I'' the two dyadic subintervals of I. We write (7.3) for I, I', I'' and get $\int_I |u(x) - u_j(x)|dx \le C|I|$. But also $\int_I |u(x) - u_{j+1}(x)| \le C|I|$. Therefore $\int_I |u_j(x) - u_{j+1}(x)|dx \le 2C|I|$.

We now observe that if $f(x)$ is affine on some interval $[a,b]$, then

$$\int_a^b |f(x)|dx \ge \frac{1}{4}(|f(b)| + |f(a)|)(b - a) \ge \frac{1}{4}(b - a) \sup_{[a,b]} |f|.$$

We apply this trivial observation both to $\int_{I'} |v_j(x)|dx$ and to $\int_{I''} |v_j(x)|dx$ to get $\sup_I |u_j| \le 16C$. But the union of the intervals I is the real line, and (7.4) is therefore proved.

Since $v_j(x)$ is continuous and piecewise linear with nodes at $k2^{-j-1}$, $k \in \mathbb{Z}$, we immediately obtain

$$\left\| \frac{d}{dx} v_j(x) \right\|_\infty \leq 32C2^j. \tag{7.5}$$

If $j_o < j_1$, we obviously have

$$u_{j_1}(x) - u_{j_1}(0) = u_{j_0}(x) - u_{j_0}(0) + \sum_{j_0 \leq j < j_1} v_j(x) - v_j(0). \tag{7.6}$$

If j_1 is kept fixed and j_0 tends to $-\infty$, the series which appears on the right-hand side of (7.6) converges uniformly on compact sets to a continuous function $s_{j_1}(x)$. Therefore $u_{j_0}(x) - u_{j_0}(0)$ also converges uniformly to a continuous function $r(x)$ which does not depend on j_1. Since $u_j(x) = a_j + b_j x$ when $0 \leq x \leq 2^{-j}$ we therefore have $r(x) = a + bx$ when $x \geq 0$. Similarly $r(x) = a' + b'x$ when $x \leq 0$ and the continuity of $r(x)$ implies $a = a'$.

Finally $r(x) = a + bx + c|x|$, and all these functions satisfy (7.3) with $u_j(x) = r(x)$.

We now subtract $r(x)$ from the given $u(x)$ which amounts to assuming that $u_j(x) - u_j(0)$ converges uniformly on compact intervals to 0 as j tends to $-\infty$. Therefore we have

$$u_{j_1}(x) - u_{j_1}(0) = \sum_{j<j_1} (v_j(x) - v_j(0)). \tag{7.7}$$

This series can be differentiated term by term and we obtain

(7.8) $$\frac{d}{dx} u_{j_1}(x) = \sum_{j<j_1} \frac{d}{dx} v_j(x).$$

Therefore $\|\frac{d}{dx} u_j(x)\|_\infty \leq 32C2^j$.

Once this information is obtained, we can easily prove that $u(x)$ belongs to BMO. Let $[a,b]$ be any interval of length 2^{-j}. Then $[a,b]$ is contained in the union J of two adjacent dyadic intervals of length 2^{-j}. Therefore $\int_{[a,b]} |u(x) - u_j(x)|dx \leq \int_J |u(x) - u_j(x)|dx \leq C2^{-j}$. We now replace $u_j(x)$ by $u_j(a)$ when $a \leq x \leq b$. The estimate we have obtained on the derivative shows that the error term will also be less than $C'2^{-j}$. We have now obtained the usual BMO condition.

Lemma 12. _If_ $u(x)$ _belongs to_ $BMO(\mathbb{R})$, _then we have_

(7.9) $$\int_{k2^{-j}}^{(k+1)2^{-j}} |u(x) - u_j(x)|^2 dx \leq C2^{-j}$$

when $u_j = E_j(u)$.

If $u(x)$ belongs to BMO, then for any interval I there exists a constant $\gamma(I)$ such that $\int_I |u(x) - \gamma(I)|^2 dx \leq C|I|$. Generally one takes for $\gamma(I)$ the mean value of $u(x)$ on I but any reasonable mean value adapted to the size of I would also work. For example, if $I = [k2^{-j}, (k+1)2^{-j}]$, one can take $\gamma(I) = \int_{-\infty}^{\infty} u(x)2^j\varphi(2^jx - k)dx = \gamma(u,I)$.

The second observation we need is the already mentioned property that $\|u_j\|_{BMO} \leq C\|u\|_{BMO}$ for a fixed constant C. The last remark comes from the fact that $\gamma(u_j,I) = \gamma(u,I)$ since $u - u_j$ is orthogonal to the linear span B_j of $\varphi(2^jx - k)$.

There are other methods to build piecewise linear approximations $u_j(x)$ to a BMO function. One can decide for example to define u_j by $u_j(k2^{-j}) = 2^j \int_{k2^{-j}}^{(k+1)2^{-j}} u(t)dt$.

Returning to the weak boundedness property of the other operator R_u, we consider the functions $f(x) = 2^m(1 - |2^m x - \ell|)^+$ and $g(x) = 2^m(1 - |2^m x - \ell'|)^+$ and we want to prove the inequality $|\langle R_u(f), g\rangle| \leq C2^m$. We first observe that $E_j(f) = f$ for $j \geq m$. Therefore $\sum_{j\geq m} D_j(u)E_j(f) = (u - E_m(u))f$, whose L^2-norm is estimated by lemma 12.

The other part is even easier. It is estimated by

$$\sum_{j<m} \int |D_j(u)||E_j(f)||g(x)|dx \leq \sum_{j<m} \|D_j u\|_\infty \|E_j(f)\|_\infty \|g\|_1 .$$

But we have $\|D_j u\|_\infty \leq C\|u\|_*$ and, for any $f \in L^1(\mathbb{R})$, $\|E_j(f)\|_\infty \leq C2^j\|f\|_1$. The whole sum is therefore bounded by $C2^m$.

Finally $R_u(1) = \sum_{-\infty}^{\infty} D_j(u)E_j(1) = \sum_{-\infty}^{\infty} D_j(u) = u$. The computation of $(R_u)^*(1)$ is similar and we obtain $(R_u)^*(1) = 0$ since the ranges of D_j and of E_j are orthogonal.

We have obtained uniform operator estimates on R_u when $u = u_N$. It is now a routine argument to pass to the limit. This amounts to proving that the truncated operators $T_N = \sum_{-N}^{N} D_j(u)E_j$ converge to R_u in the strong operator topology. We already know that the operator norms $\|T_N\|$ are uniformly bounded. Therefore it suffices to check the convergence on a total subset of $L^2(\mathbb{R})$. If, for example, $f = \psi_I$ is a wavelet and $|I| = 2^{-m}$, then $E_j(f) = 0$ unless $j \geq m + 1$ (and $E_j(f) = f$). Therefore $T_N(f) = \left[\sum_{m+1}^{N} D_j(u)\right]f$, which converges in $L^2(\mathbb{R})$ to $\left[\sum_{m+1}^{\infty} D_j(u)\right]f$.

For proving the converse implication some very mild assumptions on u are needed. We assume for example that $u(x)$ is locally square integrable on the line, together with the condition $\int_{-\infty}^{\infty} |u(x)|^2 e^{-\epsilon|x|} dx < \infty$ for all $\epsilon > 0$.

Then $D_j(u)$ makes sense as well as $\sum_{-\infty}^{\infty} D_j(u)E_j(f)$ -- at least when f is a finite linear combination of wavelets. Since the linear span of wavelets is dense in $L^2(\mathbb{R})$ it makes sense to ask when $\sum_{-\infty}^{\infty} D_j(u)E_j(f) = R_u(f)$ is bounded on L^2. If it is the case, we can write $\|R_u(\psi_I)\|_2 \leq C$, which reads

$$(7.10) \qquad \int |u - E_j(u)|^2 |\psi(2^j x - k)|^2 dx \leq C2^{-j}.$$

We would like to obtain from (7.10) the inequality

$$(7.11) \qquad \int_{k2^{-j}}^{(k+1)2^{-j}} |u - E_j(u)|^2 dx \leq C' 2^{-j} .$$

and then apply lemma 11. The only problem is the fact that ψ vanishes inside [0,1]. But the sum $|\psi(x)|^2 + |\psi(x+1)|^2 + |\psi(x-1)|^2$ does not vanish on [0,1]. If we then write (7.10) with both $k, k+1$ and $k-1$ and sum the three inequalities, that will give (7.11).

Theorem 4 is completely proved.

VII. Takafumi Murai's proof of the boundedness of the Cauchy kernel.

Let us denote by $A(x)$ a real-valued Lipschitz function on the line and by Γ_k the singular integral operator whose kernel is given by p.v. $\frac{(A(x)-A(y))^k}{(x-y)^{k+1}}$. When $k = 0$, $\Gamma_0 = \pi H$ where H is the Hilbert transform. The $T(1)$ theorem can be applied to these operators Γ_k, and a simple integration by parts yields $\Gamma_k(1) = \Gamma_{k-1}(A')$. The boundedness of all Γ_k's is then proved by induction. Since all the kernels satisfy the smoothness and size properties which are required for real variable methods and since $A' \in L^\infty$, we therefore have

$$\Gamma_0 \text{ is bounded} \Rightarrow \Gamma_1(1) = \Gamma_0(A') \in BMO$$
$$\Rightarrow \Gamma_1 \text{ is bounded} \Rightarrow \Gamma_2(1) = \Gamma_1(A') \in BMO$$
$$\Rightarrow \Gamma_2 \text{ is bounded} \Rightarrow \ldots\ldots$$

Each time we apply the $T(1)$-theorem we introduce in the estimates a multiplicative factor which comes from the fact that the $T(1)$ theorem in the case of an antisymmetric kernel does not give $\|T\|_{2,2} \leq \|T(1)\|_{BMO} + C_0C$, where C_0 is some constant and C is the constant in the kernel estimates. In fact there is a constant in front of the right hand side. Another numerical constant appears when we apply J. Peetre's theorem (a Calderón-Zygmund operator is bounded from L^∞ into BMO). Using this inductive approach, we obtain $\|\Gamma_k\|_{2,2} \leq \pi C^k \|A'\|_\infty^k$ where C is some fixed large constant.

Therefore the series $\sum_{0}^{\infty} \varphi^k \Gamma_k$ converges for $|\varphi| < C^{-1}$, which gives the L^2-boundedness of the Cauchy kernel for Lipschitz norms. This is indeed what A. P. Calderón obtained in 1977 using complex methods.

We now want to show that this estimate suffices to obtain the general case. The proof which will be given is based on a fundamental idea by Guy David to use the rising sun lemma in order to decrease the slope of the Lipschitz graph. But G. David's proof was burdened by two difficulties: he was working directly on the Cauchy kernel and the $T(1)$ theorem did not exist.

T. Murai computes $T(1) = T_A(1)$ when T_A is the operator given by the kernel

$$(1.1) \qquad E_A(x,y) = \frac{1}{x-y} \exp\left[i\frac{A(x)-A(y)}{x-y}\right].$$

From the estimates $\|\Gamma_k\|_{2,2} \le \pi C^k \|A'\|_\infty^k$ we immediately obtain $\|T_A\|_{2,2} \le \pi \exp(C\|A'\|_\infty)$.

Assuming that A is real-valued, Murai then proves

$$(1.2) \qquad \|T_A\|_{BMO} \le C'(1 + \|A'\|_\infty)^5,$$

which gives, by the $T(1)$ theorem,

$$(1.3) \qquad \|T_A\|_{2,2} \le C''(1 + \|A'\|_\infty)^5.$$

But this estimates immediately gives the boundedness of the Cauchy kernel for any Lipschitz graph since

$$(x - y - i(A(x) - A(y)))^{-1} = \int_0^\infty E_{\lambda A}(x,y)e^{-\lambda}d\lambda.$$

Therefore the Cauchy kernel is a mean value of the operators $T_{\lambda A}$, and the growth given by (1.3) yields the boundedness of the Cauchy kernel.

T. Murai's proof of (1.2) is also based on induction. What he essentially proves is the following: there exist constants C, C' such that $\sup\{\|T_A(1)\|_{BMO};\ \|A'\|_\infty \le (3/2)^{k+1}\} \le C \sup\{\|T_A(1)\|_{BMO};\ \|A'\|_\infty \le (3/2)^k\} + C'$. Then the polynomial growth follows immediately and the exponent 5 is related to the numerical value of C.

2. The rising sun lemma.

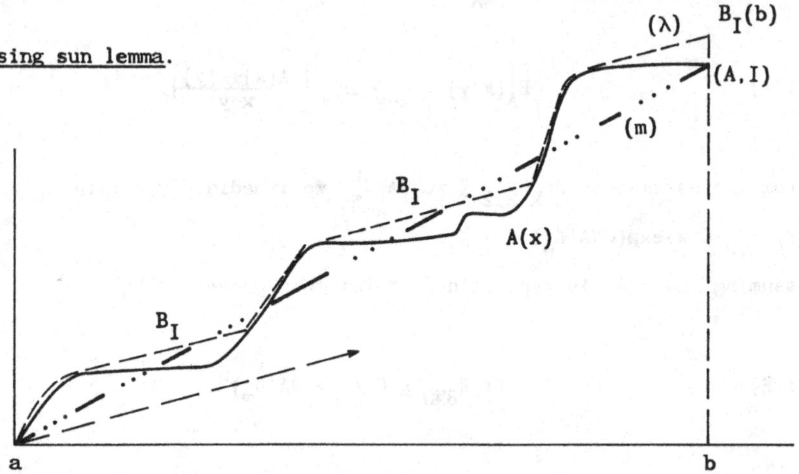

We are assuming that $A(x)$ is an increasing lipschitz function on $[a,b]$ and that $0 \le A'(x) \le M$. We denote by $m = \frac{A(b)-A(a)}{b-a}$ the average increase of A on $[a,b]$. We therefore have $0 \le m \le M$. Finally we consider λ such that $0 \le \lambda \le m \le M$. We define $B_I(x)$ to be the least continuous increasing function on $[a,b]$ such that $B_I'(x) \ge \lambda$ and $B_I(x) \ge A(x)$. The function B_I is pictured in dashes. Then we have

Lemma 1. <u>We have</u> $B_I(x) = A(x)$ <u>on a closed subset</u> E <u>of</u> $[a,b]$. <u>The measure of the complement</u> Ω <u>of</u> E <u>in</u> $[a,b]$ <u>does not exceed</u> $\frac{M-m}{M-\lambda}|I|$, <u>and</u> $B_I(x) = \lambda x + c_k$ <u>on each component</u> $]a_k,b_k[$ <u>of</u> Ω.

We have always $B_I(a) = A(a)$ but it may happen that $B_I(b) > A(b)$. Then $]a',b]$ is one of the components of Ω and $B_I(a') = A(a')$.

We observe that $|B_I(x) - A(x)| = B_I(x) - A(x) \leq M \operatorname{dist}(x,E)$.

Therefore the kernels $E_A(x,y)$ and $E_B(x,y)$ where B stands for B_I differ at most by $M(x-y)^{-2}(\operatorname{dist}(x,E) + \operatorname{dist}(y,E))$.

3. <u>An induction on the slope of the graph</u>.

We know that the operator T_A is a Calderón-Zygmund operator and we want to have a better control on its norm. We introduce the following estimator

$$\sigma(A,I) = \int_I \left| \int_I \frac{1}{x-y} \exp\left[i\frac{A(x)-A(y)}{x-y}\right] dy \right| dx.$$

In other words $\sigma(A,I) = \|T_A(\chi_I)\|_{L^1(I)}$ where χ_I is the characteristic function of I.

The second estimator will be $\sup_I \frac{\sigma(A,I)}{|I|}$ where the supremum ranges over all intervals I and will be denoted $\sigma(A)$. This definition is motivated by the following lemma.

Lemma. <u>There exists a numerical constant</u> C_0 <u>such that</u>

$$\|T_A(1)\|_{BMO} \leq \sigma(A) + C_0(1 + M)$$

<u>when</u> $\|A'\|_\infty \leq M$.

For the proof it suffices to know that $|E_A(x,y)| \le \frac{1}{|x-y|}$ and $|\partial/\partial x E_A(x,y)| \le (1+M)|x-y|^{-2}$. Then we decompose the function 1 as a sum $b_0 + b_1 + b_2 + b_3$, where $b_0 = \chi_I$, b_1 is the characteristic function of I_1, b_2 of I_2, and $I \cup I_1 \cup I_2$ is a partition of 3I, with $|I| = |I_1| = |I_2|$.

We have $\int_I \int_{I_1} |x-y|^{-1} dx dy = (\log 2)|I|$ and the same with $\int_I \int_{I_2}$, since I, I_1 and I_2 are disjoint with the same length.

Finally if x_0 is the center of I, we majorize as usual $\int_{|x_0-y|\ge(3/2)|I|} |K(x_0,y) - K(x,y)| dy$ by $C(1+M)$. Finally the only non-trivial term in checking the BMO norm of $T_A(1)$ on I is $\sigma(A,I)$.

We now introduce

(3.1) $\tau(M) = \sup\{\sigma(A),\ 0 \le A'(x) \le M\}$,

and we want to prove the fundamental inequality

(3.2) $\tau(M) \le 4\tau\left[\frac{2M}{3}\right] + C_1(1+M)$.

This inequality will provide the polynomial growth estimate we need on $\|T_A(1)\|_{BMO}$, which itself implies a polynomial growth estimate on $\|T_A\|_{2,2}$. As in any inductive process we have to start from somewhere. We need to know that $\tau(M)$ is finite for some small values of M. But this information is precisely yielded by the T(1)-theorem.

The restriction $0 \le A'(x) \le M$ can be easily removed. In fact if one adds a real constant γ to $A'(x)$, then the corresponding kernel is multiplied by the unimodular factor $e^{i\gamma}$. We also need to observe that changing the real-valued function A into −A has the effect of changing $E_A(x,y)$ into its complex conjugate. In both cases $\sigma(A,I)$ remains unchanged.

4. The proof of the main inequality.

We want to apply the rising sun lemma to the function $A(x)$ on the interval I in order to compute $\sigma(A,I)$. There are two cases. If $m = \frac{A(b)-A(a)}{b-a} \geq \frac{M}{2}$, we can apply the lemma with $\lambda = \frac{M}{3}$ and we obtain $|\Omega| \leq \frac{3}{4}|I|$. If $0 \leq m \leq \frac{M}{2}$, we do not get a good estimate for $|\Omega|$. But in that case, we first replace $A(x)$ by $Mx - A(x)$. Then $\sigma(A,I)$ remains unchanged and we can apply the rising sun lemma to the function $Mx - A(x)$ which we relabel as A.

The function B_I which is produced by the rising sun lemma has the fundamental property that $\frac{M}{3} \leq B_I'(x) \leq M$ on I. To simplify the notation we drop the index I and write B. Then $0 \leq B'(x) - \frac{M}{3} \leq 2\frac{M}{3}$ and therefore we have

$$\sigma(B,I) \leq \tau\left[\frac{2M}{3}\right]|I|. \tag{4.1}$$

We now write the open set Ω as a union of disjoint intervals I_k and want to prove

$$\sigma(A,I) \leq \sigma(B,I) + \sum_{k\geq 0} \sigma(A,I_k) + C(1+M)|I|. \tag{4.2}$$

Once (4.2) is obtained, (3.2) follows immediately.

Indeed $\sigma(A,I_k) \leq \sigma(A)|I_k|$ and $\sum |I_k| \leq \frac{3}{4}|I|$, which gives

$$\frac{\sigma(A,I)}{|I|} \leq 4\tau\left[\frac{2M}{3}\right] + 4C(1+M). \tag{4.3}$$

It now suffices to take the supremum with respect to all I and A such that $0 \leq A'(x) \leq M$.

Let us return to (4.2).

We write $E_A(x,y) = E_B(x,y) + R(x,y)$ where the error term satisfies

$$|R(x,y)| \leq M(x-y)^{-2}(\text{dist}(x,E) + \text{dist}(y,E)).$$

This bound is interesting except in one case, when x and y belong to the same I_k. Let us denote by G the complement in $I \times I$ of the union of $I_k \times I_k$. Then we have $\iint_G |R(x,y)|dx\,dy \leq 4M|I|$ by straightforward computation.

On the other hand we have $\sigma_A(I) = \int_I \left|\int_I E_A(x,y)dy\right|dx \leq \int_I \left|\int_I E_B(x,y)dy\right|dx + \int_I \left|\int_I R(x,y)dy\right|dx = \sigma_B(I) + \text{Error}$. We then write the error term as $\iint_G |R(x,y)|dy\,dx + \sum_{k\geq 0} \int_{I_k} \left|\int_{I_k} R(x,y)dy\right|dx$.

Each of these integrals is then majorized by

$$\int_{I_k} \left|\int_{I_k} E_A(x,y)dy\right|dx + \int_{I_k} \left|\int_{I_k} E_B(x,y)dy\right|dx. \tag{4.4}$$

The first integral in (4.4) is immediately rewritten $\sigma(A,I_k)$ while the second one can be explicitly computed. We know that $B(x) = \lambda x + c_k$ on each interval I_k. Therefore $E_B(x,y) = e^{i\lambda}(x-y)^{-1}$ if x and y belong to the same I_k. Therefore $\int_{I_k} \left|\int_{I_k} E_B(x,y)dy\right|dx = \pi\int_{I_k} |H\chi_{I_k}|dx \leq \pi|I_k|$ where H denotes the Hilbert transform.

REFERENCES

The existence of an orthonomal basis of smooth wavelets was announced in

[1] MEYER, Y. Principe d'incertitude, bases hilbertiennes et algèbres d'opérateurs, Séminaire Bourbaki, Feb. 1986, no. 662.

and the proofs will appear in

[2] LEMARIE, P. G. and MEYER, Y. Ondelettes et bases hilbertiennes, Revista Matematica lbero Americana Publisher A. Cordoba, Universided Autonoma de Madrid MADRID (34) SPAIN.

The results presented in the first chapters have been obtained by both G. Battle and P. G. Lemarié. G. Battle used wavelets in Constructive Quantum Field Theory. An illuminating reference is the survey paper:

[3] FEDERBUSH, P. Quantum field theory in ninety minutes. Preprint Department of Mathematics University of Michigan, Ann Arbor MI 48109. Another reference is

[4] BATTLE, G. and FEDERBUSH P. A Phase Cell Cluster Expansion for Euclidean Field Theories, Ann. of Physics, 142, 95-139 (1982).

We now list some references concerning Schauder bases or unconditional bases in classical function spaces.

[5] BOČKAREV, S.V. Existence of a basis in the space of functions analytic in a disc and some properties of the Franklin system (Russian) Mat. Sbornik 95(1974) 3-18.

[6] CIESIELSKI, Z. Bases and approximation by splines, Proc. Internat. Congr. Math. (Vancouver 1974), Vol. 2 47-51.

[7] FRAZIER, M., JAWERTH, B. Decomposition of Besov spaces (Dept. of Mathematics, Washington Univ. St. Louis, MO 63130).

[8] PEETRE J., New thoughts on Besov Spaces Duke Univ. Math. Series (1976).

[9] SJÖLIN P. and STRÖMBERG, J. O., Spline systems as bases in Hardy spaces, Dept. of Mathematics, Univ. of Stockholm, Report 1, 1982.

[10] WOJTASZCZYK, P., H^p-spaces and spline systems, Studia Math. 77 (1984) 289-320.

The basis of smooth wavelets is an unconditional basis for all Besov spaces $B_p^{s,q}$. Moreover the suitably normalized coefficients in the wavelets expansion give a realization of the isomorphism between $B_p^{s,q}$ and λ_p^q (Theorem 2, Chapter 9 of J. Peetre's book).
The Calderón-Zygmund theory for operators which are not given by convolutions has been developed in

[11] COIFMAN, R.R., and WEISS G., Analyse harmonique non-commutative sur certains espaces homogènes. L. Notes 242 (1971) SPRINGER.

At that time a criterion giving the basic L^2 estimate was missing and the theory has been fully completed by the fundamental papers

[12] DAVID G., JOURNÉ J.L. and SEMMES S., Opérateurs de Calderón-Zygmund, fonctions para-accrétives et interpolation, Revista Matematica lbero Americana (1986).

[13] DAVID G., JOURNÉ J.L., A boundedness criterion for generalized Calderón-Zygmund operators Ann. of Math. 120 (1984) 371-397.

[14] DAVID G., Opérateurs de Calderón-Zygmund, to appear in the Proceedings of the I.C.M. BERKELEY (1986).

The application of this new Calderón-Zygmund theory to solving the Dirichlet and the Neumann problems in Lipschitz domains is given in

[15] VERCHOTA, G., Layer potentials and regularity for the Dirichlet problem for Laplace's equation in Lipschitz domains, J. Funct. Anal. 59 (1984), 3, 372-611.

The symbolic calculus for the Lemarié operator algebra $\mathcal{A}$ has been investigated by Ph. Tchamitchian. Using the wavelet basis he could prove that there exists an operator T in $\mathcal{A}$ which is invertible on L^2 but whose inverse does not belong to $\mathcal{A}$. Moreover for any $p > 2$ there is an operator $T \in \mathcal{A}$ which is invertible on L^2 but not on L^p. A reference is

[16] TCHAMITCHIAN, Ph., Calcul symbolique sur les operateurs de Calderón-Zygmund et bases inconditionnelles de $L^2(\mathbb{R})$, C.R. Acad. Sc. Paris, t. 303, Série 1, 6, 1986, 215-218.

A last group of papers has been my motivation for trying to construct orthonormal bases with wavelets.

[17] GOUPILLAUD, P., GROSSMANN, A. and MORLET, J., Cycle octave and related transforms in seismic signal analysis, Geoexploration 23 (1984-1985) Elsevier Science Publishers, B. V. Amsterdam 85-102.

[18] GROSSMANN, A. and MORLET, J., Decomposition of Hardy functions into square integrable wavelets of constant shape, SIAM, J. Math. Anal. 15, 4, (1984).

[19] DAUBECHIES, I., GROSSMANN, A. and MEYER Y., Painless non-orthogonal expansions. J. Math. Phys. 27 (1986) 1271-1283.

[20] DAUBECHIES, I., Discrete sets of coherent states and their use in signal analysis, Preprint, Courant Institute of Mathematical Sciences, New York.

WAVELETS AND OPERATORS

ADDENDUM

Yves Meyer

For each integer m, Jan-Olov **Stromberg** constructed an orthonormal basis of $L^2(\mathbf{R})$ of the form $2^{j/2}w_m(2^jx-k)$, $j \in \mathbf{Z}$, $k \in \mathbf{Z}$, where $w_m(x)$ is of class C^m with an exponential decay at infinity. The reference is

J.O.Stromberg, A modified Franklin system and higher-order spline systems on $\mathbf{R}^n$ as unconditional bases for Hardy spaces, Conference in Harmonic Analysis in honor of Antoni Zygmund, Vol.II, 475-494, The University of Chicago, March 1981, edited by B.Beckner & al., Wadworth, Belmont, California.

Stromberg's wavelets are similar but not identical to our Franklin wavelets and Stromberg's construction does not extend to the C^∞ case.

My visit to the Department of Mathematics of the University of Illinois has been a very happy one. It is a pleasure to thank the organizers of the Special Year in Modern Analysis for their kindness and the participants for their patience during my too numerous lectures.

But without those lectures and without the remarkable job of typing done by Hilda Britt, these notes could not have existed.

Thanks to all of you and "au revoir"!

On the Structure of the Graph of the Franklin Analyzing Wavelet

by Earl Berkson

Notation and Terminology: We adopt the notation and terminology in [2, Chapter I]. As usual, the symbols $\mathbb{R}$, $\mathbb{Z}$, $\mathbb{N}$, $\mathbb{C}$, and $\mathbb{T}$ will denote, respectively, the real line, the set of integers, the set of positive integers, the complex plane, and the unit circle in $\mathbb{C}$. The Fourier transform of a function f (on $\mathbb{R}$ or $\mathbb{T}$) will be written $\hat{f}$. The real and imaginary parts of a complex number z will be denoted by $\mathcal{R}e\, z$ and $\mathcal{I}m\, z$, respectively. Additional notation will be introduced as needed.

1. Introduction. The Franklin analyzing wavelet $g: \mathbb{R} \longrightarrow \mathbb{R}$, has a remarkable ability to generate bases for classical function spaces (see [2, Chapter I] for an extensive treatment and development of the nature of g). By definition

$$(1.1) \qquad g(\lambda) \equiv \int_{\mathbb{R}} e^{2\pi i\lambda x}\, e^{-i\pi x}\, \omega(x)\, \frac{\sin^2 \frac{\pi x}{2}}{\left[\frac{\pi x}{2}\right]^2}\, dx\ , \text{ where}$$

$$(1.2) \qquad \omega(x) = \frac{\sin^2 \frac{\pi x}{2}}{1 - \frac{2}{3}\sin^2 \frac{\pi x}{2}} \left[\frac{\sin^4 \frac{\pi x}{2}}{1 - \frac{2}{3}\sin^2 \frac{\pi x}{2}} + \frac{\cos^4 \frac{\pi x}{2}}{1 - \frac{2}{3}\cos^2 \frac{\pi x}{2}} \right]^{-1/2}$$

$$\text{for all } x \in \mathbb{R}\ .$$

In particular, $g(2^{-1} + \lambda) = g(2^{-1} - \lambda)$ for all $\lambda \in \mathbb{R}$. The function g is known to have exponential decay at infinity, and to be piecewise linear with nodes at the points $2^{-1}m$, $m \in \mathbb{Z}$.

We shall show below that the function g has the additional properties stated in the following theorem.

(1.3) Theorem. For each non-zero real number x, let $\operatorname{sgn}(x)$ denote the sign of x, and for each real number y let $\langle y\rangle$ denote the largest $m \in \mathbb{Z}$ such that $m \leq y$. Then:

(i) $\operatorname{sgn}(g(n/2)) = (-1)^{n/2}$, for all even $n \in \mathbb{N}$;

(ii) $\operatorname{sgn}(g(n/2)) = (-1)^{\langle n/2\rangle}$, for all sufficiently large $n \in \mathbb{N}$;

(iii) $|g(\frac{n}{2})| > \left|g\left[\frac{n+1}{2}\right]\right|$, for all sufficiently large $n \in \mathbb{N}$;

(iv) $\sigma\, n^{-1/2} \beta_1^n \leq |\, g(n/2)\,| \leq \lambda\, n^{-1/2} \beta_1^n$, for all sufficiently large $n \in \mathbb{N}$, where σ and λ are positive absolute constants, and $\beta_1 = \left[\, 2 - 3^{1/2}\,\right]^{1/2}$.

REMARKS. (a) The second inequality in Theorem (1.3)-(iv), which states that $|\, g(n/2)\,| = O(\, n^{-1/2} \beta_1^n)$, reproduces the known exponential decay at infinity of $|g|$ ([2, Chapter I]). (b) The equality in (1.3)-(ii) actually holds for all positive integers n distinct from 3 (see §6 below).

The first step in obtaining Theorem (1.3) is to recast (1.1) (at the nodes of g) into the following form more convenient for our purposes.

(1.4) Theorem. For every $k \in \mathbb{Z}$,

(1.5) $$g\left[\frac{k+1}{2}\right]$$

$$= -\frac{2}{\pi}\int_0^{2\pi} \left[\sin^2(ku)\right] \frac{\sin^2 u}{1 - \frac{2}{3}\sin^2 u} \left[\frac{\sin^4 u}{1 - \frac{2}{3}\sin^2 u} + \frac{\cos^4 u}{1 - \frac{2}{3}\cos^2 u}\right]^{-1/2} du$$

$$+ g(1/2),$$

and,

(1.6) $g\left[\frac{k+1}{2}\right]$

$$= \frac{2}{\pi}\int_0^{\pi}\left[\cos(2ku)\right]\frac{\sin^2 u}{1-\frac{2}{3}\sin^2 u}\left[\frac{\sin^4 u}{1-\frac{2}{3}\sin^2 u}+\frac{\cos^4 u}{1-\frac{2}{3}\cos^2 u}\right]^{-1/2} du .$$

While the finite intervals of integration in (1.5) and (1.6), coupled with the positivity of the integrands in (1.5) and, for $k = 0$, (1.6), are convenient for numerical methods of integration, neither (1.5) nor (1.6) permits ready estimation of the sign or size of $g\left[\frac{k+1}{2}\right]$. However, by contour integration, Theorem (1.4) has the following consequence which alleviates this difficulty.

(1.7). Theorem. Let $\alpha_1 = 2 - \sqrt{3}$, $\alpha_2 = 2 + \sqrt{3}$, $\beta_1 = \sqrt{\alpha_1}$, $\beta_2 = \sqrt{\alpha_2}$. Then:

(1.8) for $n \in \mathbb{N}$, $\frac{\sqrt{2}\,\pi}{\sqrt{3}}\, g\left[\frac{n+1}{2}\right] = -\Lambda_1^{(n)} + \Lambda_2^{(n)} + 2\Lambda_3^{(n)}$,

where

$$(1.9) \quad \Lambda_1^{(n)} = (-1)^{n-1}\int_0^{\alpha_1} x^{n-1}\frac{(x+1)^2}{(\alpha_1-x)^{1/2}(\alpha_2-x)^{1/2}}\,\frac{(\alpha_1+x)^{1/2}(\alpha_2+x)^{1/2}}{(x^2+\beta_1^2)^{1/2}(x^2+\beta_2^2)^{1/2}}\,dx ,$$

$$(1.10) \quad \Lambda_2^{(n)} = \int_0^{\alpha_1} x^{n-1}\frac{(x-1)^2}{(\alpha_1+x)^{1/2}(\alpha_2+x)^{1/2}}\,\frac{(\alpha_1-x)^{1/2}(\alpha_2-x)^{1/2}}{(x^2+\beta_1^2)^{1/2}(x^2+\beta_2^2)^{1/2}}\,dx ,$$

$$(1.11) \quad \Lambda_3^{(n)} = \mathcal{Im}\int_0^{\beta_1}\frac{(iy)^{n-1}}{(\beta_1-y)^{1/2}}\,\frac{y^4+6iy^3-10y^2-6iy+1}{(y^2+\alpha_1^2)^{1/2}(y^2+\alpha_2^2)^{1/2}}\,(\beta_1+y)^{-1/2}(\beta_2^2-y^2)^{-1/2}dy.$$

As will be discussed below, the advantage of (1.8) for our purposes is that it provides the relation:

$$(1.12) \quad \frac{\pi}{\sqrt{6}}\, g\left[\frac{n+1}{2}\right] = \Lambda_3^{(n)} + O(n^{-1/2}\alpha_1^n) , \quad \text{for } n \in \mathbb{N} .$$

We shall see that, since $0 < \alpha_1 < \beta_1 < 1$, the behavior near β_1 of the function (of y) expressed by $\mathcal{Im}\left[(i)^{n-1}(y^4+6iy^3-10y^2-6iy+1)\right]$ can be used to control the right-hand side of (1.12).

2. *Proof of* Theorem (1.4). We set the stage by taking note of the following proposition, which can easily be obtained either from the theory of distributions or by direct elementary considerations.

(2.1) *Proposition. Suppose that* $F \in L^1(\mathbb{R})$, F *is continuous on* $\mathbb{R}$, *the function* $x^2 \hat{F}(x)$ *is periodic of period* 1 *on* $\mathbb{R}$, *and* F *and* $\hat{F}$ *are real-valued. Let* $G(x) = -(2\pi)^2 x^2 \hat{F}(x)$ *for* $x \in \mathbb{R}$, *and, for* $k \in \mathbb{Z}$, *define* $a_k \in \mathbb{R}$ *by putting* $a_k = \hat{G}(k) \equiv \int_0^1 G(x)e^{-2\pi ikx}dx$. *For* $n \in \mathbb{Z}$, *let*

$$c_n = \begin{cases} \sum_{k=1}^{n} a_k , & \text{if } n > 0 ; \\ 0 , & \text{if } n = 0 ; \\ -\sum_{k=n+1}^{0} a_k , & \text{if } n < 0 . \end{cases} \tag{2.2}$$

Let $Q: \mathbb{R} \to \mathbb{R}$ *be the continuous function on* $\mathbb{R}$ *such that for each* $n \in \mathbb{Z}$, Q *is linear on* $[n,n+1]$, *and*

$$Q(n) = \begin{cases} \sum_{k=0}^{n-1} c_k , & \text{if } n > 0 ; \\ 0 , & \text{if } n = 0 ; \\ -\left(\sum_{k=n}^{-1} c_k\right) , & \text{if } n < 0 . \end{cases} \tag{2.3}$$

Then F , $\hat{F}$, *and* G *are even functions on* $\mathbb{R}$, *and*

$$F(x) = Q(x) + Ax + B , \quad \text{for all } x \in \mathbb{R} , \tag{2.4}$$

where A *and* B *are real constants.*

We shall require the next proposition.

(2.5). Proposition. Under the hypotheses of Proposition (2.1), for each $k \in \mathbb{Z}$, we have

$$(2.6)\quad F(k+1) - F(0) = \frac{-1}{\pi}\int_0^{2\pi} u^2\, \hat{F}\left(\frac{u}{\pi}\right) \frac{\sin^2[(k+1)u]}{\sin^2 u}\, du\ ;$$

$$(2.7)\quad F(k) = (2\pi)^{-1}\int_0^{2\pi} u^2\, \hat{F}(u/2\pi)\, \frac{\cos(ku)}{4\sin^2(u/2)}\, du\ .$$

Proof. Since $x^2\,\hat{F}(x)$ is bounded on $\mathbb{R}$, $\hat{F} \in L^1(\mathbb{R})$ and the series

$$\sum_{k=-\infty}^{\infty} \hat{F}(x+k)$$

converges uniformly on compact intervals of $\mathbb{R}$ to a function $P(x)$ which is continuous and of period 1 on $\mathbb{R}$. With the aid of Fourier inversion on $\mathbb{R}$, a standard argument shows that

$$(2.8)\qquad F(k) = \int_0^1 P(x)e^{-2\pi ikx}dx\ , \qquad \text{for all } k \in \mathbb{Z}\ .$$

But for each $x \in \mathbb{R} \setminus \mathbb{Z}$,

$$P(x) = -\sum_{k=-\infty}^{\infty} (2\pi)^{-2}(x+k)^{-2}G(x+k) = -4^{-1}G(x)\sum_{k=-\infty}^{\infty}[\pi(x+k)]^{-2}$$

The last series is well-known ([1, §260]) to have sum equal to $1/\sin^2(\pi x)$. Hence

$$(2.9)\qquad P(x) = -G(x)\left[4\sin^2(\pi x)\right]^{-1}, \qquad \text{for } x \in \mathbb{R} \setminus \mathbb{Z}\ .$$

Using (2.9) in (2.8), we obtain

$$(2.10)\qquad F(k) = -\int_0^1 G(x)\left[4\sin^2(\pi x)\right]^{-1} e^{-2\pi ikx}dx\ , \quad \text{for all } k \in \mathbb{Z}$$

Changing the variable of integration in (2.10), substituting from the definition of G, and using the fact that $\hat{F}$ is an even function, we obtain (2.7) for arbitrary $k \in \mathbb{Z}$. The conclusion (2.6) is an immediate consequence of (2.7) and the double angle formula for the cosine.

To see that Theorem (1.4) holds, define $g_0 : \mathbb{R} \longrightarrow \mathbb{R}$ as follows:

$$g_0(\lambda) = \frac{1}{2}\, g\left[\frac{\lambda + 1}{2}\right], \quad \text{for all } \lambda \in \mathbb{R} .$$

Since the function ω in (1.2) has period 2, it is clear from (1.1) that g_0 is continuous on $\mathbb{R}$. Moreover, for all $x \in \mathbb{R}$,

$$x^2 \hat{g}_0(x) = \pi^{-2}\, \omega(2x) \sin^2 \pi x ,$$

and so $x^2 \hat{g}_0(x)$ is periodic on $\mathbb{R}$ with period 1. It is thus evident that g_0 satisfies the hypotheses of Proposition (2.1). Application of Proposition (2.5) to g_0 now completes the proof of Theorem (1.4).

3. <u>The Application of Contour Integration</u>. In this section, we shall use (1.6) to set up a suitable contour integration which, in turn, will establish Theorem (1.7). We begin with a slight reformulation of (1.6) that will expedite the calculations. From repeated use of the double angle formulas, we find that

$$(3.1) \qquad \frac{\sin^2 u}{1 - \frac{2}{3}\sin^2 u} = \frac{3}{2}\,\frac{1 - \cos(2u)}{2 + \cos(2u)} ,$$

and

$$(3.2) \qquad \frac{\sin^4 u}{1 - \frac{2}{3}\sin^2 u} + \frac{\cos^4 u}{1 - \frac{2}{3}\cos^2 u} = 3\,\frac{1 + 2\cos^2(2u)}{4 - \cos^2(2u)} .$$

Substituting into (1.6) from the equations in (3.1) and (3.2), we obtain for all $k \in \mathbb{Z}$,

$$(3.3) \qquad g\left[\frac{k+1}{2}\right]$$

$$= \frac{\sqrt{3}}{\pi} \int_0^{\pi} [\cos(2ku)]\, \frac{1 - \cos(2u)}{2 + \cos(2u)} \left[\frac{1 + 2\cos^2(2u)}{4 - \cos^2(2u)}\right]^{-1/2} du .$$

The change of variable $\theta = 2u$ on the right of (3.3) gives, for $k \in \mathbb{Z}$,

$$(3.4) \qquad g\left[\frac{k+1}{2}\right]$$

$$= \frac{\sqrt{3}}{2\pi} \int_0^{2\pi} [\cos(k\theta)]\, \frac{1 - \cos\theta}{2 + \cos\theta} \left[\frac{4 - \cos^2\theta}{1 + 2\cos^2\theta}\right]^{1/2} d\theta .$$

Since the integrand in (3.4) is an even function of period 2π , we can rewrite (3.4) as the following equation, which is thereby valid for all $k \in \mathbb{Z}$:

$$(3.5) \qquad g\left[\frac{k+1}{2}\right]$$

$$= \frac{\sqrt{3}}{\pi} \int_0^{\pi} [\cos(k\theta)]\, \frac{1 - \cos\theta}{2 + \cos\theta} \left[\frac{4 - \cos^2\theta}{1 + 2\cos^2\theta}\right]^{1/2} d\theta .$$

Because the graph of the Franklin analyzing wavelet g is symmetric about the line $x = 1/2$, we shall henceforth confine our attention to the nodes of the function g at the points $m/2$ for $m \in \mathbb{N}$. Accordingly, it will be convenient to specialize (3.5) as follows:

$$(3.6)\qquad g\left(\frac{n+1}{2}\right) = \frac{\sqrt{3}}{\pi}\int_0^{\pi} [\cos(n\theta)]\frac{1-\cos\theta}{2+\cos\theta}\left[\frac{4-\cos^2\theta}{1+2\cos^2\theta}\right]^{1/2} d\theta ,$$

for all non-negative $n \in \mathbb{Z}$.

To set the stage for contour integration, we rewrite (3.6), for $n \in \mathbb{N}$, in the form:

$$(3.7)\qquad g\left(\frac{n+1}{2}\right) = \frac{\sqrt{3}}{\pi}\,\mathcal{R}e\int_0^{\pi} e^{in\theta}\,\frac{1-\cos\theta}{2+\cos\theta}\left[\frac{4-\cos^2\theta}{1+2\cos^2\theta}\right]^{1/2} d\theta ,$$

for $n \in \mathbb{N}$.

The integral on the right of (3.7) can now be replaced with a contour integral by the change of variable $z = e^{i\theta}$. This gives us for all $n \in \mathbb{N}$:

$$(3.8)\qquad g\left(\frac{n+1}{2}\right)$$

$$= -\frac{3^{1/2}}{2^{1/2}\pi}\,\mathcal{I}m\int_{\Gamma} z^{n-1}\,\frac{(z-1)^2}{z^2+4z+1}\left[-\frac{z^4-14z^2+1}{z^4+4z^2+1}\right]^{1/2} dz ,$$

where Γ is the upper semicircle $z = e^{i\theta}$, $0 \le \theta \le \pi$, oriented in the sense of increasing θ. Here, and in what follows,

$$w^{1/2} \equiv e^{2^{-1}\mathrm{Log}\, w} ,$$

where Log denotes the principal branch of the logarithm. With α_1, α_2, β_1, and β_2 as in the statement of Theorem (1.7), it is easy to see from the quadratic formula that:

$$(3.9)\qquad f(z) \equiv -\frac{z^4 - 14z^2 + 1}{z^4 + 4z^2 + 1} = -\frac{(z-\alpha_1)(z+\alpha_1)(z-\alpha_2)(z+\alpha_2)}{(z-i\beta_1)(z+i\beta_1)(z-i\beta_2)(z+i\beta_2)},$$

and

$$(3.10)\qquad \frac{(z-1)^2}{z^2 + 4z + 1} = \frac{(z-1)^2}{(z+\alpha_1)(z+\alpha_2)}.$$

It will also be important to observe the obvious chain of inequalities:

$$(3.11)\qquad 0 < \alpha_1 < \beta_1 < 1 < \beta_2 < \alpha_2 .$$

For all sufficiently small $\epsilon > 0$, let C_ϵ be the simple, closed, counterclockwise curve comprised of: Γ ; the upper semicircles of radius ϵ centered at α_1 and $(-\alpha_1)$; the arc $\mathcal{A}_\epsilon$ of the circle $\mathcal{K}_\epsilon \equiv \{ z: |z-i\beta_1| = \epsilon \}$ obtained by adjoining to the upper semicircle of $\mathcal{K}_\epsilon$ the part of the lower semicircle of $\mathcal{K}_\epsilon$ lying outside the strip $-\frac{\epsilon}{4} < \mathcal{R}e\ z < \frac{\epsilon}{4}$; the vertical line segments joining the end-points of $\mathcal{A}_\epsilon$ to the real axis; and the closed intervals on the real axis $[-1,-\alpha_1-\epsilon]$, $[-\alpha_1+\epsilon,-\frac{\epsilon}{4}]$ together with the negatives of these last two intervals. Let z_0 be one of the complex numbers α_1 , $(-\alpha_1)$, α_2 , $(-\alpha_2)$, $i\beta_1$, $(-i\beta_1)$, $(-i\beta_2)$. We introduce an analytic logarithm for $(z - z_0)$ by deleting from $\mathbb{C}$ the vertical downwards-directed ray issuing from z_0 , and taking $(-\frac{\pi}{2}) < \arg(z - z_0) < \frac{3\pi}{2}$. Further, we consider also the case $z_0 = i\beta_2$. In this case, we introduce an analytic logarithm for $(z - z_0)$ by deleting from $\mathbb{C}$ the ray $\{ x + i\beta_2 :\ x \le 0 \}$, and then taking $(-\pi) < \arg(z - z_0) < \pi$.

It is clear that C_ϵ is contained in the simply connected domain D_ϵ , which is defined to be the intersection of the domains of definition for the analytic logarithms of $(z - z_0)$ mentioned above. For each choice of z_0 listed above, we use the corresponding analytic logarithm defined above to write $\sqrt{z - z_0} \equiv \exp\left\{ 2^{-1}\log(z - z_0) \right\}$ on D_ϵ . In terms of this notation, we put

(3.12) $$F(z) = (-i) \frac{\sqrt{z-\alpha_1}\ \sqrt{z+\alpha_1}\ \sqrt{z-\alpha_2}\ \sqrt{z+\alpha_2}}{\sqrt{z-i\beta_1}\ \sqrt{z+i\beta_1}\ \sqrt{z-i\beta_2}\ \sqrt{z+i\beta_2}}, \quad \text{for } z \in D_\epsilon .$$

We know from the way that the contour integral in (3.8) was obtained that the function $f(z)$ in (3.9) maps Γ into the open right half-plane. Hence there is a simply connected domain H_ϵ containing Γ and contained in D_ϵ such that $f(H_\epsilon)$ is contained in the open right half-plane. It follows that on H_ϵ we have either $F(z) \equiv [f(z)]^{1/2}$ or $F(z) \equiv -[f(z)]^{1/2}$. It is easy to check that $F(1) > 0$, and $[f(1)]^{1/2} = 2^{1/2} > 0$. Hence $F(z) \equiv [f(z)]^{1/2}$ on $H_\epsilon \supseteq \Gamma$, and we can rewrite (3.8) in the form

(3.13) $$g\left[\frac{n+1}{2}\right] = -\frac{3^{1/2}}{2^{1/2}\pi} \mathcal{Im} \int_\Gamma z^{n-1} \frac{(z-1)^2}{z^2+4z+1} F(z)dz ,$$

for $n \in \mathbb{N}$.

Taking note of the fact that $\int_{C_\epsilon} z^{n-1} \frac{(z-1)^2}{z^2+4z+1} F(z)dz = 0$, we decompose this contour integral into the corresponding sum of integrals along the constituent curves of C_ϵ, then take imaginary parts and let $\epsilon \to 0^+$. With due attention to the branches indicated in (3.12), we thereby obtain Theorem (1.7) from (3.13).

4. Proof of Theorem (1.3)-(i),(ii). In outline, our method of proof for Theorem (1.3) will be fashioned by showing that the term $2\Lambda_3^{(n)}$ dominates the right-hand side of (1.8), and analyzing the behavior of $\Lambda_3^{(n)}$. The following well-known scholium, which is easily established by induction, or, alternatively, by the change of variable $t = \sin^{-1}\left[\sqrt{s}/\sqrt{a}\right]$, will be instrumental for these purposes.

(4.1) Scholium. For $a > 0$, and k a non-negative integer, we have:

(4.2) $$\int_0^a \frac{s^k}{(a-s)^{1/2}}\, ds = C_k\, a^{k+(1/2)}, \quad \text{and}$$

(4.3) $$\int_0^a (a-s)^{1/2}\, s^k ds = \frac{1}{2(k+1)}\, C_{k+1}\, a^{k+(3/2)},$$

where, for $k \in \mathbb{Z}$, $k \geq 0$,

(4.4) $$C_k = \frac{2^{2k+1}\,(k!)^2}{(2k+1)!}.$$

From Scholium (4.1), (1.9) and (1.10), we obtain the following lemma.

(4.5). Lemma. For $n \in \mathbb{N}$:

(4.6) $$\operatorname{sgn}\left[\Lambda_1^{(n)}\right] = (-1)^{n-1}, \quad \text{and} \quad \Lambda_2^{(n)} < |\Lambda_1^{(n)}|,$$

(4.7) $$K_1\, C_{n-1}\, \alpha_1^{n-(1/2)} \leq |\Lambda_1^{(n)}| \leq K_2\, C_{n-1}\, \alpha_1^{n-(1/2)},$$

(4.8) $$K_3\, \frac{1}{2n}\, C_n\, \alpha_1^{n+(1/2)} \leq \Lambda_2^{(n)} \leq K_4\, \frac{1}{2n}\, C_n\, \alpha_1^{n+(1/2)},$$

where K_1, K_2, K_3, K_4 are positive constants.

Moreover, with the aid of Wallis' formula ([1, (262.5)]) in the second inequality, the following lemma is a straightforward consequence of Stirling's formula, which we use in the form ([1, (276.14),(278.9)])

$$n! = \sqrt{2\pi n}\; n^n\, e^{-n}\, e^{\theta(n)/(12n)}, \qquad \text{for } n \in \mathbb{N},$$

where $0 < \theta(n) < 1$.

(4.9) Lemma. For $n \in \mathbb{N}$,

$$\frac{11}{10}(n+1)^{-1/2} < C_n = \frac{\pi^{1/2}\,\omega_n}{n^{1/2}} < 2(n+1)^{-1/2},$$

where $\omega_n > 0$, and $\omega_n \longrightarrow 1$ as $n \longrightarrow +\infty$.

Next we observe that by virtue of (1.8), (4.7), (4.8), and (4.9), we have the following lemma.

(4.10) Lemma. For $n \in \mathbb{N}$:

(4.11) $$|\Lambda_1^{(n)}| = O\left[n^{-1/2}\, \alpha_1^n\right];$$

(4.12) $$0 < \Lambda_2^{(n)} = O\left[n^{-3/2}\, \alpha_1^n\right];$$

(4.13) $$\frac{\pi}{\sqrt{6}}\, g\left[\frac{n+1}{2}\right] = \Lambda_3^{(n)} + O\left[n^{-1/2}\, \alpha_1^n\right].$$

As an immediate consequence of (4.13) we have the following.

(4.14) Corollary. For $n \in \mathbb{N}$,

$$\frac{\pi}{\sqrt{6}} \left| g\left[\frac{n+1}{2}\right] \right| = |\Lambda_3^{(n)}| + O\left[n^{-1/2} \alpha_1^n\right] .$$

Easy estimates with (1.9) show that for $n \in \mathbb{N}$,

$$(4.15) \qquad |\Lambda_1^{(n)}| \le (\alpha_1 + 1)^2 \int_0^{\alpha_1} \frac{u^{n-1}}{(\alpha_1 - u)^{1/2}} \frac{(\alpha_1 + \alpha_2)^{1/2}}{(\alpha_2 - \alpha_1)^{1/2}} \frac{(2\alpha_1)^{1/2}}{\beta_1\beta_2} du .$$

However, $\dfrac{(\alpha_1 + \alpha_2)^{1/2}}{(\alpha_2 - \alpha_1)^{1/2}} = \dfrac{\sqrt{2}}{3^{1/4}}$, and $\beta_1\beta_2 = 1$. Using these facts in (4.15), we get:

$$(4.16) \qquad |\Lambda_1^{(n)}| \le 2 (\alpha_1 + 1)^2 \, 3^{-1/4} \alpha_1^{1/2} \int_0^{\alpha_1} \frac{u^{n-1}}{(\alpha_1 - u)^{1/2}} du ,$$

for $n \in \mathbb{N}$.

We now take up the proof of Theorem (1.3)-(i). From (1.11) we have for $m \in \mathbb{N}$, m odd:

(4.17)

$$\Lambda_3^{(m)} = 6(-1)^{(m+1)/2} \int_0^{\beta_1} \frac{y^{m-1}}{(\beta_1 - y)^{1/2}} \frac{y - y^3}{(y^2+\alpha_1^2)^{1/2} (y^2+\alpha_2^2)^{1/2}} (\beta_1+y)^{-1/2} (\beta_2^2-y^2)^{-1/2} dy,$$

and,

$$(4.18) \qquad \operatorname{sgn}\left[\Lambda_3^{(m)}\right] = (-1)^{(m+1)/2} .$$

We temporarily fix an arbitrary odd $m \in \mathbb{N}$. From (4.6) and (4.12), we have:

$$(4.19) \qquad |-\Lambda_1^{(m)} + \Lambda_2^{(m)}| = \Lambda_1^{(m)} - \Lambda_2^{(m)} < \Lambda_1^{(m)} .$$

Easy estimates with (4.17) give:

(4.20) $$|\Lambda_3^{(m)}| \geq \frac{6(1-\alpha_1)}{(2\beta_1)^{½}\ \beta_2\ (2\alpha_1)^{½}(\alpha_1+\alpha_2^2)^{½}} \int_0^{\beta_1} \frac{y^m}{(\beta_1 - y)^{½}}\, dy \,.$$

From the definitions of α_1, α_2 in the statement of Theorem (1.7), we see that $\alpha_1 + \alpha_2^2 = 3(3+\sqrt{3})$, and so $(\alpha_1 + \alpha_2^2)^{½} < 4$. Using this last inequality, together with the relations $\beta_1 = \alpha_1^{½}$, and $\beta_1\beta_2 = 1$, in (4.20), we find that

(4.21) $$|\Lambda_3^{(m)}| \geq \frac{3(1-\alpha_1)}{4\,\beta_1^{½}} \int_0^{\beta_1} \frac{y^m}{(\beta_1 - y)^{½}}\, dy \,.$$

From (4.16), (4.19), and (4.21), we get

(4.22) $$|2\Lambda_3^{(m)}| - |-\Lambda_1^{(m)} + \Lambda_2^{(m)}|$$
$$\geq \frac{3(1-\alpha_1)}{2\,\beta_1^{½}} \int_0^{\beta_1} \frac{y^m}{(\beta_1 - y)^{½}}\, dy$$
$$- 2\,(\alpha_1 + 1)^2\ 3^{-1/4}\ \alpha_1^{½} \int_0^{\alpha_1} \frac{u^{m-1}}{(\alpha_1 - u)^{½}}\, du \,.$$

By virtue of (4.2), and the fact that $(\alpha_1 + 1)^2 = 6\,\alpha_1$, we can rewrite (4.22) in the following form.

(4.23)
$$|2\Lambda_3^{(m)}| - |-\Lambda_1^{(m)} + \Lambda_2^{(m)}|$$
$$\geq 3(1-\alpha_1)\,\frac{2^{2m}\,(m!)^2}{(2m+1)!}\,\beta_1^m - 6\alpha_1 3^{-1/4}\,\frac{2^{2m}[(m-1)!]^2}{(2m-1)!}\,\alpha_1^m \,.$$

Thus, for $m \in \mathbb{N}$, m odd, in order to show that

(4.24) $$|2\Lambda_3^{(m)}| - |-\Lambda_1^{(m)} + \Lambda_2^{(m)}| > 0 \,,$$

it suffices to establish

$$(1-\alpha_1)\,\frac{m^2}{2m(2m+1)} > 2\alpha_1\ 3^{-1/4}\ \alpha_1^{m/2} \,.$$

Thus, since $\alpha_1 < 3^{-1}$, it suffices for (4.24), in the case of $m \in \mathbb{N}$ with m odd, to have

$$(4.25) \qquad \frac{m^2}{2m(2m+1)} > 3^{-(1/4)-(m/2)} .$$

But, trivially, for m odd, $m \geq 3$,

$$\frac{m^2}{2m(2m+1)} = \frac{1}{2(2+m^{-1})} > \frac{1}{6} ,$$

while

$$3^{-(1/4)-(m/2)} \leq 3^{-7/4} .$$

Hence, in view of the sufficiency of (4.25), for m odd, $m \geq 3$, we shall have (4.24) provided $6^{-1} > 3^{-7/4}$. This last inequality is trivially true, and so by (4.24) and (1.8), together with (4.18), we see that

$$(4.26) \qquad \operatorname{sgn}\left\{g\left[\frac{m+1}{2}\right]\right\} = (-1)^{(m+1)/2} , \quad \text{for } m \text{ odd}, \ m \geq 3 .$$

Moreover, $\Lambda_3^{(1)} < 0$ by (4.18), while by virtue of (4.6) $\Lambda_1^{(1)} > \Lambda_2^{(1)}$. Hence by (1.8), $g(1) < 0$. This shows that (4.26) holds for $m = 1$, and consequently (4.26) holds for all odd $m \in \mathbb{N}$, and this is precisely (1.3)-(i).

Next we proceed to the demonstration of (1.3)-(ii). Suppose that $\nu \in \mathbb{N}$ and ν is even. From (1.11) we see that

(4.27)

$$\Lambda_3^{(\nu)} = (-1)^{(\nu+2)/2} \int_0^{\beta_1} \frac{y^{\nu-1}}{(\beta_1 - y)^{1/2}} \, \frac{y^4 - 10y^2 + 1}{(y^2+\alpha_1^2)^{1/2}(y^2+\alpha_2^2)^{1/2}} (\beta_1+y)^{-1/2}(\beta_2^2-y^2)^{-1/2} dy.$$

Let $\delta_1 = 5 - 2\sqrt{6} > 0$, $\delta_2 = 5 + 2\sqrt{6}$. Thus,

$$y^4 - 10y^2 + 1 \equiv (\sqrt{\delta_1} - y)(\sqrt{\delta_1} + y)(\delta_2 - y^2) ,$$

and $0 < \sqrt{\delta_1} < \beta_1$. We can rewrite (4.27) more concisely as follows:

(4.28) $$\Lambda_3^{(\nu)} = (-1)^{(\nu+2)/2} \int_0^{\beta_1} \frac{y^{\nu-1}}{(\beta_1 - y)^{½}} (\sqrt{\delta_1} - y)\, W(y)\, dy ,$$

for all even $\nu \in \mathbb{N}$,

where W is a fixed continuous function on $[0,\beta_1]$ such that $W(y) > 0$ for all $y \in [0,\beta_1]$. Put $\tau = 2^{-1}(\sqrt{\delta_1} + \beta_1)$. We see from (4.28) that

(4.29) $$(-1)^{\nu/2} \Lambda_3^{(\nu)} = \int_\tau^{\beta_1} \frac{y^{\nu-1}}{(\beta_1 - y)^{½}} (y - \sqrt{\delta_1})\, W(y)\, dy$$

$$+ O(\nu^{-1} \tau^\nu) , \quad \text{for all even } \nu \in \mathbb{N}$$

Hence there is a positive constant A such that

(4.30) $$(-1)^{\nu/2} \Lambda_3^{(\nu)} \geq A \left\{ \int_0^{\beta_1} \frac{y^{\nu-1}}{(\beta_1 - y)^{½}} dy - \int_0^{\tau} \frac{y^{\nu-1}}{(\beta_1 - y)^{½}} dy \right\}$$

$$+ O(\nu^{-1} \tau^\nu) .$$

The second integral occurring in (4.30) is obviously $O(\nu^{-1} \tau^\nu)$. Using this fact and (4.2) in (4.30), we obtain

(4.31) $$(-1)^{\nu/2} \Lambda_3^{(\nu)} \geq A\, C_{\nu-1}\, \beta_1^{\nu-(1/2)} + O(\nu^{-1} \tau^\nu) .$$

Hence after application of Lemma (4.9) to (4.31), we get another positive constant B so that:

$$(-1)^{\nu/2} \Lambda_3^{(\nu)} \geq B\, \nu^{-½} \beta_1^\nu + O(\nu^{-1} \tau^\nu) .$$

This can be rewritten in the form

(4.32) $$(-1)^{\nu/2} \Lambda_3^{(\nu)} \geq \nu^{-½} \beta_1^\nu \left\{ B + O\left[\nu^{-½} (\tau/\beta_1)^\nu \right] \right\} ,$$

for all even $\nu \in \mathbb{N}$.

We infer from (4.32) that the positive constant B satisfies:

(4.33) $$(-1)^{\nu/2} \Lambda_3^{(\nu)} \geq \frac{B}{2} \nu^{-½} \beta_1^\nu ,$$

for all sufficiently large even $\nu \in \mathbb{N}$

It follows, in particular, from (4.33) that

(4.34) $$\operatorname{sgn}\left[\Lambda_3^{(\nu)} \right] = (-1)^{\nu/2} ,$$

for all sufficiently large even $\nu \in \mathbb{N}$.

It now follows from (4.7), (4.8), (4.9), and (4.33) that:

(4.35) $$| 2\Lambda_3^{(\nu)} | - | -\Lambda_1^{(\nu)} + \Lambda_2^{(\nu)} | \geq B \nu^{-1/2} \beta_1^{\nu} + O(\nu^{-1/2} \alpha_1^{\nu}) ,$$

for all sufficiently large even $\nu \in \mathbb{N}$.

Hence the positive constant B satisfies:

(4.36) $$| 2\Lambda_3^{(\nu)} | - | -\Lambda_1^{(\nu)} + \Lambda_2^{(\nu)} | \geq \frac{B}{2} \nu^{-1/2} \beta_1^{\nu} ,$$

for all sufficiently large even $\nu \in \mathbb{N}$.

Using (4.36) together with (1.8) and (4.34), we easily see that for all sufficiently large even $\nu \in \mathbb{N}$:

(4.37) $$\operatorname{sgn}\left[g\left[\frac{\nu + 1}{2}\right] \right] = (-1)^{\nu/2} ,$$

and

(4.38) $$\left| g\left[\frac{\nu + 1}{2}\right] \right| \geq B_1 \nu^{-1/2} \beta_1^{\nu} ,$$

where B_1 is a positive constant.

Combining (4.37) with (1.3)-(i) completes the demonstration of (1.3)-(ii).

5. Proof of Theorem (1.3)-(iii),(iv). From (4.27) and (4.34) we see that

(5.1)

$$| \Lambda_3^{(2k)} | = \int_0^{\beta_1} \frac{y^{2k-1}}{(\beta_1 - y)^{1/2}} \frac{-y^4 + 10 y^2 - 1}{(y^2+\alpha_1^2)^{1/2} (y^2+\alpha_2^2)^{1/2}} (\beta_1+y)^{-1/2} (\beta_2^2-y^2)^{-1/2} dy ,$$

for all sufficiently large $k \in \mathbb{N}$.

From (4.17) we have

(5.2)

$$| \Lambda_3^{(2k+1)} | = \int_0^{\beta_1} \frac{y^{2k-1}}{(\beta_1 - y)^{½}} \frac{6y^2 - 6y^4}{(y^2+\alpha_1^2)^{½} (y^2+\alpha_2^2)^{½}} (\beta_1+y)^{-½}(\beta_2^2-y^2)^{-½} dy,$$

for $k \in \mathbb{N}$.

Using (5.1) and (5.2) successively in the equation of Corollary (4.14), and then subtracting the second resulting equation from the first, we find that

(5.3) $$\frac{\pi}{\sqrt{6}} \left\{ \left| g\left[\frac{2k+1}{2}\right] \right| - \left| g\left[\frac{2k+2}{2}\right] \right| \right\}$$

$$= \int_0^{\beta_1} \frac{y^{2k-1}}{(\beta_1 - y)^{½}} \frac{5y^4 + 4y^2 - 1}{(y^2+\alpha_1^2)^{½} (y^2+\alpha_2^2)^{½}} (\beta_1+y)^{-½}(\beta_2^2-y^2)^{-½} dy + O\left[k^{-½}\alpha_1^{2k}\right],$$

for all sufficiently large $k \in \mathbb{N}$.

We rewrite (5.3) more concisely in the following fashion.

(5.4) $$\frac{\pi}{\sqrt{6}} \left\{ \left| g\left[\frac{2k+1}{2}\right] \right| - \left| g\left[\frac{2k+2}{2}\right] \right| \right\}$$

$$= \int_0^{\beta_1} \frac{y^{2k-1}}{(\beta_1 - y)^{½}} \phi(y)\, \Psi(y)\, dy + O\left[k^{-½}\alpha_1^{2k}\right],$$

for all sufficiently large $k \in \mathbb{N}$,

where $\Psi(y) \equiv (y^2 + \alpha_1^2)^{-½} (y^2 + \alpha_2^2)^{-½} (\beta_1 + y)^{-½} (\beta_2^2 - y^2)^{-½} > 0$ is continuous on $[0,\beta_1]$, and

$$\phi(y) \equiv 5y^4 + 4y^2 - 1 \equiv 5(y^2 + 1)(y + 5^{-½})(y - 5^{-½}) .$$

Since $0 < 5^{-½} < \beta_1$, we have:

(5.5) $\phi < 0$ on $[0,5^{-½})$, and $\phi > 0$ on $(5^{-½},\beta_1]$.

By virtue of (3.11) we can choose and fix a number η such that

(5.6) $$\max \{ \alpha_1 , 5^{-½} \} < \eta < \beta_1 .$$

From (5.4) we get

(5.7)
$$\frac{\pi}{\sqrt{6}}\left\{\left|g\left[\frac{2k+1}{2}\right]\right| - \left|g\left[\frac{2k+2}{2}\right]\right|\right\}$$
$$= \int_0^{\eta} \frac{y^{2k-1}}{(\beta_1 - y)^{1/2}}\,\phi(y)\,\Psi(y)\,dy + \int_{\eta}^{\beta_1} \frac{y^{2k-1}}{(\beta_1 - y)^{1/2}}\,\phi(y)\,\Psi(y)\,dy$$
$$+ O\left[k^{-1/2}\alpha_1^{2k}\right]$$
$$= \int_{\eta}^{\beta_1} \frac{y^{2k-1}}{(\beta_1 - y)^{1/2}}\,\phi(y)\,\Psi(y)\,dy + O\left[k^{-1}\eta^{2k}\right] + O\left[k^{-1/2}\alpha_1^{2k}\right],$$

for all sufficiently large $k \in \mathbb{N}$.

Since ϕ and Ψ are positive and continuous on $[\eta,\beta_1]$, we can utilize (5.7) to obtain a positive constant ρ such that:

(5.8)
$$\frac{\pi}{\sqrt{6}}\left\{\left|g\left[\frac{2k+1}{2}\right]\right| - \left|g\left[\frac{2k+2}{2}\right]\right|\right\}$$
$$\geq \rho\left[\int_0^{\beta_1} \frac{y^{2k-1}}{(\beta_1 - y)^{1/2}}\,dy - \int_0^{\eta} \frac{y^{2k-1}}{(\beta_1 - y)^{1/2}}\,dy\right]$$
$$+ O\left[k^{-1}\eta^{2k}\right] + O\left[k^{-1/2}\alpha_1^{2k}\right],$$

for all sufficiently large $k \in \mathbb{N}$.

Hence, by (5.8), (4.2), and Lemma (4.9), there is a positive constant r such that:

(5.9)
$$\frac{\pi}{\sqrt{6}}\left\{\left|g\left[\frac{2k+1}{2}\right]\right| - \left|g\left[\frac{2k+2}{2}\right]\right|\right\}$$
$$\geq r\,k^{-1/2}\,\beta_1^{2k} + O\left[k^{-1}\eta^{2k}\right] + O\left[k^{-1/2}\alpha_1^{2k}\right],$$

for all sufficiently large $k \in \mathbb{N}$.

In view of (5.6), it follows from (5.9) that

(5.10)
$$\left|g\left[\frac{2k+1}{2}\right]\right| > \left|g\left[\frac{2k+2}{2}\right]\right|,$$

for all sufficiently large $k \in \mathbb{N}$.

Next we replace k by (k + 1) in (5.1). Using the resulting equation together with (5.2) and Corollary (4.14), we obtain the following relation.

(5.11) $$\frac{\pi}{\sqrt{6}}\left\{\left|\, g\left[\frac{2k+2}{2}\right]\right| - \left|\, g\left[\frac{2k+3}{2}\right]\right|\right\}$$

$$= \int_0^{\beta_1} \frac{y^{2k+1}}{(\beta_1 - y)^{1/2}} \frac{y^4 - 16y^2 + 7}{(y^2+\alpha_1^2)^{1/2}\,(y^2+\alpha_2^2)^{1/2}} (\beta_1+y)^{-1/2}(\beta_2^2-y^2)^{-1/2}dy + O\left[k^{-1/2}\alpha_1^{2k}\right],$$

for all sufficiently large $k \in \mathbb{N}$.

Let $q(y) \equiv y^4 - 16y^2 + 7$. Direct calculation shows that $q(\beta_1) > 0$. Since $q(1) < 0$ and the quartic polynomial $q(y)$ is an even function tending to $+\infty$ as $y \to +\infty$, we see that q has four distinct zeroes, all situated in $\mathbb{R} \setminus [0,\beta_1]$. Hence $q(y) > 0$ for all $y \in [0,\beta_1]$. Applying this last fact to (5.11), we see that there is a positive constant μ such that

(5.12) $$\frac{\pi}{\sqrt{6}}\left\{\left|\, g\left[\frac{2k+2}{2}\right]\right| - \left|\, g\left[\frac{2k+3}{2}\right]\right|\right\}$$

$$\geq \mu \int_0^{\beta_1} \frac{y^{2k+1}}{(\beta_1 - y)^{1/2}}\, dy + O\left[k^{-1/2}\alpha_1^{2k}\right],$$

for all sufficiently large $k \in \mathbb{N}$.

In view of (3.11), application of (4.2) and (4.9) to (5.12) readily provides that

(5.13) $$\left|\, g\left[\frac{2k+2}{2}\right]\right| > \left|\, g\left[\frac{2k+3}{2}\right]\right|$$

for all sufficiently large $k \in \mathbb{N}$.

Combining (5.10) with (5.13) establishes (1.3)-(iii).

We next show (1.3)-(iv). This will complete the proof of Theorem (1.3). From (4.21) together with (4.2) and Corollary (4.14) we see that there is a positive constant R such that:

$$\left| g\left(\frac{m+1}{2}\right) \right| \geq R\, m^{-1/2} \beta_1^m + O\left(m^{-1/2} \alpha_1^m\right),$$

for all odd $m \in \mathbb{N}$.

Hence for all sufficiently large odd $m \in \mathbb{N}$,

(5.14) $$\left| g\left(\frac{m+1}{2}\right) \right| \geq \frac{R}{2}\, m^{-1/2} \beta_1^m .$$

Combining (4.38) with (5.14), we obtain (for all sufficiently large $n \in \mathbb{N}$) the lower estimate for $| g(n/2) |$ in (1.3)-(iv). From the definition of $\Lambda_3^{(n)}$ in (1.11) together with (4.2) and Lemma (4.9), we see that for $n \in \mathbb{N}$,

(5.15) $$| \Lambda_3^{(n)} | = O\left(n^{-1/2} \beta_1^n\right) .$$

Use of (5.15) in Corollary (4.14) completes the proof of (1.3)-(iv).

6. Refinements of the Computations. The theorem of this concluding section improves upon Theorem (1.3)-(ii), and will serve to illustrate that, with due attention to the constants arising, the methods of calculation in §§4,5 can be refined so as to provide additional precision in Theorem (1.3). We shall show the following.

(6.1) Theorem. *In the notation of Theorem* (1.3),

$$\operatorname{sgn}(g(n/2)) = (-1)^{\langle n/2 \rangle} ,$$

for every positive integer n *different from* 3 .

In order to achieve the precision expressed by Theorem (6.1) it will be necessary to encounter straightforward, detailed, sometimes tedious computations involving many-digit numbers. The results of such computations will be described at the appropriate stages below. However, in all cases, we do not equate any numbers with rounded-off decimal approximants to them, so that all computations and the resulting inequalities are rigorously accurate.

We remark at the outset of the demonstration of Theorem (6.1) that, in view of Theorem (1.3)-(i), we need only consider the case of odd $n \in \mathbb{N}$. Accordingly, we return to the considerations in (4.27) through (4.28). The function W on $[0,\beta_1]$ occurring in (4.28) has the form:

$$(6.2)\qquad W(y) = \frac{(\sqrt{\delta_1}+y)\,(\delta_2-y^2)}{(y^2+\alpha_1^2)^{1/2}\,(y^2+\alpha_2^2)^{1/2}}\,(\beta_1+y)^{-1/2}(\beta_2^2-y^2)^{-1/2} .$$

Elementary computations from (6.2) show that:

$$(6.3)\qquad W(y) \le \frac{2\sqrt{\delta_1}\,\delta_2}{\sqrt{\beta_1}}\,(\beta_2^2-\delta_1)^{-1/2} < (14.76)\,\sqrt{\delta_1}\,,$$

$$\text{for } 0 \le y \le \sqrt{\delta_1}\,,$$

and

$$(6.4)\qquad W(y) \ge \frac{\sqrt{2\delta_1}\,(\delta_2-\alpha_1)}{(\alpha_1+\alpha_1^2)^{1/2}\,(\alpha_1+\alpha_2^2)^{1/2}\,\sqrt{\beta_1}\,(\beta_2^2-\delta_1)^{1/2}} > (4.489)\,\sqrt{\delta_1}\,,$$

$$\text{for } \sqrt{\delta_1} \le y \le \beta_1 .$$

It follows from (4.28), (6.3), and (6.4) that

$$(6.5)\qquad (-1)^{\nu/2}\,\Lambda_3^{(\nu)} \ge -(14.76)\sqrt{\delta_1}\int_0^{\sqrt{\delta_1}} \frac{y^{\nu-1}}{(\beta_1-y)^{1/2}}\,(\sqrt{\delta_1}-y)\,dy$$

$$+\,(4.489)\sqrt{\delta_1}\int_{\sqrt{\delta_1}}^{\beta_1} \frac{y^{\nu-1}}{(\beta_1-y)^{1/2}}\,(y-\sqrt{\delta_1})\,dy,$$

$$\text{for all even } \nu \in \mathbb{N} .$$

Expressing the last integral of (6.5) in the form $\int_0^{\beta_1} - \int_0^{\sqrt{\delta_1}}$, and collecting terms, we find that

$$(6.6)\qquad (-1)^{\nu/2}\Lambda_3^{(\nu)} \ge -(10.271)\sqrt{\delta_1}\int_0^{\sqrt{\delta_1}} \frac{y^{\nu-1}}{(\beta_1 - y)^{1/2}}(\sqrt{\delta_1} - y)\,dy$$

$$+ (4.489)\sqrt{\delta_1}\int_0^{\beta_1} \frac{y^{\nu-1}}{(\beta_1 - y)^{1/2}}(y - \sqrt{\delta_1})\,dy,$$

for all even $\nu \in \mathbb{N}$.

Before proceeding further, it will be convenient to take note of the following inequalities, which can be established by elementary computations.

$$(6.7)\qquad \beta_1 \ge .515 ;$$

$$(6.8)\qquad \sqrt{\delta_1} < .318 .$$

From (6.7) and (6.8) we obtain:

$$(6.9)\qquad (\beta_1 - \sqrt{\delta_1})^{1/2} \ge .443 ;$$

$$(6.10)\qquad \frac{\sqrt{\delta_1}}{\beta_1} < \frac{.318}{.515} < .6175 .$$

From (6.9) we get:

$$(6.11)\qquad \int_0^{\sqrt{\delta_1}} \frac{y^{\nu-1}}{(\beta_1 - y)^{1/2}}(\sqrt{\delta_1} - y)\,dy$$

$$\le \frac{\sqrt{\delta_1}}{.443}\int_0^{\sqrt{\delta_1}} y^{\nu-1}dy = \frac{\sqrt{\delta_1}}{.443}\,\frac{(\sqrt{\delta_1})^{\nu}}{\nu} .$$

Using (6.11) in (6.6), we obtain:

(6.12)

$$(-1)^{\nu/2}\Lambda_3^{(\nu)}$$

$$\ge -(10.271)\,\frac{\delta_1}{.443}\,\frac{(\sqrt{\delta_1})^{\nu}}{\nu}$$

$$+ (4.489)\sqrt{\delta_1}\left\{\int_0^{\beta_1} \frac{y^{\nu}}{(\beta_1 - y)^{1/2}}dy - \sqrt{\delta_1}\int_0^{\beta_1} \frac{y^{\nu-1}}{(\beta_1 - y)^{1/2}}dy\right\} ,$$

for all even $\nu \in \mathbb{N}$.

From (4.2) and (4.4), it is easily seen that for all even $\nu \in \mathbb{N}$:

(6.13)
$$\left[\int_0^{\beta_1} \frac{y^{\nu}}{(\beta_1 - y)^{1/2}} dy \right] \left[\int_0^{\beta_1} \frac{y^{\nu-1}}{(\beta_1 - y)^{1/2}} dy \right]^{-1}$$

$$= \frac{2\nu^2}{2\nu^2 + \nu} \beta_1 \geq \frac{4}{5} \beta_1 .$$

Using (6.13) in (6.12), we find that:

(6.14)
$$(-1)^{\nu/2} \Lambda_3^{(\nu)} \geq -(10.271) \frac{\delta_1}{.443} \frac{(\sqrt{\delta_1})^{\nu}}{\nu}$$

$$+ (4.489)\sqrt{\delta_1} \left\{ \frac{4}{5} \beta_1 - \sqrt{\delta_1} \right\} \int_0^{\beta_1} \frac{y^{\nu-1}}{(\beta_1 - y)^{1/2}} dy ,$$

for all even $\nu \in \mathbb{N}$,

Using (4.2) to express the last integral, we get for all even $\nu \in \mathbb{N}$,

(6.15)
$$(-1)^{\nu/2} \Lambda_3^{(\nu)} \geq -(10.271) \frac{\delta_1}{.443} \frac{(\sqrt{\delta_1})^{\nu}}{\nu}$$

$$+ (4.489)\sqrt{\delta_1} \left\{ \frac{4}{5} \beta_1 - \sqrt{\delta_1} \right\} C_{\nu-1} \beta_1^{\nu-(1/2)} .$$

By virtue of (6.10), $\left\{ \frac{4}{5} \beta_1 - \sqrt{\delta_1} \right\} > 0$. Thus if we apply the first inequality in (4.9) to (6.15), we see that:

(6.16)
$$(-1)^{\nu/2} \Lambda_3^{(\nu)} \geq -(10.271) \frac{\delta_1}{.443} \frac{(\sqrt{\delta_1})^{\nu}}{\nu}$$

$$+ (4.489)\sqrt{\delta_1 \beta_1} \left\{ \frac{4}{5} - \frac{\sqrt{\delta_1}}{\beta_1} \right\} \frac{1.1}{\sqrt{\nu}} \beta_1^{\nu} ,$$

for all even $\nu \in \mathbb{N}$

By (6.10), $\left\{ \frac{4}{5} - \frac{\sqrt{\delta_1}}{\beta_1} \right\} > .1825$. Using this in (6.16) gives:

(6.17)
$$(-1)^{\nu/2} \Lambda_3^{(\nu)} > -(10.271) \frac{\delta_1}{.443} \frac{(\sqrt{\delta_1})^{\nu}}{\nu}$$

$$+ (4.489)\sqrt{\delta_1 \beta_1} \, (.1825) \frac{1.1}{\sqrt{\nu}} \beta_1^{\nu} ,$$

for all even $\nu \in \mathbb{N}$

Hence for even $\nu \in \mathbb{N}$, in order that $(-1)^{\nu/2} \Lambda_3^{(\nu)}$ should be positive, it is sufficent that ν satisfy the following condition:

$$(6.18) \quad [(4.489)\sqrt{\beta_1}\,(.1825)(1.1)] \geq (10.271)\frac{\sqrt{\delta_1}}{.443}\,\frac{1}{\sqrt{\nu}}\left[\frac{\sqrt{\delta_1}}{\beta_1}\right]^{\nu} .$$

It follows from (6.7) that

$$(6.19) \qquad \sqrt{\beta_1} > .7175 .$$

Application of (6.19), (6.8), and (6.10) to (6.18) shows that if ν is an even positive integer, then $(-1)^{\nu/2} \Lambda_3^{(\nu)}$ will be positive provided:

$$(6.20) \quad [(4.489)(.7175)(.1825)(1.1)] \geq (10.271)\frac{.318}{.443}\,\frac{1}{\sqrt{\nu}}\,(.6175)^{\nu} .$$

Direct calculation shows that (6.20) holds for $\nu = 4$, and hence for all even $\nu \geq 4$. Thus we have shown that

$$(6.21) \qquad \operatorname{sgn}\left[\Lambda_3^{(\nu)}\right] = (-1)^{\nu/2} , \qquad \text{for all even integers } \nu \geq 4 .$$

Since (1.9) and (1.10) insure that for even $\nu \in \mathbb{N}$, $-\Lambda_1^{(\nu)} + \Lambda_2^{(\nu)} > 0$, an application of (6.21) to (1.8) shows that if $\nu \in 4\mathbb{N}$, then $g\left[\frac{\nu + 1}{2}\right]$ is positive. Thus we have established the following fact (which will be useful later):

$$(6.22) \qquad g\left[\frac{4k + 1}{2}\right] > 0 , \qquad \text{for all } k \in \mathbb{N} .$$

In order to complete the proof of Theorem (6.1) by exploiting (6.21), we next show that for a suitable range of ν , the term $2\Lambda_3^{(\nu)}$ dominates the right-hand side of (1.8). From (6.17), (6.19), and (6.8), we see that:

(6.23)

$$\frac{\sqrt{\nu}}{\sqrt{\delta_1}}(.443)\,(-1)^{\nu/2}\,\Lambda_3^{(\nu)} > -(10.271)\,(.318)\frac{(\sqrt{\delta_1}\,)^{\nu}}{\sqrt{\nu}}$$

$$+\,(4.489)(.443)(.7175)(.1825)(1.1)\beta_1^{\nu}\,,$$

for all even $\nu \in \mathbb{N}$.

Direct calculation shows that the coefficient of β_1^{ν} in (6.23) is at least (.2864301) . Using this and the calculated value of the coefficient in (6.23) of $\dfrac{(\sqrt{\delta_1}\,)^{\nu}}{\sqrt{\nu}}$, we obtain:

(6.24)

$$\frac{\sqrt{\nu}}{\sqrt{\delta_1}}(.443)\,(-1)^{\nu/2}\,\Lambda_3^{(\nu)} > -(3.266178)\frac{(\sqrt{\delta_1}\,)^{\nu}}{\sqrt{\nu}} + (.2864301)\,\beta_1^{\nu}\,,$$

for all even $\nu \in \mathbb{N}$.

Moreover, for even $\nu \in \mathbb{N}$, $\left[\,-\Lambda_1^{(\nu)}\,\right]$ is expressed by the integral on the right of (1.9), whose integrand is dominated at each $x \in [0,\alpha_1)$ by

$$\frac{x^{\nu-1}}{(\alpha_1-x)^{1/2}}\;\frac{(\alpha_1+1)^2}{(\alpha_2-\alpha_1)^{1/2}}\;\frac{2^{1/2}\alpha_1^{1/2}\,(\alpha_2+\alpha_1)^{1/2}}{\beta_1\beta_2} = \frac{12}{3^{1/4}}\,\alpha_1^{\;3/2}\,\frac{x^{\nu-1}}{(\alpha_1-x)^{1/2}}$$

$$\le (9.12)\alpha_1^{\;3/2}\,\frac{x^{\nu-1}}{(\alpha_1-x)^{1/2}}\,.$$

Hence, with the aid of (4.2), we find that

$$0 < -\Lambda_1^{(\nu)} \le (9.12)\alpha_1^{\;3/2}\int_0^{\alpha_1}\frac{x^{\nu-1}}{(\alpha_1-x)^{1/2}}\,dx = (9.12)\alpha_1^{\;3/2}\,C_{\nu-1}\,\alpha_1^{\nu-(1/2)}\,.$$

Applying (4.9) to the last expression gives

(6.25) $$0 < -\Lambda_1^{(\nu)} < (9.12)\ 2\nu^{-½}\ \alpha_1^{\nu+1} = (18.24)\ \nu^{-½}\ \alpha_1^{\nu+1}\ ,$$

for all even $\nu \in \mathbb{N}$.

Similar considerations with (1.10) show that for even $\nu \in \mathbb{N}$,

$$0 < \Lambda_2^{(\nu)} \leq \alpha_2^{½} \int_0^{\alpha_1} x^{\nu-1}\ (\alpha_1 - x)^{½} dx = \alpha_2^{½}\ (2\nu)^{-1}\ C_\nu\ \alpha_1^{\nu+(½)} = (2\nu)^{-1} C_\nu \alpha_1^\nu\ .$$

Hence by (4.9), for even $\nu \in \mathbb{N}$,

(6.26) $$0 < \Lambda_2^{(\nu)} < \nu^{-1}\ (\nu + 1)^{-½}\ \alpha_1^\nu\ .$$

Combining (6.25) and (6.26) we get for all even $\nu \in \mathbb{N}$:

(6.27)

$$\begin{aligned} \left| -\Lambda_1^{(\nu)} + \Lambda_2^{(\nu)} \right| = -\Lambda_1^{(\nu)} + \Lambda_2^{(\nu)} &< (18.24)\nu^{-½}\ \alpha_1^{\nu+1} + \nu^{-1}(\nu + 1)^{-½}\ \alpha_1^\nu \\ &< \alpha_1^\nu\ \nu^{-½}\left\{ (18.24)\ \alpha_1 + \nu^{-1} \right\} . \end{aligned}$$

It is easy to verify that

(6.28) $$(18.24)\ \alpha_1 < 4.9\ .$$

Using (6.28) in (6.27) gives for all even $\nu \in \mathbb{N}$,:

(6.29) $$\left| -\Lambda_1^{(\nu)} + \Lambda_2^{(\nu)} \right| < \alpha_1^\nu\ \nu^{-½}\left\{ (4.9) + \nu^{-1} \right\} .$$

For even $\nu \geq 4$, (6.21) allows us to rewrite (6.24) as follows:

(6.30)

$$|\Lambda_3^{(\nu)}| > \frac{\sqrt{\delta_1}}{(.443)}\ \frac{\beta_1^\nu}{\sqrt{\nu}} \left\{ (.2864301) - (3.266178)\ \nu^{-½}\left(\sqrt{\delta_1}/\beta_1 \right)^\nu \right\} .$$

Straightforward computations show that the quantity in braces in (6.30) is positive for all even $\nu \geq 4$, and also that

$$\frac{\sqrt{\delta_1}}{(.443)} > .7174\ .$$

Applying these facts to (6.30), we get:

(6.31)

$$|\Lambda_3^{(\nu)}| > (.7174)\,\frac{\beta_1^{\nu}}{\sqrt{\nu}}\left\{ (.2864301) - (3.266178)\,\nu^{-1/2}(\sqrt{\delta_1}/\beta_1)^{\nu} \right\},$$

for all even $\nu \geq 4$

Further straightforward computations show that:

$$(.7174)\,(.2864301) > (.20548),$$

and

$$(.7174)\,(3.266178) < (2.344)\ .$$

Applying these to (6.31) gives:

(6.32) $$|\Lambda_3^{(\nu)}| > \frac{\beta_1^{\nu}}{\sqrt{\nu}}\left\{ (.20548) - (2.344)\,\nu^{-1/2}(\sqrt{\delta_1}/\beta_1)^{\nu} \right\},$$

for all even $\nu \geq 4$

Using (6.10) in (6.32), we see that:

(6.33) $$|2\Lambda_3^{(\nu)}| > \frac{\beta_1^{\nu}}{\sqrt{\nu}}\left\{ (.41096) - (4.688)\,\nu^{-1/2}(.6175)^{\nu} \right\},$$

for all even $\nu \geq 4$

Combining (6.29) and (6.33), we find that:

(6.34)

$$|2\Lambda_3^{(\nu)}| - |-\Lambda_1^{(\nu)} + \Lambda_2^{(\nu)}| > \frac{\beta_1^{\nu}}{\sqrt{\nu}}\left\{ (.41096) - (4.688)\,\nu^{-1/2}(.6175)^{\nu} \right\} - \alpha_1^{\nu}\,\nu^{-1/2}\left\{ (4.9) + \nu^{-1} \right\},$$

for all even $\nu \geq 4$

Since $\beta_1 = \sqrt{\alpha_1}$, we can rewrite the minorant in (6.34) to get:

(6.35)

$$|2\Lambda_3^{(\nu)}| - |-\Lambda_1^{(\nu)} + \Lambda_2^{(\nu)}|$$

$$> \frac{\beta_1^{\nu}}{\sqrt{\nu}}\left\{(.41096) - (4.688)\nu^{-1/2}(.6175)^{\nu} - [(4.9) + \nu^{-1}]\,\beta_1^{\nu}\right\}$$

$$\equiv \frac{\beta_1^{\nu}}{\sqrt{\nu}}\,T(\nu)\ ,$$

for all even $\nu \geq 4$

Since simple computations show that $\beta_1 < .518$, we readily see that, in the notation of (6.35),

(6.36) $$T(6) > (.41096) - (4.688)6^{-½}(.6175)^6 - [(4.9) + 6^{-1}]\ (.518)^6 .$$

Straightforward calculations with the minorant in (6.36) show that $T(6) > 0$, and hence $T(\nu) > 0$, for all even $\nu \geq 6$. Applying this fact to (6.35), we obtain:

(6.37) $$|2\Lambda_3^{(\nu)}| - |-\Lambda_1^{(\nu)} + \Lambda_2^{(\nu)}| > 0 , \qquad \text{for all even } \nu \geq 6 .$$

Hence, in view of (1.8),

(6.38) $$\operatorname{sgn}\left\{ g\left[\frac{\nu+1}{2}\right] \right\} = \operatorname{sgn}\left[\Lambda_3^{(\nu)} \right] , \qquad \text{for all even } \nu \geq 6 .$$

Applying (6.21) to (6.38), we have for all odd integers $n \geq 7$:

(6.39) $$\operatorname{sgn}\left\{ g(n/2) \right\} = (-1)^{<n/2>} .$$

Thus, by virtue of Theorem (1.3)-(i), we see that (6.39) is valid for all integers $n \geq 6$, as well as for $n = 2$ and $n = 4$. By (1.6), for $k = 0$, and (6.22) for $k = 1$, we get, respectively, $g(1/2) > 0$, and $g(5/2) > 0$. Combining these facts completes the proof of Theorem (6.1).

Remarks. Although we have not rigorously determined whether or not (6.39) fails to be true in the case $n = 3$, computerized numerical integration, employing Simpson's rule on the integral expression for $g(3/2)$ afforded by (3.6), provides an approximate value for $g(3/2)$ which is small and positive. Hence numerical approximations suggest, but do not prove, that the case $n = 3$ is an exception to the equation in Theorem (6.1) .

ACKNOWLEDGEMENT. The work of the author was supported by a grant from the National Science Foundation (U.S.A.).

REFERENCES

1. C. Carathéodory, Theory of functions of a complex variable, vol. I, Chelsea Publishing Co., New York, 1954.

2. Y. Meyer, Wavelets and operators, Proceedings of the Special Year in Modern Analysis at the University of Illinois (1986/87), Cambridge University Press, 1988.

UNIVERSITY OF ILLINOIS
DEPARTMENT OF MATHEMATICS
1409 W. GREEN STREET
URBANA, ILLINOIS 61801
U.S.A.

Boundedness of the Canonical Projection for Sobolev Spaces Generated by Finite Families of Linear Differential Operators.

A. Pelczynski.

Abstract.

Let S be a finite set of linear partial differential operators with constant coefficients containing the identity operator. We consider the orthogonal projection P_S from the Cartesian product $\oplus_S L^2(\mathbf{R}^n)$ of $\#S$ (= the number of elements of S) copies of $L^2(\mathbf{R}^n)$ in $\oplus_S L^2(\mathbf{R}^n)$ onto the image of the Sobolev space $L^2_S(\mathbf{R}^n)$ in $\oplus_S L^2(\mathbf{R}^n)$ via the map $f \to (Df)_{D\in S}$. Applying the theory of Fourier multipliers we state in terms of S various criteria for p-boundedness of P_S for $1 \le p \le \infty$, for $1 < p < \infty$ and for the weak type $(1,1)$ of P_S.

Introduction.

We study Sobolev spaces on $\mathbf{R}^n$ determined by non-empty finite families of linear partial differential operators with constant coefficients. Such families containing the identity operator are called smoothnesses. The anisotropic Sobolev spaces are special cases of our spaces when the differential operators are partial derivatives. Given a smoothness S we consider the canonical embedding $f \to (Df)_{D\in S}$ of the Sobolev space $L^p_S(\mathbf{R}^n)$ into the Cartesian product $\oplus_S L_p(\mathbf{R}^n)$ of $\#S$ (= the number of elements of S) copies of $L^p(\mathbf{R}^n)$. The main object of our study is the orthogonal projection P_S — called the canonical projection from $\oplus_S L^2(\mathbf{R}^n)$ onto the canonical image of the canonical embedding of $L^2_S(\mathbf{R}^n)$. The canonical projection plays a similar role for the scale $L^p_S(\mathbf{R}^n)$ $(1 \le p \le \infty)$ as the Riesz projection for the Hardy spaces H^p. Our principal objective is to determine for what smoothnesses the canonical projection is p-bounded for all p with $1 \le p \le \infty$, for what it is of weak type $(1,1)$, and for what it is only p-bounded for $1 < p < \infty$.

In Section 2 we give a quick reduction of the problem to a study of certain multipliers on $\mathbf{R}^n$. For a fixed smoothness S the multipliers are rational functions on $\mathbf{R}^n$ having common denominator (called the fundamental polynomial of S) $Q_S = \sum_{D\in S} |s(D)|^2$ where $s(D)$ is the symbol of the operator D.

In Section 3 we consider so called trivial smoothnesses, i.e. such that the canonical projection is p-bounded for all $p \in [1,\infty]$; equivalently it is p-bounded either for $p = 1$ or for $p = \infty$. We present a simple proof due essentially to J-P. Kahane of a result from [Si] and [P-S] (cf. also [K-Si]) that if S is a smoothness consisting of partial derivatives then S is not trivial if it has more than one maximal element with respect to the partial order:

$$\partial_1^{a(1)}\partial_2^{a(2)}\dots\partial_n^{a(n)} \le \partial_1^{b(1)}\partial_2^{b(2)}\dots\partial_n^{b(n)}$$
$$\stackrel{\text{df}}{\equiv} a(j) \le b(j) \quad \text{for } j = 1,2,\dots,n$$

We also show the triviality of smoothnesses $\{\mathrm{Id}, \Delta\}$ where Δ is the Laplacian in $\mathbf{R}^n$ $(n = 1,2,\dots)$ and the 2 dimensional $\overline{\partial}$ smoothness $\{\mathrm{Id}, \partial_1 + i\partial_2\}$

Section 4 is devoted to the study of smoothnesses for which the canonical projection is of weak type $(1,1)$. The main result of this section says that if the fundamental polynomial Q_S is h-elliptic with respect to some h homogeneity, in particular Q_S is elliptic, then P_S is of weak type $(1,1)$. This result heavily depends on the Fabes-Riviére [F-R] generalization of the Hörmander-Mihlin multiplier theorem. We also construct simple examples of smoothnesses whose canonical projections are not of weak type $(1,1)$.

In Section 5 we give a criterion of p-boundedness for $1 < p < \infty$ of the canonical projection. The criterion is based on the multidimensional Marcinkiewicz multiplier theorem (cf. [St]). It generalizes the observation in [P-S] that canonical projections of all smoothnesses consisting of partial derivatives are p-bounded for $1 < p < \infty$.

Acknowledgement.

Most of the research in this paper has been done when the author was participating in the Special Year on Modern Analysis at the University of Illinois. The author would like to express his gratitude to the Department of Mathematics of the University of Illinois at Urbana for its hospitality and support.

1. Preliminaries.

The letters Z, R, C, I, T stand for the integers, the real line, the complex plane, the interval $\{t \in \mathsf{R} : |t| \le 1\}$, the unit circle (usually represented as the interval $\pi\mathsf{I}$ with the endpoints identified) respectively. By X^n we denote the n-th cartesian power of a set X. Given $a = (a(j))$ and $b = (b(j))$ in Z^n we write

$$a \le b \text{ iff } a(j) \le b(j) \text{ for } j = 1,2,\dots,n.$$

An $a \in \mathsf{Z}^n$ with $a \ge 0$ is called a multindex. Given a multindex $a \in \mathsf{Z}^n$ and $x = (x(j))$, $y = (y(j))$ in $\mathbf{R}^n$, we write $x^a = \prod_{j=1}^n x(j)^{a(j)}$, $|a| = \sum_{j=1}^n a(j)$, $|x|_2 = (\sum_{j=1}^n |x(j)|^2)^{1/2}$, $\langle x, y\rangle = \sum_{j=1}^n x(j)y(j)$. By ∂^a we denote the operator of partial derivative corresponding to multindex a; precisely $\partial^a = \partial_1^{a(1)}.\partial_2^{a(2)}\dots\partial_n^{a(n)}$ where $\partial_j = \partial/\partial x(j)$. To the identity operator Id we assign the multindex 0, i.e we admit $\mathrm{Id} = \partial^{(0,0,\dots,0)} = \partial_j^0$ for $j = 1,2,\dots,n$. A linear partial differential operator is an operator $D = \sum_{0 \le a \in \mathsf{Z}^n} c_a \partial^a$ (c_a-complex eventually zero). The symbol of D is the polynomial

$$s(D) = \sum_{0 \le a \in \mathsf{Z}^n} (-i)^{|a|} c_a \xi^a, \quad (\xi \in \mathbf{R}^n).$$

An *n-dimensional smoothness* is a finite set of linear differential operators of n variables which contains the identity operator.

Function spaces on the groups $\mathbf{R}^n$ *and* $\mathbf{T}^n$.

For $1 \le p \le \infty$, by $L^p(\mathbf{R}^n)$ (resp. $L^p(\mathbf{T}^n)$) we denote the usual L^p space on $\mathbf{R}^n$ (resp. on $\mathbf{T}^n$) with respect to the n-dimensional Lebesgue measure λ_n normalized so that $\lambda_n(\mathbf{I}^n) = 2^n$ (resp. the Lebesgue measure on $\mathbf{T}^n$ normalized so that $\lambda_n(\mathbf{T}^n) = (2\pi)^n$). The norms of these spaces are usually denoted by $\|\cdot\|_p$; sometimes we write $\|\cdot\|_{L^p(\mathbf{R}^n)}$ (resp. $\|\cdot\|_{L^p(\mathbf{T}^n)}$). By $C(\mathbf{R}^n)$ we denote the Banach space of all continuous complex valued functions vanishing at infinity; $C(\mathbf{T}^n)$ has the usual meaning.

Recall that the Schwartz class $\mathcal{S}(\mathbf{R}^n)$ consists of all infinitely differentiable functions $f : \mathbf{R}^n \to \mathbf{C}$ such that

$$\sup_{x \in \mathbf{R}^n} |x^a \partial^b f(x)| < \infty \text{ for all multindices } a, b \in \mathbf{Z}^n.$$

Given an n-dimensional smoothness S the space $C_S(\mathbf{R}^n)$ is the completion of $\mathcal{S}(\mathbf{R}^n)$ in the norm

$$\|f\|_{S,\infty} = \max_{D \in S} \|Df\|_\infty.$$

For $1 \le p < \infty$ the Sobolev space $L^p_S(\mathbf{R}^n)$ is the completion of $\mathcal{S}(\mathbf{R}^n)$ in the norm

$$\|f\|_{S,p} = \left(\sum_{D \in S} \|Df\|_p^p \right)^{\frac{1}{p}}.$$

The Sobolev space $L^\infty_S(\mathbf{R}^n)$ is defined to be the space of all $f \in L^\infty(\mathbf{R}^n)$ such that there exists a sequence (f_m) in $\mathcal{S}(\mathbf{R}^n)$ with the property that the limits $\lim_m \int_{\mathbf{R}^n} Df_m g\, dx$ exist for all $D \in S$ and all $g \in L^1(\mathbf{R}^n)$ and $\lim_m \int_{\mathbf{R}^n} f_m \cdot g\, dx = \int fg\, dx$. Given $f \in L^\infty_S(\mathbf{R}^n)$ for each $D \in S$ there exists a unique element in $L^\infty(\mathbf{R}^n)$ denoted by Df such that $\lim_m \int_{\mathbf{R}^n} Df_m g\, dx = \int_{\mathbf{R}^n} Dfg\, dx$ for $g \in L^1(\mathbf{R}^n)$. We admit $\|f\|_{S,\infty} = \max_{D \in S} \|Df\|_\infty$. Also for $1 \le p < \infty$ the elements of $L^p_S(\mathbf{R}^n)$ can be identified with functions in $L^p(\mathbf{R}^n)$; if f is such a function then for every $D \in S$, Df exists in the weak sense (cf. [St], chap V, § 2) and belongs to $L^p(\mathbf{R}^n)$. Moreover

$$\|f\|_{S,p} = \left(\sum_{D \in S} \int_{\mathbf{R}^n} |Df(x)|^p\, dx \right)^{\frac{1}{p}}.$$

The definition of Sobolev spaces on $\mathbf{T}^n$ determined by a smoothness S, i.e. the spaces $C_S(\mathbf{T}^n)$ and $L^p_S(\mathbf{T}^n)$ for $1 \le p \le \infty$ is similar; the role of $\mathcal{S}(\mathbf{R}^n)$ is played by the space Trig$(\mathbf{T}^n)$ of all trigonometric polynomials in n variables, i.e. finite linear combinations of the exponents $x \to e^{i\langle x,m\rangle}$ for $m \in \mathbf{Z}^n$ regarded as functions of x on $\mathbf{T}^n$. We write L^p, C, L^p_S, C_S if we do not specify whether we consider function spaces on $\mathbf{R}^n$ or on $\mathbf{T}^n$.

The Fourier transform.

By $\mathcal{F}$ we denote the operator of the Fourier transform and by $\mathcal{F}^{-1}$ the operator of the inverse Fourier transform. We also write $\mathcal{F}(f) = \hat{f}$. Precisely for $f \in L^1(\mathbf{R}^n)$

$$\hat{f}(\xi) = (2\pi)^{-\frac{n}{2}} \int_{\mathbf{R}^n} f(x) e^{-i\langle x,\xi\rangle}\, dx \quad \text{for } \xi \in \mathbf{R}^n,$$

and for $f \in L^1(\mathbf{T}^n)$

$$\hat{f}(m) = (2\pi)^{-\frac{n}{2}} \int_{\mathbf{T}^n} f(x) e^{-i\langle x,m\rangle}\, dx \quad \text{for } m \in \mathbf{Z}^n$$

Recall that the Fourier transform of a finite Borel measure μ on $\mathbf{R}^n$ (resp. on $\mathbf{T}^n$) is the function $\hat{\mu}$ on $\mathbf{R}^n$ (resp. on $\mathbf{Z}^n$) defined by

$$\hat{\mu}(\xi) = (2\pi)^{-\frac{n}{2}} \int_{\mathbf{R}^n} e^{-i\langle x,\xi\rangle} \mu(dx) \quad \text{for } \xi \in \mathbf{R}^n$$

$$(\text{resp. } \quad \hat{\mu}(m) = (2\pi)^{-\frac{n}{2}} \int_{\mathbf{T}^n} e^{-i\langle x,m\rangle} \mu(dx) \quad \text{for } m \in \mathbf{Z}^n\ ).$$

The p-boundedness of operators; the multipliers.

Let X_1, X_2 be linear spaces of measurable functions on measure spaces (Ω_1, μ_1) and (Ω_2, μ_2) respectively and let $U : X_1 \to X_2$ be a homogeneous and additive operator. Recall that U is said to be (p_1,p_2)-bounded for some p_i with $1 \le p_i \le \infty$ $(i = 1, 2)$ provided there is a $K > 0$ such that

$$\|U(f)\|_{L^{p_2}(\mu_2)} \le K\|f\|_{L^{p_1}(\mu_1)} \quad \text{for } f \in X_1 \cup L^{p_1}(\mu_1).$$

If $p = p_1 = p_2$ we say that U is p-bounded. U is said to be of weak type (p_1, p_2) for some p_i with $1 \le p_i \le \infty$ $(i = 1, 2)$ provided there is a $K > 0$ such that for every $a > 0$

$$(\mu_2\{|U(f)| > a\})^{\frac{1}{p_2}} \le K\|f\|_{L^{p_1}(\mu_1)} a^{-1} \quad \text{for } f \in X_1 \cup L^{p_1}(\mu_1).$$

Let $U : X_1 \to X_2$ be a linear operator acting between linear topological spaces of measurable functions on $\mathbf{R}^n$ (resp. on $\mathbf{T}^n$) and let $f : \mathbf{R}^n \to \mathbf{C}$ (resp. $f : \mathbf{Z}^n \to \mathbf{C}$) be a measurable function. We say that the operator U is determined by the multiplier f or vice versa f is determined by U if for g in a dense subset of X

$$\mathcal{F}(U(g))(\xi) = f(\xi)\mathcal{F}(g)(\xi) \quad (\xi \text{ almost everywhere}).$$

2. The canonical projection and related multipliers.

Let (Ω, μ) be a measure space, S a finite set of cardinality $\#S > 0$. Denote by $\oplus_S(\Omega, \mu) = \bigcup_{s\in S}\Omega_s, \oplus_S\, \mu)$ the measure space of $\#S$ copies of Ω; a $\oplus_S\,\mu$-measurable subset of $\bigcup_{s\in S}\Omega_s$ is the disjoint union $\bigcup_{s\in S} A_s$ of μ-measurable subsets of Ω; we define $\oplus_S\,\mu(\bigcup_{s\in S} A_s) = \sum_{s\in S}\mu(A_s)$. If X is a space of functions on Ω then $\oplus_S X$ denotes the $\#S$-th cartesian power of X regarded as the space of functions on $\bigcup_{s\in S}\Omega_s$ via the natural identification $F \to (f_s)_{s\in S}$ where $f_s = F\,|_{\Omega_s}$ for $s \in S$. For $\oplus_S\,\mu$-measurable functions $F = (f_s)_{s\in S}$ and $G = (g_s)_{s\in S}$ with $F\overline{G}$ being $\oplus_S\,\mu$-integrable we put

$$\langle F, G\rangle = \int_{\bigcup_s \Omega_s} F\overline{G}\, d(\oplus_S\,\mu) = \sum_{s\in S}\int_\Omega f_s\overline{g}_s\, d\mu.$$

Now we are ready to define the canonical projection which is the main concept of the present paper.

Given an n-dimensional smoothness S, the *canonical embedding* $J_S : L^p_S \to \oplus_S L^p$ (resp. $C_S \to \oplus_S C$) is defined by

$$J_S(f) = (Df)_{D\in S}$$

The *canonical projection* P_S is the orthogonal projection from $\oplus_S L^2$ onto $J_S(L^2_S)$. If P_S is p-bounded we use the same name and the same symbol for the unique canonical projection from $\oplus_S L^p$ onto $J_S(L^p_S)$ which coincides with the orthogonal one on $\oplus_S L^2 \cap \oplus_S L^p$. We put

$$a_S(p) = \|P_S : \oplus_S L^p \to J_S(L^p_S)\|.$$

To distinguish between the "$\mathbf{R}^n$ model" and the "$\mathbf{T}^n$ model" we shall write $J^{\mathbf{R}}_S$, $P^{\mathbf{R}}_S$, $a^{\mathbf{R}}_S, \ldots$, and $J^{\mathbf{T}}_S$, $P^{\mathbf{T}}_S$, $a^{\mathbf{T}}_S, \ldots$

The polynomial

$$Q_S = \sum_{D\in S} |s(D)|^2$$

is called the *fundamental polynomial* of the smoothness S. Since $\mathrm{Id} \in S$ and $s(\mathrm{Id}) = 1$, it follows that $Q_S \geq 1$.

Our first result reduces the study of canonical projections to the theory of multipliers.

Proposition 2.1. *The canonical projection P_S is given by the formula*

$$P_S((f_E)_{E\in S}) = \left(\sum_{E\in S} T_{D,E}(f_E)\right)_{D\in S} \tag{2.1}$$

where $T_D : L^2 \to L^2$ is determined by the multiplier $m_{D,E} = s(D)\overline{s(E)}Q_S^{-1}$.

Proof. The following criterion easily yields that P_S determined by (2.1) is an orthogonal projection:

Let S be a k-element set, $k > 0$. Let H be a Hilbert space. Assume that an operator $P : H^k \to H^k$ has the matrix representation $(T_{D,E})_{D\in S, E\in S}$ where $T_{D,E} : H \to H$ are bounded linear operators such that

$$T^*_{D_1,D_2} = T_{D_2,D_1} \text{ and } \sum_{E\in S} T_{D_1,E}T_{E,D_2} = T_{D_1,D_2} \quad (D_1, D_2 \in S). \tag{2.2}$$

Then P satisfies $P^* = P = P^2$, hence it is an orthogonal projection.

Observe that in our case $T_{D,E}$ is bounded because it is determined by a bounded multiplier (in fact $|m_{D,E}| \leq 1$). Furthermore the identities $m_{D_1,D_2} = \overline{m_{D_2,D_1}}$ and $\sum_{E\in S} m_{D_1 E} m_{E,D_2} = m_{D_1,D_2}$ imply (2.2).

To complete the proof we shall show that $P_S(\oplus_S L^2_S) = J_S(L^2_S)$. Let X denote either $\mathcal{S}(\mathbf{R}^n)$ or $\mathrm{Trig}(\mathbf{T}^n)$. Pick $(f_E)_{E\in S}$ in $\oplus_S X$ and define f by $\mathcal{F}(f) =$

$\sum_{E\in S}\overline{s(E)}Q_S^{-1}\cdot\mathcal{F}(f_E)$. Since $\mathcal{F}(X)=X$, it follows that $f\in X$. Thus, for $D\in S$,

$$\begin{aligned}\mathcal{F}(Df)&=s(D)\mathcal{F}(f)\\&=\sum_{E\in S}s(D)\overline{s(E)}Q_S^{-1}\mathcal{F}(f_E)\\&=\mathcal{F}\left(\sum_{E\in S}T_{D,E}f_E\right).\end{aligned}$$

Hence, if P_S is defined by(2.1), then

$$P_S((f_E)_{E\in S})=(Df)_{D\in S}\subset J_S(L_S^2)\quad\text{for }(f_E)_{E\in S}\in\oplus_S X.$$

A similar argument shows that if $(f_E)_{E\in S}\in J_S(X)$, i.e. $(f_D)_{D\in S}=(Df)_{D\in S}$ for some $f\in X$ then $P_S((f_D)_{D\in S})=(f_D)_{D\in S}$. Taking into account that $\oplus_S X$ is dense in $\oplus_S L^2$ and that $J_S(X)$ is dense in $J_S(L_S^2)$ and that P_S is a bounded linear operator we get the desired conclusion. ∎

The matrix $(m_{D,E})_{D\in S,E\in S}$ is called the multiplier matrix of P_S. An immediate consequence of Proposition 2.1 is

Corollary 2.1. *The canonical projection P_S is (p,q)-bounded (resp. of weak type (p,q)) iff all the operators $T_{D,E}$ for $D\in S$ and $E\in S$ have the same property. In particular P_S is p-bounded iff there is an absolute constant C (independent of p) such that*

$$C^{-1}a_S(p)\le\sum_{D\in S}\sum_{E\in S}\|T_{D,E}:L^p\to L^p\|\le Ca_S(p).$$

∎

Corollary 2.2. *Let S be an n-dimensional smoothness. Assume that for some p with $1\le p\le\infty$ the canonical projection $P_S^{\mathbf{R}}$ is p-bounded. Then the canonical projection $P_S^{\mathbf{T}}$ is also p-bounded; moreover there is $C>0$ such that $a_S^{\mathbf{T}}(p)\le Ca_S^{\mathbf{R}}$.*

Proof. Combine Corollary 2.1 with the following classical transference theorem (cf. [St-W], chap. VII, § 3, Theorem 3.8): let $U:L^2(\mathbf{R}^n)\to L^2(\mathbf{R}^n)$ be a p-bounded operator for some p with $1\le p\le\infty$. Assume that U is determined by a multiplier $f:\mathbf{R}^n\to\mathbf{C}$ such that f is continuous at each point $a\in\mathbf{Z}^n$. Then $(f(a))_{a\in\mathbf{Z}^n}$ determines an operator, say $U^{\mathbf{T}};L^2(\mathbf{T}^m)\to L^2(\mathbf{T}^n)$, which is p-bounded for the same p. Moreover

$$\|U^{\mathbf{T}}:L^p(\mathbf{T}^m)\to L^p(\mathbf{T}^n)\|\le\|U:L^p(\mathbf{R}^n)\to L^p(\mathbf{T}^n)\|.$$

∎

For $p\in[1,\infty]$ denote by $[p,p^*]$ the closed interval with the endpoints p and p^* where $p^*=p(p-1)^{-1}$ for $1<p<\infty$, $p^*=\infty$ for $p=1$ and $p^*=1$ for $p=\infty$.

Clearly the canonical projection P_S, being self adjoint, satisfies $\langle P_S(F),G\rangle=\langle F,P_S(G)\rangle$ whenever at least one side of the identity makes sense. Thus using the duality between $\oplus_S L^p$ and $\oplus_S L^{p^*}$ and the classical interpolation theorem we get

Corollary 2.3. *Let S be a smoothness such that for some p with $1 \le p \le \infty$ the canonical projection P_S is p-bounded. Then P_S is q-bounded for every $q \in [p, p^*]$; moreover*

$$a_S(q) \le a_S(p) = a_S(p^*).$$

■

3. Trivial Smoothnesses.

A smoothness S is called trivial if $P_S^{\mathbf{R}}$ is p-bounded for all $p \in [1, \infty]$. In view of Corollary 2.3, S is trivial iff $P_S^{\mathbf{R}}$ is either 1-bounded or ∞-bounded.

For both the "$\mathbf{R}^n$-model" and the "$\mathbf{T}^n$-model" we have

Corollary 3.1. *For every smoothness S the following conditions are equivalent*

(i) *P_S is 1-bounded;*

(ii) *P_S is ∞-bounded;*

(iii) *P_S extends to a bounded projection from $\oplus_S C$ onto $J_S(C_S)$;*

(iv) *for every $D \in S$ and for every $E \in S$ the multiplier $m_{D,E} = s(D)\overline{s(E)}Q_S^{-1}$ is the Fourier transform of a (complex valued) finite Borel measure;*

(v) $\sup_{1<p<\infty} a_S(p) < \infty$.

For the proof combine Corollaries 2.1 – 2.3 with the classical result (cf. [St-W], chap. I, Theorem 3.19) that a multiplier determines a 1-bounded operator iff it is a Fourier transform of a finite Borel measure. To prove the implication (ii) $\Rightarrow$ (iii) we also use the observation that if X denotes either $\mathcal{S}(\mathbf{R}^n)$ or $\mathrm{Trig}(\mathbf{T}^n)$ then $P_S(\oplus X) \subset J_S(X)$ and $J_S(X)$ is dense in the sup norm in $J_S(C_S)$. ■

Next we present a criterion for non-triviality of a smoothness which was suggested to us by J.-P. Kahane. We begin with auxiliary notation.

A parellelopiped in $\mathbf{R}^n$ is any set of the form

$$I(a, r) = \{\xi \in \mathbf{R}^n : |\xi(j) - a(j)| \le r(j) \text{ for } j = 1, 2, \ldots, n\}$$

where $a = (a(j)) \in \mathbf{R}^n$ and $r = (r(j)) \in \mathbf{R}^n$ with $r(j) \ge 0$ for $j = 1, 2, \ldots, n$. A parellelopiped in $\mathbf{Z}^n$ is the intersection of a parellelopiped $I(a, r) \subset \mathbf{R}^n$ with $\mathbf{Z}^n$ provided a and r are in $\mathbf{Z}^n$.

We put $d(\xi) = \min_{i \le j \le n} |\xi(j)|$ for $\xi \in \mathbf{R}^n$.

A sequence $(I_k) = (I(a_k, r_k))$ of parellelopipeds is called admissable provided $\lim_k d(r_k) = \infty$.

Given $f : \mathbf{R}^n \to \mathbf{C}$ (resp. $f : \mathbf{Z}^n \to \mathbf{C}$) and a parellelopiped $I = I(a, r)$ in $\mathbf{R}^n$ (resp. in $\mathbf{Z}^n$) we put

$$\mathrm{Av}(f; I) = \lambda_n(I)^{-1} \int_I f(\xi)\, d\xi$$

$$(\text{resp. } \mathrm{Av}(f; I) = (\#I)^{-1} \sum_{b \in I} f(b)).$$

Note that if $I = I(a, r)$ is a parellelopiped in $\mathbf{Z}^n$, then $\#I = \prod_{j=1}^n (2r(j)+1)$.

An n-dimensional smoothness S has the property $(*)_\mathbf{R}$ (resp. $(*)_\mathbf{T}$) if for every $D \in S$ there is an admissable sequence (I_k) of parellelopipeds in $\mathbf{R}^n$ (resp. in $\mathbf{Z}^n$) such that $\lim_k \sup_{\xi \in I_k} m_{D,D}(\xi) = 0$.

Now we are ready to state

Theorem 3.1. *If S has $(*)_\mathbf{T}$ then $P_S^\mathbf{T}$ is 1-unbounded, hence (by Corollary 2.2) S is non-trivial. If S satisfies $(*)_\mathbf{R}$ then S is non-trivial.*

To prove Theorem 3.1 we need the following observation due to Wiener (cf. [K], p. 42 for one-dimensional case).

Proposition 3.1. *If $f : \mathbf{R}^n \to \mathbf{C}$ (resp. $f : \mathbf{Z}^n \to \mathbf{C}$) is a Fourier transform of a finite Borel measure μ on $\mathbf{R}^n$ (resp. on $\mathbf{T}^n$) then*

$$(2\pi)^{-\frac{n}{2}}\mu(\{0\}) = \lim_k \mathrm{Av}(f; I_k)$$

for every admissable sequence of parellelopipeds (I_k) in $\mathbf{R}^n$ (resp. in $\mathbf{Z}^n$).

Proof. We consider only the case of $\mathbf{Z}^n$; the argument in the case of $\mathbf{R}^n$ is similar.

Let $I_k = I(a_k, r_k)$ be a fixed parellelopiped in $\mathbf{Z}^n$. Let us put

$$g_k(x) = (2\pi)^{-\frac{n}{2}}(\#I_k)^{-1}\sum_{b \in I_k} e^{-i\langle b,x\rangle} \qquad (x \in \mathbf{T}^n).$$

Then, remembering that $f(b) = (2\pi)^{-n/2}\int_{\mathbf{T}^n} e^{-i\langle b,x\rangle}\mu(dx)$, we have

$$\int_{\mathbf{T}^n} g_k(x)\mu(dx) = \mathrm{Av}(f; I_k).$$

Clearly $g_k(0) = (2\pi)^{-n/2}$ and $|g_k(x)| \le (2\pi)^{-n/2}$ for $x \in \mathbf{T}^n$. A simple calculation shows that for $0 \ne x \in \mathbf{T}^n$,

$$g_k(x) = (-\sqrt{2\pi})^{-n} e^{-i\langle a_k,x\rangle} \prod_{\{j:x(j)\ne 0\}} \frac{\sin[(r_k(j)+2^{-1})x(j)]}{(2r_k(j)+1)\sin[2^{-1}x(j)]}.$$

Thus $|g(x)| \le [d(r_k)C(x)]^{-1}$, where $C(x) = (2\pi)^{-n/2}\prod_{\{j:x(j)\ne 0\}} \sin[2^{-1}x(j)]$. Hence, if (I_k) is an admissable sequence of parellelopipeds in $\mathbf{Z}^n$ then $\lim_k g_k(x) = 0$ for $0 \ne x \in \mathbf{T}^n$ and $\lim_k g_k(0) = (2\pi)^{-n/2}$. Thus the desired conclusion follows from the Lebesgue domination convergence theorem. ∎

Proof of Theorem 3.1. If P_S were 1-bounded then, by Corollary 3.1 (iv) and by Proposition 3.1, there would exist the limit, $\lim_k \mathrm{Av}(m_{D,D}; I_k)$ for every $D \in S$ and every admissable sequence of parellelopipeds; moreover thelimit would be independent of a particular choice of the admissable sequence. Since $\sum_{D\in S} m_{D,D} \equiv 1$, the limit would be positive for some D_0 in S. On the other hand, by $(*)_\mathbf{R}$ (resp. by $(*)_\mathbf{T}$) there exists an admissable sequence (I_k) such that $\lim_k \mathrm{Av}(m_{D_0,D_0}; I_k) = 0$, a contradiction. ∎

The condition $(*)$ in Theorem 3.1 can be replaced by a formally weaker one in the case where the symbols of elements of the smoothness satisfy some growth condition.

Call a polynomial $W : \mathbf{R}^n \to \mathbf{C}$ *slightly elliptic* if there are $\alpha > 0$ and $d > 0$ such that for every $a \in \operatorname{supp} W$

$$|W(\xi)|^2 \geq \alpha \xi^{2a} \text{ for } \xi \in \mathbf{R}^n \text{ with } d(\xi) > d$$

where $\operatorname{supp} W$ is the minimal subset A of $\mathbf{Z}^n$ such that $W(\xi) = \sum_{a \in A} c_a \xi^a$. W is *strongly slightly elliptic* if $|W(\xi)|^2 \geq \alpha \xi^{2a}$ for all $\xi \in \mathbf{R}^n$ and for every $a \in \operatorname{supp} W$.

Theorem 3.2. *Let S be such a smoothness that for every $D \in S$ the symbol $s(D)$ is slightly elliptic. Assume that for every $D \in S$ there is a sequence (ξ_k) such that*

$$\lim_k m_{D,D}(\xi_k) = 0; \quad \lim_k d(\xi_k) = \infty. \tag{3.1}$$

Then S satisfies $(*)_{\mathbf{T}}$ and $(*)_{\mathbf{R}}$, hence S is not trivial.

Proof. Let us put $\operatorname{supp} S = \bigcup_{D \in S} \operatorname{supp} s(D)$. Since $\operatorname{supp} S$ is a finite set, it follows from slight ellipticity of the symbols of all the elements of S that there are $\alpha_S > 0$ and $d_S > 0$ such that

$$|s(D)(\xi)|^2 \geq \alpha_S \sum_{a \in \operatorname{supp} s(D)} \xi^{2a}$$

for $D \in S$ and for $\xi \in \mathbf{R}^n$ with $d(\xi) > d_S$. On the other hand the Schwarz inequality and again the finiteness of $\operatorname{supp} S$ yield the existence of a $\beta_S > 0$ such that

$$|s(D)(\xi)|^2 \leq \beta_S \sum_{a \in \operatorname{supp} s(D)} \xi^{2a} \tag{3.2}$$

for $D \in S$ and for $\xi \in \mathbf{R}^n$. Now fix $D \in S$ and pick (ξ_k) as in (3.1). Then, remembering that $m_{D,D} = |s(D)|^2 \left(\sum_{E \in S} |s(E)|^2\right)^{-1}$, it follows from (3.2) that

$$\lim_k |s(D)(\xi_k)|^2 \left(\sum_{b \in \operatorname{supp} S} \xi_k^{2b} \right)^{-1} = 0$$

Thus there is a multindex $b \in \operatorname{supp} S$ such that

$$\lim_k |s(D)(\xi_k)| \xi_k^{2b} = 0.$$

Hence, replacing (ξ_k) by an appropriate subsequence and taking into account the estimate from below for $|s(D)|^2$ we can assume without loss of generality that

$$d(\xi_k) > 2^k d_S \text{ and } \xi_k^{2a} \xi_k^{-2b} < 2^{-k} \text{ for } k = 1, 2, \ldots \quad (a \in \operatorname{supp} s(D)).$$

Let us put, for $k = 1, 2, \ldots$

$$I_k = \{\xi' \in \mathbf{R}^n : |\xi_k(j) - \xi'(j)| < k + 1 \text{ for } j = 1, 2, \ldots, n\}$$

$I_k^{\mathbf{T}}$ – the largest parellelopiped in $\mathbf{Z}^n$ contained in I_k.

Note that the condition $d(\xi_k) > 2^k d_S$ for $k = 1, 2, \ldots$ yields

$$\lim_k [\sup_{\xi' \in I_k} \xi'(j)\xi_k^{-1}(j)] = 1 \quad \text{for } j = 1, 2, \ldots, n.$$

Thus for every $a \in \operatorname{supp} s(D)$,

$$\lim_k [\sup_{\xi' \in I_k} (\xi'_k)^{2(a-b)}] = \lim_k \xi_k^{2(a-b)} = 0.$$

Hence, in view of (3.2), the I_k's (resp. the $I_k^{\mathbf{T}}$'s) form an admissable sequence of parellelopipeds in $\mathbf{R}^n$ (resp. in $\mathbf{Z}^n$) such that

$$\lim_k \operatorname{Av}(m_{D,D}; I_k) = \lim_k \operatorname{Av}(m_{D,D}; I_k^{\mathbf{T}}) = 0.$$

Thus the desired conclusion follows from Theorem 3.1. ∎

As an application of Theorem 3.2 we obtain

Corollary 3.2. *Let S be a smoothness in $\mathbf{Z}^n$ consisting of partial derivatives. Assume that S contains more than one maximal element (with respect to the ordering $\partial^a \leq \partial^b \stackrel{\text{df}}{\equiv} a \leq b$). Then S is not trivial.*

Proof. The symbol of every partial derivative is obviously slightly elliptic. Since S has more than one maximal element, for every $\partial^a \in S$ there is a $\partial^b \in S$ such that $b(j) > a(j)$ for some j with $1 \leq j \leq n$. Thus there is a sequence $(\xi_k) \subset \mathbf{R}^n$ such that $\lim_k d(\xi_k) = \infty$ and $\lim_k \xi_k^{2a}\xi_k^{-2b} = 0$. Therefore

$$\lim_k |m_{\partial^a,\partial^a}(\xi_k)| = \lim_k \xi_k^{2a} \left(\sum_{\partial^b \in S} \xi_k^{2b} \right)^{-1} = 0.$$

Thus by Theorem 3.2, S is not trivial. ∎

In [P-S], [Si] (cf. also [K-Si]) Corollary 3.2 has been derived from a general Sobolev embedding theorem.

Theorems 3.1 and 3.2 show that triviality of a smoothness is a very rare property. They can be used for instance to show that S is not trivial if it contains more than on maximal element in the Hörmander's ordering (cf. [H II], p. 34):

$$D_1 \prec\!\!\prec D_2 \stackrel{\text{df}}{\equiv} \lim_{t=\infty} \sup_{\xi \in \mathbf{R}^n} \widetilde{s(D_1)}(\xi, t)/\widetilde{s(D_2)}(\xi, t) = 0$$

where for a polynomial W,

$$\widetilde{W}(\xi, t) = \left(\sum_{a \in \mathbf{Z}^n} |(\partial^a W)(\xi)|^2 t^{2|a|} \right)^{\frac{1}{2}} \qquad (\xi \in \mathbf{R}^n, t \in \mathbf{R})$$

There are however some interesting trivial smoothnesses which we shall discuss next.

Let $D \neq \mathrm{Id}$ be an arbitary linear partial differential operator. First we consider the smoothness

$$S_D = \{\mathrm{Id}, D\}.$$

We put $P_D = P_{S_D}$, $Q_D = Q_{S_D} = 1 + |s(D)|^2$. The multiplier matrix of P_D is

$$Q_D^{-1} \begin{pmatrix} 1 & s(D) \\ \overline{s(D)} & |s(D)|^2 \end{pmatrix}.$$

The following observation is easy but useful.

Proposition 3.2. *If $s(D)$ is real-valued, i.e. $s(D)(\xi) \in \mathbf{R}$ for $\xi \in \mathbf{R}^n$, then S_D is trivial iff $(1 + is(D))^{-1}$ is the Fourier transform of a finite Borel measure on $\mathbf{R}^n$.*

Proof. The triviality of S_D yields that the entries $m_{\mathrm{Id},\mathrm{Id}}$ and $m_{\mathrm{Id},D}$ of the multiplier matrix of P_D are the Fourier transforms of finite Borel measures. Hence so is $(1 + is(D))^{-1} = m_{\mathrm{Id},\mathrm{Id}} - im_{\mathrm{Id},D}$ (because if $s(D)$ is real valued then $Q_D = 1 + s(D)^2$).

Conversely, let $\mathcal{F}(\mu) = (1 - is(D))^{-1}$ for some finite Borel measure μ on $\mathbf{R}^n$. Define the measure $\tilde{\mu}$ by $\tilde{\mu}(B) = \overline{\mu(-B)}$ ($B \subset \mathbf{R}^n$ Borel). Then $\mathcal{F}(\tilde{\mu}) = \overline{\mathcal{F}(\mu)} = (1 - is(D))^{-1}$ because $s(D)$ is real valued. Hence $m_{\mathrm{Id},\mathrm{Id}} = \mathcal{F}(2^{-1}(\mu + \tilde{\mu}))$; $m_{\mathrm{Id},D} = m_{D,\mathrm{Id}} = \mathcal{F}(2^{-1}i(\mu + \tilde{\mu}))$; $m_{D,D} = \mathcal{F}((2\pi)^{n/2}\delta - 2^{-1}(\mu + \tilde{\mu}))$ where δ stands for the point mass at zero. Thus all entries of the multiplier matrix of P_D are the Fourier transforms of finite Borel measures. ∎

Remark. Obviously a version of Proposition 3.2 can be used in the case where $\alpha s(D)$ is real-valued for some $\alpha \in \mathbf{C}\setminus\{0\}$; this is the case of D being a partial derivative.

Recall that a polynomial $W = \sum_{j=0}^{m} \sum_{|a|=j} c_a \xi^a$ of degree m is called elliptic provided there is an $\alpha > 0$ such that $|\sum_{|a|=n} c_a \xi^a| \geq \alpha |\xi|_2^m$ for $\xi \in \mathbf{R}^n$.

For D with $s(D)$ elliptic and non negative the triviality of S_D is expressd in terms of the fundemental solution of the equation $(\mathrm{Id} + D)u = u$.

Proposition 3.3. *If $s(D) \geq 0$ and $s(D)$ is elliptic and if $(1 + s(D))^{-1}$ is a Fourier transform of a finite Borel measure on $\mathbf{R}^n$, then S_D is trivial.*

For the proof we need

Fact I. (cf. [HI], Corollary 7.9.4) Let n be a positive integer and r a real number with $2r > n$. Assume that a function $h : \mathbf{R}^n \to \mathbf{C}$ satisfies $h(1 + |\cdot|_2)^r \in L^2(\mathbf{R}^n)$. Then $\mathcal{F}^{-1}(h) \in L^1(\mathbf{R}^n)$.

Proof of Proposition 3.3. One has the identity

$$(1 + it)^{-1} = -i \sum_{k=0}^{m-1} (1 + t)^{k-m}(1 + i)^{m-k-1} + (1 + i)^m (1 - t)^{-m}(1 + it)^{-1}$$

for $t \geq 0$ and $m = 1, 2, \ldots$ Putting $t = s(D)$ we infer that $1 + is(D)$ is the Fourier transform of a finite Borel measure. Indeed using the formula $\mathcal{F}(\mu * \nu) = (2\pi)^{n/2}\mathcal{F}(\mu)\mathcal{F}(\nu)$ (where $\mu * \nu$ denotes the convolution of measures μ and ν) we infer that if $(1 + s(D))^{-1}$ is the Fourier transform of a finite Borel measure so is $(1 + s(D))^{k-m}$ for $k = 0, 1, \ldots, m - 1$ and therefore the sum "$\sum_{k=0}^{m-1}$" in the right hand side of the identity has the same property. On the other hand the ellipticity and the positivity of $s(D)$ yields that as we choose m large enough then the term $(1 + i)^m(1 + t)^{-m}(1 + it)^{-1}$ satisfies the assumptions of Fact I. ∎

Remark. The analysis of the proof shows that the assertion of Proposition 3.3 remains valid if we replace the ellipticity of $s(D)$ by a weaker condition that there exists $\beta > 0$ such that $|s(D)(\xi)| \geq |\xi|_2^{-\beta}$ for $\xi \in \mathbf{R}^n$ with $|\xi|_2$ large enough. Obviously non negativity of $s(D)$ can be replaced by non positivity.

The Smoothness of the Laplacian.

Let $\Delta = \sum_{j=1}^{n} \partial_j^2$ be the n dimensional Laplacian. We have $s(\Delta)(\xi) = -|\xi|_2 = -\sum_{j=1}^{n} \xi(j)^2$ for $\xi \in \mathbf{R}^n$. Thus $s(\Delta)$ is non positive and elliptic.

Proposition 3.4. *The smoothness S_Δ is trivial.*

Proof. Combine Proposition 3.3 with the following

Fact II. (cf. [St] pp. 131–132) For $x \in \mathbf{R}^n$ let

$$G(x) = (2\pi)^{\frac{n}{2}-1} \int_0^\infty e^{\frac{-\pi|x|_2^2}{t}} e^{\frac{-t}{4\pi}} t^{-\frac{n}{2}+1} \frac{dt}{2t}.$$

Then $G \in L^1(\mathbf{R}^n)$ and $\mathcal{F}(G) = (1 + |\xi|_2^2)^{-1}$. (Note that our formula for G differs slightly from that in [St] because our definition of the Fourier transform is different.) ∎

The $\overline{\partial}$ smoothness.

Let $\overline{\partial} = \partial_1 - i\partial_2$. We have $s(\overline{\partial})(\xi) = i\xi(1) + \xi(2)$ for $\xi \in \mathbf{R}^2$. The symbol $s(\overline{\partial})$ is elliptic but not real valued. Nevertheless we have

Proposition 3.5. *The 2-dimensional smoothness $S_{\overline{\partial}}$ is trivial.*

Proof. Clearly $Q_{\overline{\partial}} = 1 + |\xi|_2^2$. Thus, by Fact II, the entries of the multiplier matrix of $P_{\overline{\partial}}$ on the main diagonal are the Fourier transforms of finite Borel measures on $\mathbf{R}^2$. It remains to examine the entry $m_{\mathrm{Id},\overline{\partial}} = m_{\overline{\partial},\mathrm{Id}}$. We have

$$m_{\mathrm{Id},\overline{\partial}}(\xi) = (i\xi(1) + \xi(2))(1 + |\xi|_2^2)^{-1}.$$

Thus it is enough to show that the function $\xi \to \xi(1)(1 + |\xi|_2^2)^{-1}$ is the Fourier transform of a function in $L^1(\mathbf{R}^2)$. By Fact II this reduces to verifying that $\partial_1 G \in L^1(\mathbf{R}^2)$. We have

$$\partial_1 G(x) = -2\pi \int_0^\infty x(1) e^{\frac{-\pi|x|_2^2}{t}} e^{\frac{-t}{4\pi}} \frac{dt}{2t^2} \quad \text{for } x \in \mathbf{R}^2.$$

Note that $\partial_1 G(x(1), x(2)) = -\partial_1 G(-x(1), x(2)) > 0$ for $x(1) > 0$. Thus using the Fubini Theorem we get

$$\begin{aligned} \int_{\mathbf{R}^2} |\partial_1 G(x)|\, dx &= 2\pi \int_0^\infty \left(\int_0^\infty x(1) e^{\frac{-\pi x(1)^2}{t}}\, dx(1) \right. \\ &\qquad \times \left. \int_{-\infty}^{+\infty} e^{\frac{-\pi x(2)^2}{t}}\, dx(2) \right) e^{\frac{-t}{4\pi}} \frac{dt}{t^2} \\ &= (2\pi)^{\frac{1}{2}} \int_0^\infty e^{\frac{-t}{4\pi}} \frac{dt}{\sqrt{t}} \\ &< +\infty \end{aligned}$$

∎

4 Smoothnesses with canonical projections of weak type $(1,1)$.

First we recall the classical Hörmander-Mihlin criterion for weak $(1,1)$ multipliers (cf. [St] pp. 96-99; and [HI] Theorem 7.9.5).

(H-M). *Let $f : \mathbf{R} \to \mathbf{C}$ be a bounded function and let $s > n/2$ be an integer. Assume that for every multindex a with $|a| \le s$ the partial derivative $\partial^a f$ exists and is continuous on $\mathbf{R}^n \setminus \{0\}$, and*

$$\sup_{0<r<\infty} r^{2|a|-n} \int_{\frac{r}{2}\le|\xi|_2\le 2r} |\partial^a f(\xi)|^2\, d\xi < \infty \tag{4.1}$$

Then the operator $T : L^2(\mathbf{R}^n) \to L^2(\mathbf{R}^n)$ defined by $T(g) = \mathcal{F}^{-1}(f\cdot\mathcal{F}(g))$ for $g \in \mathcal{S}(\mathbf{R}^n)$ is of weak type $(1,1)$.

We apply (H-M) to the classical smoothness of k-times continuously differentiable functions on $\mathbf{R}^n$. Let k and n be positive integers. Let

$$S(k,n) = \{\partial^a : 0 \le a \in \mathbf{Z}^n \text{ and } |a| \le k\}.$$

The following fact is essentially known (cf. e.g. [Kw-P]).

Proposition 4.1. *$P^{\mathbf{R}}_{S(k.n)}$ is of weak type $(1,1)$.*

Proof. The fundamental polynomial

$$Q_{S(k,n)} = \sum_{j=0}^{k} \sum_{|a|=j} \xi^{2a}$$

satisfies

$$|Q_{S(k,n)}(\xi)| \ge C_{k,n}|\xi|_2^{2k} \quad \text{for } \xi \in \mathbf{R}^n \tag{4.2}$$

(here $C, C', C'', \ldots$ with subscripts denote positive constants). The entries of the multiplier matrix of $P^{\mathbf{R}}_{S(k,n)}$ are, up to a factor being a power of i, of the form

$$f(\xi) = \xi^{b+c} Q^{-1}_{S(k,n)} \quad (\partial^b, \partial^c \in S(k,n)).$$

Using the rules of differentiation, by an easy induction with respect to $|a|$ (= the order of ∂^a) we get $\partial^a(f) = LQ^{-|a|-1}_{S(k,n)}$ where L is a polynomial in n variables of degree $v = |b+c| + (2k-1)|a|$. Thus, in view of (4.2) there is an $r_0 > 0$ and a constant $C_{k,n,a} > 0$ such that

$$|\partial^a(f)(\xi)| \le C_{k,n,a}|\xi|_2^{v-2k(|a|+1)} \quad \text{for } |\xi|_2 > r_0.$$

Hence for $r > r_0$ we have

$$\int_{\frac{r}{2}\le|\xi|_2\le 2r} |\partial^a(f)(\xi)|^2\, d\xi \le C'_{k,n,a} r^{n+2(|b+c|-2k-|a|)}.$$

Taking into account that $|b+c| \le |b| + |c| \le 2k$, we get $n + 2(|b+c| - 2k - |a|) \le n - 2|a|$. Thus

$$r^{2|a|-n} \int_{\frac{r}{2}\le|\xi|_2\le 2r} |\partial^a(f)(\xi)|^2\, d\xi \le C'_{k,n,a} \quad \text{for } r > r_0.$$

The latter inequality implies (4.1) because f is continuously infinitely many times differentiable for $0 \le |\xi|_2 \le r_0$. Thus, by (H-M), each entry of the multiplier matrix of $P^{\mathbf{R}}_{S(k,n)}$ is of weak type $(1,1)$ ∎

Proposition 4.1 is a special case of a more general fact which is based on the concept of mixed homogeneity.

A vector $h = (h(j)) \in \mathbf{R}^n$ with $1 \le h(1) \le h(2) \ldots \le h(n)$ is called a *mixed homogeneity*. We put $|h| = \sum_{j=1}^n h(j)$. For $\xi \in \mathbf{R}^n$ and fixed homogeneity h we define the modular ρ_h by: $\rho_h(0) = 0$, $\rho_h(\xi)$ is the unique positive solution of the equation $\sum_{j=1}^n \xi(j)^2 \rho^{-2h(j)} = 1$ for $\xi \in \mathbf{R}^n \backslash \{0\}$. For a monomial ξ^a, $0 \le a \in \mathbf{Z}^n$, put $\deg_h \xi^a = \deg_h a = \sum_{j=1}^n h(j)a(j)$, and for a polynomial $W(\xi) = \sum c_a \xi^a$ put $\deg W = \max\{\deg a : a \in \operatorname{supp} W\}$. Call W *h-elliptic* if there are $r > 0$ and $\alpha > 0$ such that $|W(\xi)| > \alpha \rho_h(\xi)^{\deg_h W}$ for $\rho_h(\xi) > r$. Note that for every polynomial L there is $C > 0$ such that for large ξ one has $|L(\xi)| \le C\rho_h(\xi)^{\deg_h L}$.

Now we are ready to state the Fabes-Riviere criterion (cf. [F-R], p. 28 and [R]).

(F-R). *Let h be a mixed homogeneity in $\mathbf{R}^n$ and let s be an integer with $s > |h|/2$. Let $f : \mathbf{R}^n \to \mathbf{C}$ be a bounded function. Asssume that f is s-times continuously differentiable outside the origin and for $0 \le a \in \mathbf{Z}^n$ with $|a| \le s$*

$$(4.3). \qquad \sup_{r>0} r^{2\deg_h a - |h|} \int_{\frac{r}{2} \le \rho_h(\xi) \le 2r} |\partial^a f(\xi)|^2 \, d\xi < \infty$$

Then the operator $T : L^2(\mathbf{R}^n) \to L^2(\mathbf{R}^n)$ defined by $T(g) = \mathcal{F}^{-1}(f\mathcal{F}(g))$ for $g \in \mathcal{S}(\mathbf{R}^n)$) is of weak type $(1,1)$.

A simple consequence of (F-R) is the following main result of this section:

Theorem 4.1. *Let S be an n-dimensional smoothness such that the fundamental polynomial Q_S is h-elliptic for some mixed homogeneity h in $\mathbf{R}^n$. Then $P_S^{\mathbf{R}}$ is of weak type $(1,1)$.*

Proof. One has to check that for every D and E in S the function $f = s(D)\overline{s(E)}Q_S^{-1}$ satisfies the hypothesis of (F-R). Observe that using the rules of differentiation and an easy induction one can show that

$$\partial^a f = LQ^{-|a|-1}$$

where L is a polynomial with

$$\deg_h L \le v = \deg_h(s(D)\overline{s(E)}) + |a| \deg_h Q - \deg_h a.$$

Thus there are $C_0 > 0$ and $r_0 > 0$ such that

$$|L(\xi)| \le C_0 \rho_h(\xi)^v \quad \text{for } \rho_h(\xi) > r_0.$$

It follows now from the h-ellipticity of Q_S that there are $C_1 > 0$ and $r_1 > 0$ such that

$$|\partial^a f(\xi)| \le C' \rho_h(\xi)^{v-(|a|+1)\deg_h Q_S} \quad \text{for } \rho_h(\xi) > r_1.$$

Note that $\deg_h(s(D)\overline{s(E)}) \le \deg_h s(D) + \deg_h s(E) \le \deg_h Q_S$. Hence

$$|\partial^a f(\xi)| \le C'' \rho(\xi)^{-\deg_h a} \quad \text{for } \rho_h(\xi) > r_2.$$

Passing to the "h-modified" polar coordinates (cf. [F-R] p. 20) we get

$$\int_{\frac{r}{2}\leq\rho_h(\xi)\leq 2r} d\xi \leq C'''r^{|h|} \quad \text{for } r>0.$$

Thus there are $r_3 > 0$ and $C > 0$ such that for $\rho_n(\xi) > r_3$

$$\int_{\frac{r}{2}\leq\rho_h(\xi)\leq 2r} |\partial^a f(\xi)|^2\, d\xi \leq Cr^{|h|-2\deg_h a}.$$

The latter inequality combined with the fact that f is continuously infinitely many times differentiable for $0 \leq \rho_h(\xi) \leq r_3$ yields (4.3) ∎

Corollary 4.1. *Let* $m = (m(1), m(2), \ldots, m(n)) \in \mathbf{Z}^n$ *satisfy* $1 \leq m(j)$ *for* $j = 1, 2, \ldots, n$. *Let*

$$S(m) = \{\mathrm{Id}, \partial_1^{m(1)}, \partial_2^{m(2)}, \ldots, \partial_n^{m(n)}\}.$$

Then $P_{S(m)}$ *is of weak type* $(1,1)$.

Proof. Without loss of generality one can assume that $m(1) \geq m(2) \geq \cdots \geq m(n) \geq 1$. Put $h(j) = m(1)/m(j)$ for $j = 1, 2, \ldots, n$. Then $h = (h(j))$ is a mixed homogeneity in $\mathbf{R}^n$ such that $Q_{S(m)}$ is h-elliptic with $\deg_h Q_{S(m)} = 2m(1)$. ∎

Remark. The assumption that $\inf m(j) = 1$ is not essential; in the case where some $m(j)$'s are zeros the problem reduces to a smaller number of variables.

Example of smoothnesses whose canonical projections are not of weak type $(1,1)$. The analysis of the proof of the Marcinkiewicz interpolation theorem (cf. [S-W], p. 183) combined with Corollary 2.3 yields

Corollary 4.2. *If the canonical projection of a smoothness is of weak type* $(1,1)$ *then there is* $\tilde{C} > 0$ *such that* $a_S(p) \leq \tilde{C}\max(p, p/(p-1))$ *for* $1 < p < \infty$.

We shall next prove

Proposition 4.2. *For every positive integer* m *there is a* $2m$*-dimensional smoothness* S_m *consisting of partial derivatives such that for some* $C_m > 0$

$$a_{S_m}(p) > C_m[\max(p, p/(p-1))]^m \quad \text{for } 1 < p < \infty.$$

The "soft" part of the proof of Proposition 4.2 is based upon a standard tensoring argument.

Write $\mathbf{Z}^n = \mathbf{Z}^{n_1} \times \mathbf{Z}^{n_2}$ with $n = n_1 + n_2$. For each n_i pick an n_i-dimensional smoothness S_i $(i = 1, 2)$. Define a smoothness S by

$$S = \{D : D = D_1 \otimes D_2 \text{ for } D_1 \in S_1 \text{ and for } D_2 \in S_2\},$$

where for differential operators $D_1 = \sum_a c_a \partial_{(1)}^a$ and $D_2 = \sum_b d_b \partial_{(2)}^b$ we put $D_1 \otimes D_2 = \sum_{a,b} c_a d_b \partial_{(1)}^a \partial_{(2)}^b$; here $\partial_{(1)}^a$ and $\partial_{(2)}^b$ denote partial derivatives with respect to the variables $x^{(1)} \in \mathbf{R}^{n_1}$ and $x^{(2)} \in \mathbf{R}^{n_2}$ respectively. Thus $s(D_1 \otimes D_2) = s(D_1) \otimes s(D_2)$

and $Q_S = Q_{S_1} \otimes Q_{S_2}$ where for functions $f_i : \mathbf{R}^{n_i} \to \mathbf{C}$ $(i = 1, 2)$ we define $f = f_1 \otimes f_2 : \mathbf{R}^{n_1} \otimes \mathbf{R}^{n_2} \to \mathbf{C}$ by $f(\xi) = f_1(\xi^{(1)}) f_2(\xi^{(2)})$ for $\xi = (\xi^{(1)}, \xi^{(2)}) \in \mathbf{R}^n = \mathbf{R}^{n_1} \times \mathbf{R}^{n_2}$. Thus, in view of Proposition 2.1, we infer that the canonical projection P_S is the tensor product of the canonical projections P_{S_1} and P_{S_2}, i.e. P_S has the following property: if

$$F = (f_{(D_1,D_2)})_{D_1 \in S_1, D_2 \in S_2} \in \oplus_S L^2(\mathbf{R}^n)$$

satisfies $f_{(D_1,D_2)} = f_{D_1} \oplus f_{D_2}$ for all $D_1 \in S$ and $D_2 \in S_2$ then

$$P_S(F) = P_{S_1}((f_{D_1})_{D_1 \in S_1}) \otimes P_{S_2}((f_{D_2})_{D_2 \in S_2}).$$

Hence $a_S(p)$, being the norm of P_S regarded as an operator from $\oplus_S L^p(\mathbf{R}^n)$ into $\oplus_S L^p(\mathbf{R}^n)$, satisfies $a_S(p) = a_{S_1}(p) \cdot a_{S_2}(p)$ for $1 < p < \infty$ (cf. e.g. [F-I-P]). Thus to complete the proof of Proposition 4.2 it is enough to show

Proposition 4.3. *The 2-dimensional smoothness* $S(1,2) = \{I, \partial_1, \partial_2\}$ *satisfies*

$$a_{S(1,2)}(p) > C \max(p, p/(p-1)) \quad \text{for } 1 < p < \infty,$$

where C is a numerical constant.

Proof. By Corollaries 2.1, 2.2 and 2.3 it is enough to show that there is $C > 0$ such that

$$\|M\|_p = \|M : L^p(\mathbf{T}^2) \to L^p(\mathbf{T}^2)\| > Cp \quad \text{for } p > 2$$

where M is the operator induced by the entry $m^{\mathbf{T}}_{\partial_1,\partial_2}$ of the multiplier matrix of $P^{\mathbf{T}}_{S(1,2)}$; we have

$$m^{\mathbf{T}}_{\partial_1,\partial_2}(k) = k(1)k(2)(1 + k(1)^2 + k(2)^2)^{-1} \quad \text{for } k \in \mathbf{Z}^2.$$

Let (f_n) be the sequence of trigonometric polynomials on $\mathbf{T}^2$ defined by

$$f_n(\xi) = \varphi_n(\xi(1))\varphi_n(\xi(2)) \quad \text{for } \xi = (\xi(1), \xi(2)) \in \mathbf{T}^2$$

where

$$\varphi_n(t) = \sum_{j=1}^{n} 2ij^{-1} \sin jt = \sum_{0<|j|\le n} j^{-1} e^{ijt} \quad (-\pi < t < \pi).$$

Lemma 4.1. *There are positive constants C_1, C_2, C_3 such that for $n = 1, 2, \ldots$, and for $p > 2$*

(4.4) $\qquad \|f_n\|_\infty \le C_1,$

(4.5) $\qquad \|M(f_n)\|_\infty \ge C_2 \log n,$

(4.6) $\qquad \|M(f_n)\|_p \ge C_3 n^{-\frac{2}{p}} \|M(f_n)\|_\infty.$

Assuming Lemma 4.1 we complete the proof of Proposition 4.3 (hence also the proof of Proposition 4.2) as follows. Since $\|f_n\|_p \le \|f_n\|_\infty$, it follows from (4.4)–(4.6) that for all $n = 1, 2, \ldots$ and for $p > 2$

$$\|M\|_p \ge \|f_n\|_p^{-1} \|M(f_n)\|_p \ge C_1^{-1} C_2 C_3 n^{-\frac{2}{p}} \log n.$$

Now fix $p > 2$ and pick n so that $e^p \le n < e^{p+1}$. Then with $C = C_1^{-1} C_2 C_3 e^{-3}$ we have

$$\|M\|_p \ge C_3 C_2 C_1^{-1} e^{-\frac{2(p+1)}{p}} p \ge Cp.$$

∎

Proof of Lemma 4.1. We have $\lim_n \varphi_n(t) = it$ for $|t| < \pi$. Thus, by the so-called Gibbs phenomenon (cf. [Z] vol I, chap. II, § 2; cf. also [N] for an elementary treatment) there is $C_1 > 0$ such that $\|\varphi_n\|_\infty \le C_1^{1/2}$, hence $\|f_n\|_\infty \le C_1$ for $n = 1, 2, \ldots$. This proves (4.4).

Next we have

$$\|M_n(f)\|_\infty \ge |M(f_n)(0)| = \sum_{1\le k(1)\le n} \sum_{1\le k(2)\le n} (1 + k(1)^2 + k(2)^2)^{-1} \ge C_2 \log n.$$

Finally let $V_n^{(2)}(x) = V_n(x(1))V_n(x(2))$ for $x \in \mathbb{T}^2$, where $V_n(t) = \sum_{|j|\le 2n} c_j e^{ijt}$ is the n-th de la Valle Pousin's kernel, i.e. $c_j = 1/2\pi$ for $|j| \le n$, $c_{2n} = c_{-2n} = 0$ and c_j is a linear function of j for $n \le j \le 2n$. Put $M(f_n) = g_n$. We have

$$g_n(x) = \sum_{0<k(1)\le n} \sum_{0<k(2)\le n} (1 + k(1)^2 + k(2)^2)^{-1} e^{ik(1)x(1)+ik(2)x(2)}.$$

Thus g_n is a trigonometric polynomial of degree n in each variable. Hence $V_n^{(2)} * g_n = g_n$ $(n = 1, 2, \ldots)$. Thus, by the Young inequality ([St-W], p. 178)

$$\|g_n\|_\infty \le \|V_n^{(2)}\|_{p^*} \|g_n\|_p = \|V_n\|_{p^*}^2 \|g_n\|_p.$$

The logarithmic convexity of the function $p \to \|\cdot\|_p$ implies that

$$\|V_n\|_{p^*} \le \|V_n\|_\infty^{1/p} \|V_n\|_1^{1/p^*}.$$

We have $\|V_n\|_\infty \le \sum_{|j|\le 2n} |c_j| \le 3n/2\pi$ and $\|V_n\|_1 \le 2$ (because $V_n = F_{2n} + 2^{-1}(e^{inx} + e^{-inx})F_n$ where F_n and F_{2n} are the Fejer kernels). Thus

$$\|V_n\|_{p^*} \le \frac{3^{\frac{1}{p}} 2^{\frac{1}{p^*}}}{(2\pi)^{\frac{1}{p}}} n^{\frac{1}{p}}.$$

Hence

$$\|g_n\|_p \ge C_3 n^{-\frac{2}{p}} \|g_n\|_\infty \quad \text{with } C_3 \ge (2\pi)^{\frac{2}{p}} 3^{-2}.$$

■

Corollary 4.3. *There are $C > 0$ and $\tilde{C} > 0$ such that*

$$C \max(p, p/(p-1)) \le a_{S(1,2)}(p) \le \tilde{C} \max(p, p/(p-1)) \quad (1 < p < \infty). \tag{4.7}$$

Remark. Observe that the fundamental polynomials $Q_{S(1,2)}$ and $Q_{\overline{\partial}}$ are equal while in contrast with (7.7) one has $\sup_{1<p<\infty} a_{S_{\overline{\partial}}}(p) < \infty$ because, by Proposition 3.5,the smoothness $S_{\overline{\partial}}$ is trivial.

Next we state a few open problems.

Problem 1. Is it true that for every n-dimensional smoothness S consisting of partial derivatives there is $C > 0$ such that

$$a_S(p) \leq C \max(p, p/(p-1))^{\frac{n}{2}} \quad (1 < p < \infty)?$$

Problem 2. Is it true that for every smoothness S consisting of partial derivatives there is a non-negative integer k such that for some $C_1, C_2 > 0$

$$C_1 \max(p, p/(p-1))^k \leq a_{S(1,2)}(p) \leq C_2 \max(p, p/(p-1))^k \quad (1 < p < \infty)?$$

Problem 3. Characterize smoothnesses consisting of partial derivatives with the canonical projection of weak type $(1, 1)$.

Problem 4. Is it true that if for some smoothness S the canonical projection $P_S^{\mathbf{R}}$ is of weak type $(1, 1)$ then so is $P_S^{\mathbf{T}}$?

The answer is "yes" in many special cases cf. e.g. [Kw-P] and [K-Si]. Problem 4 is a particular case of the following

Problem 5 (the transference Problem). Let $f : \mathbf{R}^n \to \mathbf{C}$ belong to $L^\infty(\mathbf{R}^n)$. Assume that f is continuous at each $k \in \mathbf{Z}^n$. Is it true that if the multiplier f determines a weak type $(1, 1)$ operator on $L^2(\mathbf{R}^n)$ then the multiplier $(f(k))_{k \in \mathbf{Z}^n}$ determines the weak $(1, 1)$ operator on $L^2(\mathbf{T}^n)$?

5. A criterion for p-boundedness of canonical projections.

Clearly whenever the canonical projection is of weak type $(1, 1)$ it is p-bounded for all p with $1 < p < \infty$. In this section we prove yet another criterion for p-boundedness of canonical projections based on the concept of slight ellipticity introduced before Theorem 3.2.

Theorem 5.1. *Let a smoothness S have the property that the symbol $s(D)$ is strongly slightly elliptic for every $D \in S$. Then the canonical projection P_S is p-bounded for $1 < p < \infty$.*

The proof of Theorem 5.1 requires some notation. Let $\mathcal{E} \in \mathbf{Z}^n$ denote the set of the characteristic functions of all subsets of the set $\{1, 2, \ldots, n\}$ identified with the set $\{\xi(1), \xi(2), \ldots, \xi(n)\}$ of the variables. If $e \in \mathcal{E}$ is the characteristic function of the set $\{i_1, i_2, \ldots, i_k\}$, $m : \mathbf{R}^n \to \mathbf{C}$ is a locally integrable function and ρ a finite k-dimensional cube regarded as the cartesian product of intervals being subsets of the axes of the variables $\xi(i_1), \xi(i_2), \ldots, \xi(i_k)$ then

$$\int_\rho m \, d\xi_{(e)} = \int_\rho m \, d\xi(i_1) \, d\xi(i_2) \ldots d\xi(i_k)$$

is the function of the variables belonging to the set $\{1, 2, \ldots, n\} \backslash \{i_1, i_2, \ldots, i_k\}$.

A k-dimensional cube is called dyadic if it is the cartesian product of intervals each of which has the property: if a and b are the endpoints of the interval, say $a \leq b$, then $ab \neq 0$ and either $ab^{-1} = 2^{-1}$ or $ab^{-1} = 2$.

The proof of Theorem 5.1 is based upon the multidimensional Marcinkieiwicz multiplier theorem (cf. [St], chap. VI, § 6, Theorem 6′).

(MMMT). *Let* $m : \mathbf{R}^n \to \mathbf{C}$ *be a bounded function which has outside the origin continuous partial derivatives of order* $\leq n$. *Assume that there exists a constant* K *such that*

$$\int_\rho |\partial^e m| \, d\xi_{(e)} \leq K \tag{5.1}$$

for all $e \in \mathcal{E}$ *with* $|e| \geq 1$ *and all* $|e|$*-dimensional dyadic cubes* ρ. *(The inequalitiy (3.1) meant the inequality between functions.) Then* m *is a p-bounded multiplier for* $1 < p < \infty$.

Proof of Theorem 5.1. We show that for every $D \in S$ and $E \in S$ the multiplier $m_{D,E} = s(D)\overline{s(E)}Q_S^{-1}$ satisfies (5.1). Fix $e \in \mathcal{E}$ with $|e| \geq 1$. Applying the rules of differentiation of products and quotients of functions, we get

$$\partial^e_{m_{D,E}} = \sum \partial^{e_0}(s(D))\partial^{e'_0}(s(\overline{E})) \prod_{j=1}^{r} \partial^{e_j}(s(D_j))\partial^{e'_j}(\overline{s(D_j)})Q_S^{-r-1} \tag{5.2}$$

where the sum extends over all sequences $D_1, D_2, \ldots, D_r$ of not necessarily different elements of S and all sequences $e_0, e'_0, e_1, e'_1, \ldots, e_r, e'_r$ such that $e = e_0 + e'_0 + e_1 + e'_1 + \cdots + e_r + e'_r$, $e_j \in \mathcal{E}$, $e'_j \in \mathcal{E}$, $|e_j + e'_j| \geq 1$ for $j = 0, 1, \ldots, r$ and for $r = 0, 1, \ldots, |e|$ (for $r = 0$ we admit $\prod_{j=1}^{0} = 1$).

Next we need the following fact which will be proved later on.

Lemma 5.1. *Assume that polynomials* W_1 *and* W_2 *are strongly slightly elliptic. Then there exists a* $K = K(W_1, W_2) > 0$ *such that for every* $e \in \mathcal{E}$ *with* $|e| \geq 1$ *and every dyadic cube being a product of intervals which belong to axes in* e *and every* e' *and* e'' *in* $\mathcal{E}$ *with* $e' \leq e$, $e'' \leq e$ *and* $|e' + e''| \geq 1$ *and every* $\xi \in \rho$,

$$|(\partial^{e'}(W_1)\partial^{e'}(W_2))(\xi)| \leq K(|W_1|^2 + |W_2|^2)(\xi)|\xi^{e'+e''}|^{-1}. \tag{5.3}$$

Fix a dyadic cube ρ. Let $e, e_0, e'_0, e_1, e'_1, \ldots, e_r, e'_r$ be as in (5.2). Combining (5.3) with the definition of Q_S we infer that there exists a $K_1 \geq 1$ such that, for $j = 1, 2, \ldots, n$

$$|(\partial^{e_0}(s(D))\partial^{e'_0}(\overline{s(E)})(\xi)| \leq K_1|Q_S(\xi)|\,|\xi^{e_0+e'_0}|^{-1},$$

$$|\partial^{e_j}(s(D_j))\partial^{e'_j}(\overline{s(D_j)})(\xi)| \leq K_1|Q_S(\xi)|\,|\xi^{e_j+e'_j}|^{-1}.$$

Thus, by (5.2),

$$|\partial^e m_{D,E})(\xi)| \leq K_2K_1^{n+1}|\xi^e|^{-1} \quad (\xi \in \rho)$$

where K_2 is another constant depending only on n (independent of S and the cube ρ). Hence, taking account that $\xi^e = \xi(i_1)\xi(i_2)\ldots\xi(i_k)$ for e being the characteristic function of the set $\{i_1, i_2, \ldots, i_k\}$, we get

$$\int_\rho |\partial^e m_{D,E}(\xi)| \, d\xi_{(e)} \leq K_2K_1^{n+1}|\log 2|^{|e|} \leq K_2K_1^{n+1}$$

This completes the verification of (5.1). ∎

Proof of Lemma 5.1. Let $W_1(\xi) = \sum c'_a \xi^a$ and let $W_2(\xi) = \sum c''_b \xi^b$. Then

$$(\partial^{e'}(W_1)\partial^{e''}(W_2))(\xi) = \sum{}^* c'_a c''_b \xi^{a-e'} \xi^{b-e''}$$

where the sum $\sum^*$ extends over all pairs (a, b) such that $c'_a c''_b \neq 0$ and $\partial^{e'} \xi^a \partial^{e''} \xi^b$ is not identically zero. Let $\xi \in \rho$ where ρ is a dyadic cube. Then

$$\xi^{e'+e''} = \prod_{\{j:(e'+e'')(j)=1\}} \xi_j \neq 0.$$

Using the inequality $\xi^{a+b} \leq 2^{-1}(\xi^{2a} + \xi^{2b})$ and the assumption that W_1 and W_2 are strictly slightly elliptic we get

$$\begin{aligned} |\partial^{e'}(W_1)\partial^{e''}(W_2)(\xi)| &\leq \sum{}^* |c'_a c''_b| 2^{-1}(\xi^{2a} + \xi^{2b}) |\xi^{e'+e''}|^{-1} \\ &\leq K(|W_1(\xi)|^2 + |W_2(\xi)|^2) |\xi^{e'+e''}|^{-1} \end{aligned}$$

with $K = 2^{-1} \max_{(a,b)} |c'_a c''_b| \cdot \alpha$ where the maximum extends over the same set of pairs (a, b) as the sum $\sum^*$ and α is any positive constant such that W_1 and W_2 both satisfy the strict slight ellipticity condition with this constant. ∎

Obviously the symbol of a partial derivative is strictly slightly elliptic. Thus (cf. [P-S], Proposition 6.2)

Corollary 5.1. *The canonical projection of any smoothness consisting of partial derivatives is p-bounded for* $1 < p < \infty$. ∎

References.

[F-R] E. B. Fabes and N. M Rivière, 'Singular integrals with mixed homogeneity', *Studia Math.* 27 (1966) pp. 19-38

[F-I-P] T. Figiel, T. Iwaniec and A. Pelczynski, 'Computing norms and critical exponents of some operators in L^p-spaces', *Studia Math.* 79 (1984), pp. 227-279

[H-I], [H-II] Lars Hörmander, *The analysis of linear partial differential operators* I, II, (Springer-Verlag, Berlin, Heidelberg New York, Tokyo, 1983)

[Ka] Y. Katznelson, *An introduction to harmonic analysis*, (John Wiley and sons, New York, 1968)

[K-Si] S. V. Kisliakov and N. G. Sidorenko, 'Anisotropic spaces of smooth functions without local unconditional structure', *LOMI preprint*, E-2-86, Leningrad 1986

[Kw-P] S. Kwapien and A. Pelczynski, 'Absolutely summing operators and translation invariant spaces of functions on compact abelian groups', *Math. Nachr.* 94 (1980) pp. 303-340

[N] I. P. Natanson *Construction function theory*, vol. I, (Ungar Publ. Co, New York, 1969)

[P-S] A. Pelczynski and K. Senator, 'On isomorphism of anisotropic Sobolev spaces with "classical Banach spaces" and a Sobolev type embedding theorem', *Studia Math.* 84 (1986), pp. 169-215

[R] N. M. Rivière, 'Singular integrals and multiplier operators', *Arkiv. Math.* 9 (1971), pp. 243-278

[Si] N. G. Sidorenko, 'Nonisomorphism of some Banach spaces of smooth functions with the space of continuous functions', *Funkc. Analiz i Priložen.* 21 No. 4 (1987), pp. 91-93 (Russian)

[St] E. Stein, *Singular integrals and differential properties of functions*, (Princeton University Press, Princeton, New Jersey, 1970)

[St-W] E. Stein and G. Weiss, Introduction to Fourier analysis on euclidian spaces, (Princeton University Press, Princeton, New Jersey, 1971)

[Z] A. Zygmund, *Trigonometric series* I, II, (second edition, Cambridge University Press, Cambridge, 1986)

Institute of Mathematics, Polish Academy of Spaces, Sniadeckich 8, Ip., 00-950 Warszawa, Poland
and
Department of Mathematics, University of Illinois at Urbana Champaign, 1409 W. Green St. Urbana, Il. 61801, USA

Remarks on L^2 restriction theorems for Riemannian manifolds

C.D. Sogge*

Around ten years ago, P. Tomas and E.M. Stein [14] showed that for $n \geq 2$ and $1 \leq p \leq 2(n+1)/(n+3)$ the Fourier transform of an $L^p(\mathbf{R}^n)$ function restricts to the unit sphere as an element of $L^2(\mathbf{R}^n)$. That is, if $d\sigma$ denotes the induced Lebesgue measure on $\mathbf{S}^{n-1}$, one has the inequality

(1)
$$\left(\int_{\mathbf{S}^{n-1}} |\hat{f}(\xi)|^2 \, d\sigma(\xi)\right)^{1/2} \leq C\|f\|_{L^p(\mathbf{R}^n)},$$

for f belonging to a dense class of functions. Earlier, A. Knapp (see [14]) had shown that such an L^2 restriction theorem cannot hold for any exponent $p > 2(n+1)/(n+3)$.

The purposes of this paper are to go over some recent work concerning generalizations of this result to the setting of compact Riemannian manifolds and to present some new results and proofs. First, though, let us set the notation. We shall be dealing with C^∞ compact boundaryless manifolds $\mathbf{M}$ of dimension ≥ 2 with Riemannian densities dx and Laplace-Beltrami operators Δ. Recall that $L^2(\mathbf{M}, dx)$ admits a complete orthogonal direct sum decomposition with respect to the eigenspaces of $-\Delta$. That is, one can write

$$L^2(\mathbf{M}) = \sum_{j=0}^{\infty} E_j,$$

where E_j is the jth eigenspace corresponding to the eigenvalue λ_j The eigenvalues are counted with multiplicity, assumed to be non-negative, and are arranged in increasing order, i.e. $0 \leq \lambda_0 \leq \lambda_1 \leq \cdots$. Also, e_j will denote the projection onto the jth eigenspace E_j. Thus, an L^2 function f can be written as

$$f = \sum_{j=0}^{\infty} e_j(f),$$

where the partial sums converge in the L^2 norm.

In order to generalize (1) let us define, for a given $1 \leq p \leq \infty$, the "critical exponent"

(2)
$$\delta(p) = \max\,(n \cdot |1/p - 1/2| - 1/2, 0).$$

Then a straightforward calculation involving Plancherel's theorem for $\mathbf{R}^n$ shows that if we define projection operators P_t as follows

(3)
$$P_t f(x) = \int_{|\xi| \in [t,t+1]} \hat{f}(\xi) e^{i<x,\xi>} \, d\xi,$$

then (if p is such that $\delta(p) > 0$) (1) is equivalent to a uniform inequality of the following form:

(1′)
$$\|P_t f\|_{L^2(\mathbf{R}^n)} \leq C(1+t)^{\delta(p)} \|f\|_{L^p(\mathbf{R}^n)}, \quad t > 0.$$

We remark that $\delta(p)$ is the critical exponent for Riesz summation for the Laplacian on $L^p(\mathbf{R}^n)$, and thus it is not surprising that a uniform inequality of the type (1′) implies a sharp theorem regarding summation on $L^p(\mathbf{R}^n)$. This connection was made by C. Fefferman and E.M. Stein (see [4]).

It is the equivalent version (1′), rather than (1), which leads itself to generalizations in the setting of Riemannian manifolds. To this end, let us define the *spectral projection operators* on $\mathbf{M}$, χ_t, which generalize those in (3), as follows:

(4)
$$\chi_t f = \sum_{\sqrt{\lambda_j} \in [t,t+1]} e_j(f).$$

Then, in analogy to (1′), we have the following result.

*Supported by an NSF postdoctoral fellowship.

THEOREM. *If $t > 0$, and if $\delta(p)$ is defined as in (2), then*

(5) $$\|\chi_t f\|_{L^2(M)} \le C(1+t)^{\delta(p)}\|f\|_{L^p(M)}, \qquad 1 \le p \le 2(n+1)/(n+3),$$

(6) $$\|\chi_t f\|_{L^2(M)} \le C(1+t)^{(n-1)(2-p)/4p}\|f\|_{L^p(M)}, \qquad 2(n+1)/(n+3) \le p \le 2.$$

Furthermore, these estimates are always sharp.

Remark. As in the Euclidean case, the inequality (5) implies sharp Riesz summation results for $L^p(M)$ and the same range of exponents. Also, one can show that (5) implies the L^2 restriction theorem for $\mathbf{R}^n$. See [3], [11].

The inequalities (5) and (6) as well as the sharpness of (5) were proved in [10]. In the present work, we shall give a new proof of the inequalities, as well as establishing the sharpness of (6). The latter is new and will be demonstrated by adapting the counterexample of Knapp mentioned before.

Before doing this, though, let us go over the main points of the previous proofs in [10]. In this work, the author proved (5) and (6) by noticing that one could generalize certain "Sobolev inequalities" for the Laplacian in $\mathbf{R}^n$ of Kenig, Ruiz and the author [8] to the setting of Riemannian manifolds, and, in fact, the following uniform differential inequality holds:

(7) $$\|u\|_{L^q(M)} \le C(1+t)^{\delta(q)-1}\|(\Delta+t^2)u\|_{L^2(M)} + C(1+t)^{\delta(q)}\|u\|_{L^2(M)}, \quad q = 2(n+1)/(n-1).$$

This inequality was proved by using the Hadamard parametrix to compute an "approximate inverse" for $\Delta + t^2$ along with an error term, both of which could be dealt with by invoking oscillatory integral theorems of Carleson and Sjölin [2] and Stein [13]. Furthermore, this inequality implies (5) and (6) since, if we take $u = \chi_t f$, where χ_t, as in (4), is the spectral projection operator associated to $\sqrt{-\Delta}$, then a simple argument involving orthogonality shows that if q is as in (7) then

$$\|\chi_t f\|_q \le C(1+t)^{\delta(q)}\|f\|_2,$$

which, by duality, implies (5) in the special case where $p = 2(n+1)/(n+3)$, which is the exponent dual to q. The other inequalities follow from interpolating this special case with

(8) $$\begin{aligned}\|\chi_t f\|_2 &\le C(1+t)^{(n-1)/2}\|f\|_1\\ \|\chi_t f\|_2 &\le \|f\|_2.\end{aligned}$$

The second inequality follows from orthogonality, while the first is a consequence of the sharp form of the Weyl formula for $\mathbf{M}$. (See [11].)

Let us now sketch a more direct proof of (5) and (6) which is modeled after the Weyl formula calculus developed by Hörmander and others. (See [5] for further references.) As we shall see, this proof will also have the virtue that it can be modified to show that (6) is sharp.

As above, by the M. Riesz interpolation theorem and (8), it suffices to prove the special case

(9) $$\|\chi_t f\|_2 \le C(1+t)^{\delta(p)}\|f\|_p, \quad p = 2(n+1)/(n+3).$$

To prove this estimate we shall also use the Hadamard parametrix, except, this time, we shall use a parametrix for the wave operator associated to Δ, namely:

$$(\partial/\partial\tau)^2 - \Delta,$$

where τ denotes a real variable.

To be able to apply the Hadamard parametrix, we shall have to study "approximate spectral projection operators," $\tilde{\chi}_t$, which we shall define now. For a nonzero Schwartz class function $\chi(t)$, to be specified later, we let $\tilde{\chi}_t$ be defined as follows:

$$\tilde{\chi}_t f = \sum_j \chi(t - \sqrt{\lambda_j})\, e_j(f). \tag{10}$$

Given any such χ, it is not difficult to see that (9) would be a consequence of the following estimate for these new operators

$$\|\tilde{\chi}_t f\|_2 \le C(1+t)^{\delta(p)}\|f\|_p, \quad p = 2(n+1)/(n+3). \tag{9'}$$

The Hadamard parametrix for the Cauchy problem with initial data f,

$$\begin{cases} [(\partial/\partial\tau)^2 - \Delta]\, u(x,\tau) = 0, & \tau > 0, \\ u(x,0) = f(x) \\ (\partial/\partial\tau)\, u(x,0) = 0, \end{cases} \tag{11}$$

only provides a good approximation to $u(x,\tau)$ when the time variable τ is small. For this reason, as will become more clear later, it will be convenient to also require that the function χ in (10) also has the property that

$$\hat{\chi}(\tau) = 0, \quad \text{when } |\tau| > \varepsilon, \tag{12}$$

where ε is to be specified later. Also, for later use, it will be convenient to in addition require that $\chi(t)$ is non-negative and $\chi(0) \ne 0$. We note that such a function exists, for if $\rho \in \mathcal{S}$ has the property that $\hat{\rho}$ is even and compactly supported, then ρ^2 is a non-negative function with compactly supported Fourier transform.

Having set things up, let us now proceed with the proof of (9′). To do so, let us first take a partial Fourier transform and rewrite the operators in (10) as follows,

$$\tilde{\chi}_t f = \int_{-\infty}^{\infty} \hat{\chi}(\tau) \sum_j e^{i\tau\sqrt{\lambda_j}}\, e_j(f)\, e^{-i\tau t}\, d\tau. \tag{13}$$

Next, we recall that the function

$$u(x,\tau) = \sum_j \cos(\tau\sqrt{\lambda_j}\,)\, e_j(f), \tag{14}$$

is the solution to the Cauchy problem (11). For this reason, it is more convenient to instead study the operators

$$\tilde{\tilde{\chi}}_t f = \int_{-\infty}^{\infty} \hat{\chi}(\tau) \sum_j \cos(\tau\sqrt{\lambda_j})\, e_j(f) e^{-i\tau t}\, d\tau. \tag{15}$$

But,

$$\begin{aligned} 2\,\tilde{\tilde{\chi}}_t f &= \tilde{\chi}_t f + \int_{-\infty}^{\infty} \hat{\chi}(\tau) \sum_j e^{-i\tau\sqrt{\lambda_j}}\, e_j(f)\, e^{-i\tau t}\, d\tau \\ &= \tilde{\chi}_t f + \sum_j \chi(t + \sqrt{\lambda_j})\, e_j(f). \end{aligned}$$

Furthermore, since $\chi \in \mathcal{S}$, it is clear from (8) that

$$\|\sum_j \chi(t+\sqrt{\lambda_j})\, e_j(f)\|_2 \le C(1+t)^{-N}\, \|f\|_p,$$

for all N, and thus it suffices to show that the operators in (15) satisfy the bounds (9′).

To do this, we shall now use the Hadamard parametrix (see Hörmander [7, §17.4] for details), which for sufficiently small times τ allows us to write the function $u(x,\tau)$ in (14) as follows

$$u(x,\tau) = \int_M \int_{\mathbf{R}^n} e^{i\Phi(x,y,\xi)} \cos(\tau|\xi|)\, \eta(x,y) f(y)\, d\xi\, dy + R_\tau f(x). \tag{16}$$

Here R_τ is also a Fourier integral operator, but it is of one order lower, and $\eta(x,y)$ is a C^∞ function which equals one on the diagonal and can be assumed to be supported in an arbitrarily small neighborhood of the diagonal as long as τ is sufficiently small. Also,

$$\Phi(x,y,\xi) = < x - \tilde{y}, \xi > . \tag{17}$$

where, for a given x, $\tilde{y}$ denotes the geodesic normal coordinates of y around x with respect to the Riemannian metric associated to Δ. This phase function is always well defined in a small enough neighborhood of the diagonal.

Thus, if we choose the number ε in (12) to be sufficiently small, it follows from (14)-(16) that, modulo an operator which the arguments below would show has an (L^p, L^2) norm that is $O((1+t)^{-1})$ better, $\tilde{\tilde{\chi}}_t$ has kernel

$$K_t(x,y) = \eta(x,y) \iint e^{i\Phi(x,y,\xi)} \cos(\tau|\xi|)\, \hat{\chi}(\tau) e^{-i\tau t}\, d\tau\, d\xi.$$

However, it is easy to check that the kernels

$$\eta(x,y) \iint e^{i[\Phi(x,y,\xi)-\tau|\xi|]}\, \hat{\chi}(\tau) e^{-i\tau t}\, d\tau\, d\xi = \eta(x,y) \int e^{i\Phi(x,y,\xi)}\, \chi(t+|\xi|)\, d\xi$$

give rise to operators with rapidly decreasing (L^p, L^2) norm, which, in turn, implies that we need only show that the operators with kernels

$$\begin{aligned} \tilde{K}_t(x,y) &= \eta(x,y) \iint e^{i[\Phi(x,y,\xi)+\tau|\xi|]}\, \hat{\chi}(\tau) e^{-i\tau t}\, d\tau\, d\xi \\ &= \eta(x,y) \int e^{i\Phi(x,y,\xi)}\, \chi(t-|\xi|)\, d\xi \end{aligned} \tag{18}$$

satisfy (9′). However, if we let $|x-y|$ denote the Riemannian distance between x and y with respect to the metric corresponding to the phase function in (17), then (17) and straightforward computations involving stationary phase (cf. [13], [10]) show that $\tilde{K}_t(x,y)$ is essentially a C^∞ function times

$$t^{(n-1)/2}\, e^{it|x-y|}\, |x-y|^{-(n-1)/2}$$

(plus a similar term where $e^{it|x-y|}$ is replaced by $e^{-it|x-y|}$). On account of this, one can argue as in [10] and use the Carleson-Sjölin method and oscillatory integral theorems of Carleson-Sjölin [2] and Stein [13] to show that the operators with kernels as in (18) satisfy the desired estimates. We shall omit the details.

Let us now turn to proof of the sharpness of (6). To motivate things let us review the counterexamp of Knapp (see [14]) that (1) cannot hold for any $p > 2(n+1)/(n+3)$. To do so we shall only conside exponents p such that $1 \le p < 2n/(n+1)$ since these are the ones for which $\delta(p) > 0$. (Clearly, there no loss of generality in doing this, since, if (1) held for some larger exponent, then interpolation with (1 for $p = 1$ would imply that (1) also held for the exponents $p < 2$ for which $\delta(p) > 0$.) To show this fix $C_0^\infty(\mathbf{R})$ function ψ, with $\hat{\psi}(0) \neq 0$, and, for a given $t > 0$, define f by setting

$$f(x) = e^{itx_1}\psi(x_1)\prod_{j=2}^{n}\psi(t^{1/2}x_j),$$

Then

$$\|f\|_p \approx t^{-(n-1)/2p}. \tag{19}$$

But, on the other hand, if we let $\xi' = (\xi_2, \ldots, \xi_n)$, then $\hat{f}(\xi)$ is essentially supported on the set

$$\{\xi : \xi_1 \in [t-1, t+1], |\xi'| \in [0, t^{1/2}]\},$$

which in turn is essentially contained in the annulus $\{\xi : |\xi| \in [t-1, t+1]\}$. Consequently, if one keeps these things in mind, it is not difficult to see that if P_t is as in (3) then,

$$\|P_t f\|_2 \approx \|f\|_2 \approx t^{-(n-1)/4}. \tag{20}$$

Thus, (19) and (20) imply that if (1′) held for a given p then one would have to have

$$(n-1)/2 \cdot [1/p - 1/2] = (n-1)(2-p)/4p \le \delta(p),$$

where $\delta(p)$ is as in (2). However, since it is easy to check that this forces $p \le 2(n+1)/(n+3)$, it follows that (1′) can only hold for this range of exponents. Finally, since, as we noted before, (1) holds if and only if (1′) holds for a given p with $\delta(p) > 0$, it follows that (1) cannot hold for $p > 2(n+1)/(n+3)$.

Remark. A similar construction was used by Stanton and Weinstein [12] to show that

$$\|\chi_t f\|_{L^2(\mathbf{S}^2)} \le C(1+t)^{1/8}\|f\|_{L^{4/3}(\mathbf{S}^2)},$$

is the best possible result, when χ_t is the spectral projection operator corresponding to the usual Laplace-Beltrami operator on $\mathbf{S}^2$. To do this, they realized that if one restricts the functions

$$f_k(x) = (x_1 + ix_2)^k$$

to the unit sphere, then the resulting functions are spherical harmonics of degree k and satisfy

$$\|f_k\|_2 / \|f_k\|_{4/3} \approx k^{1/8}.$$

Let us now prove the sharpness of (6). To do so we note that, by the proof of the theorem, we need only show that the operators with kernels as in (18) cannot satisfy better estimates than those in (6) as $t \to +\infty$. To do so, let S_t denote this operator. Then we shall fix a coordinate patch $\Omega \in \mathbf{M}$ and construct, for large t, a function f supported in Ω for which

$$\|S_t f\|_{L^2(\Omega)} / \|f\|_{L^p(\Omega)} \ge Ct^{(n-1)(2-p)/4p}, \tag{21}$$

for some positive constant C which is independent of t.

The key to this construction will be to make a judicious choice of coordinates on Ω. Given an arbitrary set of coordinates, the principal part of Δ equals

$$\sum_{j,k} g^{jk}(x)\,\frac{\partial^2}{\partial x_j \partial x_k},$$

where the g^{jk} are real and the matrix with entries g^{jk} is symmetric and positive definite. Fortunately for us though, if Ω is small enough, we can choose our coordinates so that the principal part is diagonalized with respect to, say, the first variable. That is, we can assume that in Ω,

$$\Delta = \frac{\partial^2}{\partial x_1^2} + \sum_{j,k>1} g^{jk}(x)\,\frac{\partial^2}{\partial x_j \partial x_k} + \text{lower order terms}. \tag{22}$$

For a proof of this, see Hörmander [7, pgs. 500-502].

The reason we have chosen such coordinates on Ω is that now the phase function in (17) can be taken to be of the form

$$\Phi(x,y,\xi) = <(x_1 - y_1, \gamma(x,y)),\, \xi>, \tag{23}$$

where γ is a smooth $(n-1)$-dimensional valued function having the property that when $x' = y' = 0$,

$$\gamma(x_1, 0, y_1, 0) = 0.$$

Thus, if $x' = 0$,

$$|(\partial/\partial x_1)^l\,(\partial/\partial y_1)^m < \gamma(x_1, 0, y), \xi' > | = O(|y'|\,|\xi'|). \tag{24}$$

Having made these observations, we now argue as in the Euclidean case. First of all, if, as we may assume, $0 \in \Omega$, we note that $\eta(0,0) = 1$. Thus, if now $\psi_0 \in C_0^\infty(\mathbf{R})$ has small enough support we can always find a smooth function f so that

$$\eta(0,y)f(y) = \psi_0(y). \tag{25}$$

For our purposes, though we shall take

$$\psi_0(y) = e^{ity_1}\,\psi(y_1)\prod_{j=2}^{n}\psi(\varepsilon^{-1}t^{1/2}y_j), \tag{26}$$

where ψ has the property that $\hat{\psi}$ is a non-negative function with $\hat{\psi}(0) \neq 0$, and ε is to be specified later.

For this choice of f one has

$$||f||_{L^p(\Omega)} \approx t^{-(n-1)/2p}. \tag{27}$$

On the other hand,

$$S_t f(x) = \iint e^{i\Phi(x,y,\xi)}\eta(x,y)\,f(y)\,\chi(t - |\xi|)\,dyd\xi, \tag{28}$$

and, since f is given by (25) and (26), a straightforward integration by parts argument using (24) will show that when $x' = 0$, the ξ integrand is essentially supported in the set

$$\{\xi : |\xi| \in [t-1, t+1],\ \xi_1 \in [t-1, t+1]\}.$$

Using this fact it is not difficult to see that when $|x|$ is small enough and $x' = 0$ one has

$$|(\partial/\partial x_1)e^{-itx_1}S_tf(x)| \le C \qquad (29)$$
$$|\nabla_{x'} S_tf(x)| \le Ct^{1/2},$$

where C depends only on ε.

Next, we notice that (25) and (28) give that

$$S_tf(0) = \iint \psi_0(y)e^{-iy_1\xi_1}\chi(t-|\xi|)\,dy\,d\xi - \iint \psi_0(y)e^{-iy_1\xi_1}\left(1-e^{i<\gamma(0,y),\xi'>}\right)\chi(t-|\xi|)\,dy\,d\xi. \qquad (30)$$

Furthermore, by (26), the first summand equals

$$(\varepsilon t^{-1/2}\hat{\psi}(0))^{n-1}\int \hat{\psi}(t-\xi_1)\chi(t-|\xi|)\,d\xi, \qquad (31)$$

and since we have chosen things so that $\chi, \hat{\psi} > 0$ and $\chi(0), \hat{\psi}(0) > 0$, it is not difficult to see that the expression in (31) is greater than $c\varepsilon^{n-1}$ for some positive constant c. Also, the fact that (24) implies that when $|y'| \le C\varepsilon t^{-1/2}$ one has

$$\left|\left(1-e^{i<\gamma(0,y),\xi'>}\right)\right| \le C\varepsilon\,|\xi'/t^{1/2}|,$$

can be used to show that the second integral in (30) is $O(\varepsilon^n)$. Consequently, if we choose ε to be small enough, it follows that there is a positive constant c_0 independent of t for which

$$S_tf(0) > c_0.$$

This, along with (29), implies that in some fixed dilate of the "rectangle" $\{x : |x_1| \le 1, |x'| \le t^{-1/2}\}$ one has that $S_tf(x)$ is uniformly bounded below, which, in turn, implies that

$$\|S_tf\|_{L^2(\Omega)} \ge Ct^{-(n-1)/4}.$$

Finally, this inequality along with (27) establishes (21) which completes the proof.

References

1. A. Bonami and J.L. Clerc, *Sommes de Cesáro et multiplicateurs des developpements en harmonics sphèriques*, Trans. Amer. Math. Soc. **183** (1973), 223–263.
2. L. Carleson and P. Sjölin, *Oscillatory integrals and a multiplier problem for the disc*, Studia Math. **44** (1972), 287–299.
3. F.M. Christ and C.D. Sogge, *The weak type L^1 convergence of eigenfunction expansions for pseudodifferential operators*, Inv. Math. (to appear).
4. C. Fefferman, *A note on spherical summation multipliers*, Israel J. Math. **15** (1973), 44–52.
5. L. Hörmander, *The spectral function of an elliptic operator*, Acta Math. **88** (1968), 341–370.
6. L. Hörmander, *Oscillatory integrals and multipliers on FL^p*, Ark. Mat. **11** (1971), 1–11.
7. L. Hörmander, "The Analysis of Linear Partial Differential Operators III," Springer-Verlag, New York, 1985.
8. C.E. Kenig, A. Ruiz and C.D. Sogge, *Uniform Sobolev inequalities and unique continuation for second order constant coefficient differential operators*, Duke Math. J. **55** (1987), 329–349.
9. C.D. Sogge, *Oscillatory integrals and spherical harmonics,*, Duke Math. J. **53** (1986), 43–65.
10. C.D. Sogge, *Concerning the L^p norm of spectral clusters for second order elliptic operators on compact manifolds*, J. Funct. Anal. **77** (1988), 123–134.
11. C.D. Sogge, *On the convergence of Riesz means on compact manifolds*, Annals of Math. **126**, 439–447.
12. R. Stanton and A. Weinstein, *On the L^4 norm of spherical harmonics*, Math. Proc. Camb. Phil. Proc. **89** (1981), 343–358.
13. E.M. Stein, *Oscillatory integrals in Fourier analysis*, in "Beijing Lectures in Harmonic Analysis," Princeton Univ. Press, Princeton NJ, 1986, pp. 307–356.
14. P.A. Tomas, *Restriction theorems for the Fourier transform*, Bull. Amer. Math. Soc. **81** (1975), 477–478.

Printed in the United States
By Bookmasters